INTRODUCTION TO FINITE ELEMENT METHOD
With Interdisciplinary Examples

INTRODUCTION TO FINITE ELEMENT METHOD
With Interdisciplinary Examples

Prof. Dr. N. S. V. Kameswara Rao
Former Professor,
Indian Institute of Technology Kanpur, India,
and Universiti Malaysia Sabah, Kota Kinabalu, Malaysia

INVOCATION

मातृ देवो भव।
पितृ देवो भव।
आचार्य देवो भव।
अतिथि देवो भव॥

OM OM OM

Maathru Devo Bhava, Pithru Devo Bhava, Aacharya Devo Bhava,
Athidhi Devo Bhava.

Pray thy Mother as God. Pray thy Father as God. Pray thy Teacher as God. Pray thy
Guest as God.

Be one to whom a Mother is as God, Be one to whom a Father is as God, Be one to
whom a Teacher is as God, Be one to whom a Guest is as God.
-Taittiriya Upanishad [तैत्तिरीय उपनिषद्]

Dedicated with Devotion

To

Divinity

All Pervading

Contents

Preface   **vi**
About the Author   **viii**
Acknowledgements   **ix**

List of Figures   **x**
List of Symbols   **xvi**
List of Tables   **xxi**
List of Abbreviations   **xxii**

1   Introduction   1
    1.1 Continuous and Discrete Systems   1
    1.2 Analysis and Synthesis   1
    1.3 Finite Element Method   2
    1.4 Reverse Engineering, Simulation and Virtual Testing   3
        1.4.1 Reverse Engineering   3
        1.4.2 Simulation   4
        1.4.3 Virtual Testing   4
        1.4.4 Other Applications   4

2   Finite Element Method – General Concepts   7
    2.1 General Philosophy   7
    2.2 Finite Element Procedure   9
        2.2.1 Finite Element Deformation Patterns   10
        2.2.2 Transformation of Co-ordinates   11
        2.2.3 System/ Global Stiffness Matrix, Assembly and
               System Equilibrium Equations   12
    Appendix 2A Commonly Used Finite Element Shapes   16
        2A.1 Some details of Commonly Used Finite Elements   16
        2A.2 Errors inherent in the FEM procedure   17
    Appendix 2B Choice of Displacement Distribution Functions –
            Example of Constant Strain Triangular (CST) Element   19
        2B.1 Choosing the Polynomials for Displacement / Variable
            Distribution Functions   19
        2B.2 Nodal Degrees of Freedom   20
        2B.3 Example of CST Elements – For Applications to plane problems
            (plain stress, plane strain, two-dimensional seepage, etc.)   21
        2B.4 Adopting CST elements for problems of flow through porous
            media   23
        2B.5 Isoparametric Elements   25

3   Finite Element Characteristics – Direct Stiffness Analysis   26
    3.1 Formulation of Finite Element Characteristics using
        Direct Stiffness Analysis   26

3.2 Element Stiffness Characteristics – Plane stress and
    Plane Strain Solid Element   26
3.3 Evaluation of Equivalent Nodal Forces using Principle of
    Virtual Displacement   29
3.4 System / Assemblage Equilibrium Equations   31
3.5 Displacement Method – Equivalent to Minimization of
    Potential Energy   32
3.6 Alternate Formulations of FEM   33
    3.6.1 Formulation of FEM as an Extremum Problem   33
    3.6.2 FEM Formulation using Weighted Residuals Method 34
    3.6.3 General Comments   36

**4 FEM for the Analysis of Rods, Trusses, Beams, Frames
and Elastic Solids**   **37**
4.1 Rod Element/ One Dimensional Truss Element   37
    4.1.1 Stiffness Matrix from General Direct Stiffness Analysis (FEM) 39
    4.1.2 System consisting of Several Rod Element –
        System Stiffness matrix   40
    4.1.3 Application of Distributed Loads   42
    4.1.4 Applications to Thermal Strain, Initial Strain /
        Initial Lack of Fitness / Residual Stress   44
    4.1.5 Examples   46
4.2 The Two and Three Dimensional Truss Elements   47
    4.2.1 The Two Dimensional Truss Element   47
    4.2.2 The Three Dimensional Truss Element (Space Truss Element) 49
    4.2.3 Distributed Loads, Thermal Effects, Lack of Fitness 50
    4.2.4 Examples   50
4.3 Beam Elements   51
    4.3.1 Planar Beam Elements   51
    4.3.2 System / Global Stiffness Matrix, Assembly and
        System Equilibrium Equation   54
    4.3.3 Distributed Loads   55
    4.3.4 Beams in Two-plane Bending   57
    4.3.5 Examples   58
4.4 The Frame Element   59
    4.4.1 Three Dimensional Frame Element – Space Frame Element 60
    4.4.2 Buckling of Beams and Other Problems   60
4.5 Elastic Solids   61
    4.5.1 Two Dimensional Elastic Solids – Plane Stress and
        Plane Strain Problems   61
    4.5.2 Element Characteristics   61
    4.5.3 Stress-strain Relations (Constitutive Equations) 63
    4.5.4 Stiffness Matrix of the Plane Elastic Solid Element (k matrix) 64
    4.5.5 Nodal Forces due to Initial Strain   65
    4.5.6 Equivalent Nodal Forces due to Distributed Body Forces or
        Boundary Forces   65

4.5.7 Examples 67
4.6 Summary of FEM Worked Out Examples for Rods, Trusses, Beams
and Solids 68
4.6.1 Examples of Rods 68
4.6.2 Examples of Trusses 91
4.6.3 Examples of Beams 106
4.6.4 Examples of Elastic Solids 112
4.7 Assignment Problems 126
4.7.1 Rods 126
4.7.2 Trusses 129
4.7.3 Beams 132
4.7.4 Elastic Solids 134

5   FEM for Foundation Analysis and Design   136
5.1 Soils, Foundations, and Superstructures 136
5.2 Classification of Foundations 137
5.3 Modelling, Parameters, and Analysis 139
5.4 Analysis of Shallow Foundations and Deep Foundations 140
5.5 Conventional Design and Rational Design 141
5.6 Procedures for the Analysis and Design of Footings 143
5.7 Conventional Design of Footings 143
5.8 Modelling Soil Structure Interaction for Rational Design of
Foundations 143
5.8.1 Soil – Structure Interaction Equations 144
5.8.2 Beams on Elastic Foundations (BEF) 145
5.8.3 Plates on Elastic Foundations (PEF) 146
5.8.4 Soil Reaction $q(x)/q(x,y)$ – Contact Pressure 149
5.9 Elastic Foundations 150
5.9.1 Brief Review of the Foundation Models 150
5.9.2 Winkler's Model 155
5.9.3 Evaluation of Spring Constant k, in Winkler's Soil Model
and Elastic Properties E and v of Soil 155
5.9.4 Evaluation of k_x / C_u using Plate Load Test In-situ 156
5.10 Foundation Analysis using FEM 156
5.10.1 Review of FEM Procedure 157
5.10.2 FEM Analysis of Foundations using BEF Model 160
5.10.3 Beam Elements 160
5.10.4 System / Global Stiffness Matrix, Assembly and System
Equilibrium Equations 163
5.10.5 Incorporating Soil Contact Pressure in Stiffness Characteristics
of Beam Elements 165
5.11 FEM Analysis of Foundations using PEF Model 168
5.11.1 FEM Analysis of PEF using Plate Bending Elements and
Winkler's Soil Model 168
5.11.2 Additional Details of Incorporating Soil Reaction for
PEF Analysis 170

5.11.3 Circular, Ring Shaped and Plates of General Shapes 171
5.11.4 General Comments on FEM 172
5.12 Examples of FEM Analysis for Foundations using BEF Model 172
5.12.1 Example 1 172
5.12.2 Further Illustration / Examples 183
5.12.3 Footing for Columns Carrying Loads and (or) Moments 183
5.12.4 Footing with one end resting on rigid well / curtain well 183
5.12.5 Different Sizes of Foundations 184
5.12.6 Non-homogeneous and Non-uniform Soils 184
5.12.7 Combined footings, footings with more than one load / moment
at different points 185
5.12.8 Distributed Loads 185
5.12.9 Examples of Beams without Foundation Soil 186
5.12.10 Simply Supported Beam (SSB) 187
5.12.11 Element Nodal Forces, Shear Force Diagram (SFD) and
Bending Moment Diagram (BMD) 188
5.13 Assignment Problems 191
Appendix 5A Evaluation of Modulus of Subgrade Reaction, k_x and
Elastic Properties of Soil, E_x, v_s using Plate Load Test 194
5A.1 Evaluation of Modulus of Subgrade Reaction of the Soil, k_s –
Plate Load Test 194
5A.2 Size of Contact Area 198
5A.3 Winkler's Soil Medium with or without Tension 198
5A.4 Sensitivity of Responses on k_s 198
5A.5 Modulus of Subgrade Reaction for different Plate Sizes
and Shapes 198
5A.6 Poisson's Ration v_s of the Soil Medium 200
5A.7 Evaluation of Young's Modulus of Elasticity of Soil , E_x 200
5A.8 Evaluation of k_s for Foundations subjected to Dynamic Loads 201
5A.9 Cyclic Plate Load Test 202
5A.10 Summary 203

6 **FEM for the Analysis of Field Problems – Seepage / Flow Through
Porous Media** **205**
6.1 Introduction 205
6.2 FEM for the Analysis of Seepage / Flow through Porous Media 205
6.3 General Formulation using Extremum Principle 206
6.3.1 Discretization of the Region / Continuum 208
6.3.2 Element Characteristics 208
6.3.3 Assembling System Equations governing Flow in the Region /
Continuum 213
6.3.4 Boundary Conditions 214
6.4 Examples of Seepage / Flow through Porous Media 216
6.5 Assignment Problems 229
Appendix 6A Seepage / Flow through Porous Media 233
6A.1 Introduction 233

6A.2 Flow Equations   233
6A.3 Darcy's Law   239
6A.4 General Equations of Flow   245
6A.5 Steady State Flow   246
6A.6 Two-Dimensional Flow / Laplace's Equation   247
6A.7 Methods of Solution   249

References   **250**

Subject Index   **256**

Author Index   **259**

Preface

The Finite Element Method (FEM) is one of the most widely used and popular application since 1960s for solving complex problems in Engineering, Science, Applied Mathematics and related areas. Since the publication of the first book by Zienkiewicz and Taylor (1967), the literature has grown very fast and vast in terms of large number of books, journals, reports, etc. While there are many books on the subject, there are only few books which present the method with worked out examples for easy understanding.

The present Book is focused on presenting the subject in a simple and easily understandable way with essential theory and large number of manually worked out examples in interdisciplinary areas. The FEM procedure, steps, logic, solutions and alternate formulations are presented in the first 3 chapters of the book. The above procedures are illustrated through FEM analysis of structures such as rods, trusses, beams, frames, and elastic solids with a large number of manually worked out examples. Further, FEM for analysis of few interdisciplinary areas such as Foundation Analysis and Design and Flow through Soil Media / Porous Media are presented to bring out its versatility and vast scopes.

The chapters are compiled to make them as much self-sufficient as possible. The emphasis is to present the physical implementation of FEM procedure for a thorough understanding of the applications. Several examples have been manually worked out step by step for the students to understand the subject easily and practice it for various applications. The parameters needed as inputs for FEM computations have also been discussed to clarify the ambiguities that may exist in their choice.

This book material and worked out examples have evolved from the FEM related courses offered by the Author formerly in IIT Kanpur (1966 – 2004) and subsequently in Universiti Malaysia Sabah, Kota Kinabalu, Malaysia (2004 – 2018). It is earnestly hoped that this book course will help all interested learners in understanding FEM using simple examples which are manually worked out for easy comprehension.

The book contains 6 chapters and a list of references. The chapter wise contents are as follows.

Chapter 1 presents introduction and a brief evolution of FEM and its applications in various areas of research, design and development.

General concepts and versatility of FEM and its applications in several areas of engineering, science and related areas are summarised in Chapter 2. The FEM procedures and steps involved such as discretization, evaluation of element characteristics, assembling system equations and solution, post processing, etc. are discussed in detail.

In Chapter 3, the general formulation for the evaluation of element characteristics is presented with the example of structural elements using direct stiffness analysis. The equivalence of this method as a variational approach for the solution of extremum problem of minimizing potential energy is explained. The vast scope of expanding the extremum principle as a means of solving a wide variety of field problems is brought out. Alternative formulations of FEM such as variational methods using extremum principle, weighted residual methods such as Galerkin's method, etc. are discussed.

FEM for the analysis of simple structures such as rods, trusses, frames, beams, and elastic solids is explained in detail in Chapter 4. The effect of temperature, initial strains, loads and boundary conditions, etc. are illustrated using manually worked out examples. The flexibility, modularity and versatility of FEM are demonstrated through several examples.

Chapter 5 presents the application of FEM to Foundation analysis and design problems in detail. The method is applied as an extension of the direct stiffness analysis using structural elements and adding foundation soil pressure as an applied load on the structural elements. Several examples are worked out for easy understanding of the application of FEM. The information on the required inputs such as modulus of subgrade reaction and spring constant etc. for FEM computations is also discussed in detail.

In Chapter 6, application of FEM to field problems such as seepage / flow through porous media and expanding its scope as a broad variational method of solution using extremum principle are discussed in detail. In particular, the problem of seepage of water through porous media is presented. Several examples illustrating the detailed steps involved in the method are worked out. The details of equations and factors governing flow through porous media are also presented for ready reference.

About the Author

Dr. N. S. V. Kameswara Rao was a former Professor at Indian Institute of Technology, Kanpur, India (1966-2004). Subsequently he joined as a Professor in the Faculty of Engineering, Universiti Malaysia Sabah, Kota Kinabalu, Malaysia (2004-2018). He also worked as a Consulting Design engineer in M/S M.N. Dastur & Co. (1964-1965) for about 2 years before joining IIT Kanpur. During his vast experience in academics and industry he focused on integrating Engineering analysis and, design and development, for academic research and solution of complex practical problems. He developed several inter disciplinary courses in the areas of Computational Methods and Design, Foundation Analysis and Design, Finite Element Method, Vibrations, Mechanics, Numerical Methods and Algorithms besides other courses in Engineering in general and Civil Engineering in particular.

Professor Kameswara Rao has published a large number of research papers in national and international journals of repute. He has guided several Ph.D. and M.Tech / M.Sc. students besides supervising a large number of B. Tech projects. His areas of teaching, research and consultancy include Foundation Design, Computational Mechanics, Vibrations, Finite Element Analysis, Soil-Structure Interaction, Foundation Dynamics and related areas. He was a consultant to a large number of public and private organizations such as ISRO, Defense, Railways, HAL, and KOEL etc.

Prof. Rao is a fellow of several professional organizations and was a member of the editorial board of several research journals. He received numerous awards for his academic contributions. Professor Kameswara Rao also contributed immensely towards a harmonious and purposeful academic administration by holding several senior administrative positions during his active service of more than five decades.

Professor Rao has authored four text books so far including the recent one on Foundation Design by John Wiley & Sons in 2011. He also co-authored chapters in a few books in the above areas of engineering. The present book on Introduction to Finite Element Method is the fifth book being authored by him.

Acknowledgements

I am happy to bring out this book entitled **Introduction to Finite Element Method – with Interdisciplinary Examples** after teaching several FEM related courses in IIT Kanpur, India and Faculty of Engineering, Universiti Malaysia Sabah, Kota Kinabalu, Malaysia. I express my gratitude to all my teachers, colleagues, students and authorities in both these Institutions for their cooperation during my service there. I offer my grateful thanks to my former doctoral student Dr. Ms. Chong Chee Siang, for her extensive and enthusiastic help in all aspects in bringing out this book successfully. We wish her all the best for a very bright future.

I am delighted to express my thanks to my wife Ravi Janaki, and all our family members - our children and their spouses, Sree Sai Rajeswai, Ravi, Siva, Sarada, Krishna and Kalyani and our grandchildren Shreyas, Raaghavi, Aditya, Ananya and Harish for their enthusiastic support during the preparation of this book. I offer my blessings and best wishes to all of them for their affection and help during the preparation of this book and inspiring me to continue to pursue such creative activities.

Finally, I offer my grateful salutations to my parents for their blessings for successful completion of this book.

N.S.V. Kameswara Rao
Hyderabad, Telangana, India
January, 2021

List of Figures

Figure 1.1   Deformation of Unnotched steel beams with 8x8x55mm tested   5
              using Charpy impact test and Virtual Charpy impact test

Figure 1.2   Deflection of Simply supported reinforced concrete beam with   6
              200mmx200mm cross-section and 2m span due to
              impaction of 0.60m height tested using Hammer Drop test
              and Virtual Hammer Drop test

Figure 2.1   A typical structural assemblage using interconnected finite   8
              elements

Figure 2B.1   Pascal Triangle   19

Figure 2B.2   Constant Strain Triangular Element (CST Element)   21

Figure 2B.3   A plane element of a continuum   24

Figure 3.1   Finite element discretization of a plane stress solid   27

Figure 4.1.1   37

Figure 4.1.2 Axially loaded structure with two elements   40

Figure 4.1.3 Free-body diagram of the two elements   40

Figure 4.1.4 System with applied forces at the nodes   40

Figure 4.1.5 Distributed loads on the element   42

Figure 4.1.6 Effect of Initial Stress   46

Figure 4.2.1 2D- truss element   47

Figure 4.2.2   49

Figure 4.3.1   51

Figure 4.3.2 Positive Shear Force and Bending Moments as per Bending   51
              Theory

Figure 4.3.3 Beam with General Loadings   54

Figure 4.3.4   Distributed loads on the beam   56

Figure 4.3.5   Equivalent nodal loads due to uniformly distributed load   57
                (u.d.l)

Figure 4.5.1   Plane Elastic Triangular Element   61

Figure 4.6.1   68

Figure 4.6.2   71

Figure 4.6.3   74

Figure 4.6.4   77

Figure 4.6.5   81

Figure 4.6.6   85

Figure 4.6.7   88

Figure 4.6.8   91

Figure 4.6.9   94

Figure 4.6.10   97

Figure 4.6.11   100

Figure 4.6.12   103

Figure 4.6.13   106

Figure 4.6.14   108

Figure 4.6.15   110

Figure 4.6.16   112

Figure 4.6.17   115

Figure 4.6.18   118

Figure 4.6.19   120

Figure 4.6.20   123

Figure 4.7.1   126

Figure 4.7.2   126

Figure 4.7.3   127

Figure 4.7.4   127

Figure 4.7.5   128

Figure 4.7.6   129

Figure 4.7.7   129

Figure 4.7.8   130

Figure 4.7.9   130

Figure 4.7.10   131

Figure 4.7.11   131

Figure 4.7.12   131

Figure 4.7.13   132

Figure 4.7.14   134

Figure 4.7.15   134

Figure 4.7.16   135

Figure 4.7.17   135

Figure 4.7.18   135

Figure 5.1   Building with spread foundations   136

Figure 5.2   Superstructure with pile foundations   136

Figure 5.3   Shallow and deep foundations   138

Figure 5.4   Common types of shallow foundations   138

Figure 5.5   Soil contact pressure in conventional design   141

Figure 5.6   Typical distribution of immediate settlement and contact pressure in soils   142

Figure 5.7   Beam on an elastic foundation   145

Figure 5.8   Plate on an elastic foundation   145

Figure 5.9   Convention sketch for bending theory of beams   146

Figure 5.10   Convention sketch for plate bending   148

Figure 5.11   Convention sketch for circular and annular plates in polar coordinates   148

Figure 5.12   Load on Winkler's foundation   151

Figure 5.13   Deformation of actual foundation   151

Figure 5.14   Convention sketch showing various foundation models   152

Figure 5.15   Pasternak's modified foundation model   153

Figure 5.16   161

Figure 5.17   Positive Shear Force and Bending Moments as per Bending Theory   161

Figure 5.12.1   Spread Footing   173

Figure 5.12.2   174

Figure 5.12.3   180

Figure 5.12.4   182

Figure 5.12.5   Footings subjected to Loads and Moments   183

Figure 5.12.6   Footing with one end simply supported   184

Figure 5.12.7   Footing discretized with different element length   184

Figure 5.12.8   Foundations on soils with Non uniform properties   185

Figure 5.12.9   Combined footing   185

Figure 5.12.10   Footings subjected to Distributed loads   186

Figure 5.12.11   Simply supported beam   187

Figure 5.12.12   Nodal forces for Element 1   188

Figure 5.12.13   Nodal forces for Element 2   189

Figure 5.12.14   SFD and BMD of beam   189

Figure 5.13.1   Convention sketch of spread / individual footing with   191
                  concentrated loads and moments

Figure 5.13.2   Sketch of a combined footing   193

Figure 5A.1   Plate load test setup   194
Figure 5A.2   The q-w curve obtained from plate load test   195

Figure 5A.3   Variation of subgrade reaction with plate diameter   196

Figure 5A.4   Empirical relationship between k_x and CBR value   197

Figure 5A.5   Bearing pressure – settlement curve for cyclic plate load test 202

Figure 5A.6   Determination of $k_x(C_u)$ from cyclic plate load test data 203

Figure 6.1   Discretization of a two-dimensional region into triangular   208
                  elements

Figure 6.2   Plane triangular element of the continuum   209

Figure 6.4.1a   216

Figure 6.4.1b   216

Figure 6.4.1c   218

Figure 6.4.2a   219

Figure 6.4.2b   219

Figure 6.4.3   224

Figure 6.4.4   Permeameter   226

Figure 6.5.1   229

Figure 6.5.2   230

Figure 6.5.3   230

Figure 6.5.4   231

Figure 6.5.5   232

Figure 6A.1   Element of soil volume   235

Figure 6A.2   Earth dam with boundary conditions (Numbers indicate   249
                boundary conditions as mentioned in the above paragraphs)

List of Symbols

Δ	Area of the Constant Strain Triangular (CST) element ijm
α	The coefficient of thermal expansion of the material of the element
μ	The shear modulus of the foundation material / the shear layer
β	The soil parameter, between 0.5 and 2.5 (0.5 for clayey soils and 1 to 2.5 for sandy soil)
ϕ	The fluid potential or the total head
φ	Total head at that point
β	The compressibility of the fluid
μ	The viscosity of the pore fluid
$\psi(z)$	The assumed distribution function of vertical displacement with depth
∇^2	The Laplace operator
∇^4	The bi-harmonic operator
θ_i, θ_j	The slopes at two nodes respectively
$\delta_i, \delta_j, \delta_m$	The displacement vectors of the nodes
θ_{zi}, θ_{zj}	Torsional ratations at i and j
α_1 to α_6	The six constants represent the displacements in two linear polynomials
γ_{go}	The unit weight of the gas at atmosphere pressure p_a and the appropriate temperature
ΔL	The extension of the element
ε_o	Initial strain
σ_o	The initial stresses
γ_w	The unit weight of water at the center
$\theta_x, \theta_y, \theta_z$	The angles the local coordinate axis of element makes with global axes x, y, z axes respectively
$[\varepsilon]^e$	The strains in the element
$[A]$	A square matrix
$[A]^{-1}$	A displacement transformation matrix
$[B]^e$	The strain matrix of the element
$[D]$	The usual isotropic stress-strain matrix relationship
$[D]$	The flexural rigidity of the prismatic beams or plate

$[D]$	Matrix involving element material properties
$[H_i]$	The i row of the global permeability matrix
$[H]$	The global permeability matrix
$[K]$	The total assemblage nodal stiffness matrix
$[k]$	Element Stiffness Matrix in Global Coordinates
$[K]^a$	The element stiffness matrix of the element (a)
$[k]^e$	The element stiffness matrix
$[N]$	Functions of components of position, can be obtained from the assumed distribution functions
$[R]$	The matrix of coefficients of permeability of porous medium / region
$[S]$	Stress matrix
$[S]^a$	Element stress matrix of the element (a)
$[T]$	The global transformation matrix (or the direction cosine matrix) of the local coordinates with references to the global coordinates
$\{\delta'\}$	Nodal displacement matrix in local coordinates
$\{\phi\}$	The unknown function / variables describing the phenomenon being analyzed
$\{\delta\}$	Nodal displacement matrix in Global coordinates
$\{\phi\}$	The total head / potential matrix which lists the values of the total head at all the nodes of the system
$\{\delta\}^a$	Nodal displacement at the nodes of element (a)
$\{\delta\}^e$	The listing of nodal displacements of the particular element to ensure unique displacement within the element
$\{\delta\}^e$	Element displacement matrix
$\{F'\}$	Nodal force matrix in local coordinates
$\{F\}$	Nodal force in Global coordinates
$\{F\}^a$	Forces at the nodes of element (a)
$\{F\}^a_{\varepsilon_o}$	The nodal forces required to balance any initial strains acting on the element (a)
$\{F\}^a_{other}$	The forces due to any other internal causes that the element (a) may be subjected to
$\{F\}^a_p$	The nodal forces required to balance any distributed loads acting on the element (a)
$\{F\}^e_{\sigma_o}$	Equivalent nodal forces due to initial stresses in the element
$\{F\}^e_s$	Equivalent nodal forces due to distributed surface forces
$\{P\}$	The concentrated forces applied intermittently, or the equivalent

	external nodal forces
$\{q\}$	The distributed forces on the boundary surface
$\{R\}$	The externally applied nodal forces
$\{R\}^a$	Equivalent nodal forces due to externally applied loads on the element (a)
$\{R\}_{system}$	Total nodal force matrix of externally applied forces on the system
$\{\sigma\}$	Stresses at any specified point or points of the element
A	The area of cross-section of the element
a, b	Dimensions of the rectangular plate element along x and y axes
a_1 to a_6	The arbitrary constants to be solved in terms of the nodal coordinates
B	Width of the foundation (least dimension)
C_1	The values of dimensionless relative stiffness constant
c_f, c_r	Shape constants depending on the flexibility or rigidity of the test plate used
C_u	Coefficient of elastic uniform compression / modulus of subgrade reaction
d	Diameter of soil grain
D_f	The depth of foundation below ground level (GL)
E	Young's modulus of elasticity of the material
f	The stress at any point (depth) of the beam
$f(x, y)$	The displacements of any point within the element
$F_{x, y, z}$	The body forces per unit volume due to gravity in the x-, y-, z-directions
H	The torque at the plate edges
h	Thickness of the plate
h	A total head
h_e	The potential energy per unit mass of the fluid with respect to the datum
h_p	Fluid pressure head
I	Area moment of inertia of beam cross section about the neutral axis of bending
i	Hydraulic gradient at that point
i, j, m	Numbering of nodes
J	Polar moment of inertial of the frame element. Second moment of the area of cross-section about the axial direction
k	The element stiffness matrices in global coordinates

k	The spring constant of the soil or the foundation modulus
k	Coefficient of permeability of the porous medium / soil medium
k'	The element stiffness matrices in local coordinates
k_s	The modulus of subgrade reaction
k_{s1}, k_{s2}	The value of modulus of subgrade reaction corresponding to a bearing plate of area A_1 and A_2
k_x, k_y, k_z	The coefficients of permeabilities of the flow region along x, y and z directions
L	The matrix of direction cosines or an orthonormal matrix
L	Length of the element
l, m, n	Components of the global transformation matrix / direction cosine matrix
$L\{\phi\}$	A set of governing equations to be solved over a domain / region / continuum, V
l_x, l_y	Direction cosines of outer normal 'n' with respect to x and y
M	Bending moment at that cross-section of the beam
M_i, M_j	Moments at the two nodes respectively
n	The total number of nodes for the element being considered
n	Outer normal showing direction of flow
n	The porosity of an element of soil of cross-sectional area A and height H
N'_l, N'_j, N'_m	Interpolations functions
N_i to N_j	Functions of position of the nodes
n_s	The number of grains of diameter d_s occurring in the soil
N_x, N_y	The shearing forces
p	The distributed load
p_g	The pressure in the gas measured above atmospheric pressure
q	The contact pressure at the interface between soil and foundation
q	The distributed vertical load applied on the surface of the soil
Q	The externally applied flow / seepage flux i.e., rate of flow generated
q	The rate of discharge of water through a section of gross area of soil
$q(x)$	The soil reaction or contact pressure
Q_x, Q_y	The reduced shearing forces
R	Radius of curvature of the beam due to bending
R_e	the dimensionless Reynolds number
T	The change in temperature in °C

t	Thickness of the plane elastic element in case of plane stress
T_{xi}, T_{xj}	Torsional moments at nodes i and j
U, V, W	Nodal forces in a common coordinate system
u, v, w	The displacement components along the x, y, z coordinate system
v	Poisson's ratio of the material
V	The shear forces at any cross section of the beam
v	The superficial velocity of fluid flow in soil
V_i, V_j	Vertical Forces at the two nodes respectively
v_i, v_j	The vertical displacement at two nodes respectively
v_s	The average seepage velocity
v_x	The superficial velocity at the center (at the water pressure existing at the element center)
W	Weighting function, is any function of the coordinates
w	The vertical deflection of the surface
w_{cp}	The mean settlement of the circular plate
w_{sp}	The mean settlement of the square plate
X, Y	The components of body forces
x, y, z	The global coordinates system
x', y', z'	The local coordinates system
y	The y coordinate of the point measured in the y direction from the neutral axis of the beam

List of Tables

Table 2A.1   Common Shapes of Finite Elements   16

Table 5A.1   Coefficients c_f and c_r 199

Table 5A.2   Average Values of k_s for Different Soils (Corresponding to a   200
              plate area of 10 m^2)

Table 5A.3   Values of k_s and μ_s for different soil categories (assuming $v_s =$ 201
              0.3)

Table 6A.1   Approximate values of permeability of soils   244

List of Abbreviations

FEM	Finite Element Method
AI	Artificial Intelligence
RHS	Right Hand Side
CST	Constant Strain Triangular
Det	Determinant
Grad	Gradients
LHS	Left Hand Side
BMD	Bending Moment Diagram
SFD	Shear Force Diagram
u.d.l	Uniformly Distributed Loads
2D	Two-dimensional
3D	Three-dimensional
FE	Finite Element
C	Compression
T	Tension
RCC	Reinforced Cement Concrete
BEF	Beams on Elastic Foundations
PEF	Plates on Elastic Foundations
BM	Bending Moment
SF	Shear Force
LSD	Limit State Design Method
LF	Load Factor
SSB	Simply Supported Beams
SS	Simply Supported
CBR	California Bearing Ratio
C.G.	Center of Gravity
GL	Ground Level

Chapter 1

Introduction

1.1 Continuous and Discrete Systems

Nature is infinite consisting of systems which are continuous in themselves as well as in union with all others marching along with time. Human beings are a part of it and are fortunate to be blessed with varying intellectual capabilities and still may not be able to grasp the continuous and infinite spirit of nature functioning in total coordination with all its constituents / systems. They also would like to comprehend, explore these systems and develop new ones though with varying degrees of success – by intuition, research, design, and development through their creativity, innovation and curiosity to understand these systems as much as possible.

In view of the limited capabilities of human intellect, it is not always easy to comprehend the apparent complexities of the systems that one may be interested to use / operate / modify / innovate / rehabilitate / reinvent /design / enlarge its applications and scope, mainly due to the unknown behavior of the total system as a single unit. Hence it is quite logical to consider subdividing the system into their smaller and easily understandable basic components whose behavior can be studied and understood which is referred to as **analysis** and then reassemble them to recreate the original system (as closely as possible) which is referred to as **synthesis. Synthesis is an inverse operation of analysis.** In this process one may be able to at least comprehend the behavior of the total system by synthesizing the behavior of all its constituent components which may be referred to as **elements.** This **process** of subdividing the complex **continuous system** into its basic components (which are presumably easy to study and understand using simple models) amounts to converting it into an equivalent **discrete system**. Theoretically a continuous system has infinite components which are difficult to analyze while an equivalent discrete system has finite number of discrete elements by choice and is expected to be relatively easy to analyze. If the process of subdivision is continued up to the limit of infinitesimal elements which are infinite in number, the analysis leads to mathematical equations such as differential equations or equivalents representing the phenomena and in the limit the system regains its original attributes as a continuous system.

1.2 Analysis and Synthesis

Thus, the choice is either to study / analyze the total system as a continuum or to study / analyze it as an equivalent discrete system and then synthesize to understand its behavior. It is obviously very difficult to analyze the whole system as a continuum due to limitations of mathematical / analytical methods available. An obvious approach to overcome the limitations of analytical methods in analyzing the mathematical equations representing the required phenomena in continuous systems is to discretize the mathematical equations using approximate numerical techniques such as finite difference methods, weighted residual methods and other approximate methods. In these approaches, the discretization is mathematical in nature which has limitations in attainable accuracies.

Alternatively, it is relatively easy to analyze the equivalent discrete system using simple and representative models even if the discrete system comprises of very large number of elements in view of the availability of powerful computers to handle the large number of variables necessary to study the discrete systems. In these approaches the approximations are purely physical in nature in the sense that the physical system which is continuous, is represented by an approximate discrete system using large number (as large in number as the accuracy demands) of small elements of proper size, shape and dimensionality. These large numbers of variables arising out of this physical discretization process can be easily analyzed using computers. Then the results can be synthesized to study the behavior of the original system which is represented by an equivalent discrete system. Though the discretized region is approximately equivalent to the original continuum, the elements can be mathematically analyzed as accurately as required. Thus the **'Finite Element Method'** has evolved in the last several decades. The first book on the subject is published probably by Zienkiewicz and Taylor (1967) though the method in its several primitive mathematical forms existed much before that. Since then, the literature swelled in leaps and bounds in terms of large number of Books, Journals, Reports, Software packages, and various other publications. The applications also spread to almost all areas of engineering, sciences, and related disciplines.

As can be seen from above, the Finite Element Method has several advantages compared to numerical methods (Zienkiewicz and Taylor, 1989). Both approaches use discretization for approximate analysis if the continuous system cannot be analyzed exactly using available analytical methods. The first one by directly discretizing the mathematical equation governing the phenomenon using numerical approximation (a mathematical discretization) and the other one is by discretizing the physical system using finite elements and analyze the equivalent system (physically discretized / discrete) which is referred to as Finite Element Method (FEM). However, both the methods need accurate input parameters of the system such as material properties etc. that define the phenomenon to be analyzed.

1.3 Finite Element Method

The finite element method was originally developed in aircraft industry to facilitate a refined (approximate) analysis of complex airframe structures. Though the procedure was developed as a concept of structural analysis, the wider basis of the method makes it applicable to a variety of field problems such as solid mechanics, soil structure interaction, elasticity, structural analysis, heat conduction, fluid flow, electrical networks etc. In general, this method is applicable to almost all problems in engineering, sciences including social sciences, and wide variety of applied areas. It may not be an over statement to state that FEM can be a versatile approach to solve all problems in systems that can be formulated as an extremum problem i.e., minimization of a functional involving the parameters / variables governing phenomena in that system.

The method offers a unified approach for a wide range of problems and standard computational procedures irrespective of the area of application, or whether the phenomenon has any physical meaning or not, or the phenomenon may be purely hypothetical or mathematical or even abstract. Over the years the method has evolved as a standard computational procedure for solving continuum problems posed by mathematically defined statements (which are not solvable by available analytical methods) by converting them into equivalent discrete systems, analyzing them by simple models, synthesizing them back to represent the original equivalent continuum, and evaluate the behavior by interpreting the results (post processing).The above procedures are the same irrespective of the area of application which makes FEM very popular, preferable, easily adaptable, flexible, modular and innovative. More details are discussed in Chapters 2 and 3.

It may be noted that FEM can be formulated for the analysis of systems to achieve the required accuracies by increasingly finer discretization of the region and (or) higher order elements (Chapter 2), while the input parameters should be equally accurate and reliable. The models used should be simple enough so that the physical parameters needed for computations can be accurately and reliably determined using appropriate test procedures. In view of the practical limitations of available testing techniques and costs involved, a proper and judicious balance of the refinement of the method of analysis used and the testing technique adopted to get the input parameters needed for computation is essential to achieve optimum accuracy.

1.4 Reverse Engineering, Simulation and Virtual Testing

1.4.1 Reverse Engineering

FEM is very useful for carrying out reverse engineering studies that may be needed to understand the attributes of systems whose history is not known or documented either fully or partially. There are many reasons for performing reverse engineering in various fields. Reverse engineering has its origins in the analysis of hardware for commercial or military advantage. However, the reverse engineering process, as such, is not concerned with creating a copy or changing the object in any way. It is only an analysis to deduce design features from products with little or no additional knowledge about the procedures involved in their original production and design.

In some cases, the objective of the reverse engineering process may simply be a **redocumentation** of the concerned systems. Even when the reverse-engineered product is that of a competitor, the goal may not be to copy it but to perform exploratory analysis. Reverse engineering may also be used to create interoperable objects / products. FEM may also be effectively used to analyze objects to assess their soundness and serviceability, modify and innovate, rehabilitate them if they are distressed, assess their behavior in different operating conditions and environments, quality control and a variety of other studies that may be needed during their service life.

1.4.2 Simulation

FEM is one of the most preferred and efficient methods for simulation studies of systems. Simulation is an approximate imitation of the operation of a process or system that represents its operation over time. Simulation is also used with scientific modelling of natural systems or human systems to gain insight into their functioning, as in economics. If simulation is done for an existing and functional system, it can also be a reverse engineering study. Prototype testing or nondestructive testing of an existing or a new system can be carried out using simulation and is very cost effective besides giving exhaustive data under different testing environments as the tests can be repeated without much effort and cost. It is very helpful for carrying out innovative designs and assessing their behavior. Simulation may even be used for validation of input parameters, quality control, damage control and rehabilitation, modification / improvement of a product as well as futuristic design of new products.

1.4.3 Virtual Testing

FEM is best suited for **Virtual testing** which is a process of testing a virtual product using virtual testing procedure for evaluating various characteristics, such as drop resistance, failure behavior or deformations, or standard laboratory or field testing. Numerical models including FEM are used to verify the performance of the product. Most of the **tests** performed on a physical product can be performed by using a **virtual** product. Even laboratory tests can be carried out using virtual testing techniques by creating virtual environment with FEM. Such methods are increasingly becoming popular in educational institutions, laboratory testing facilities and manufacturing processes etc.

Some examples of virtual testing carried out by Chong Chee Siang (2013, 2015, 2018) with Charpy impact test and Hammer drop test machines in Impact engineering studies are presented as Figures 1.1 and 1.2 for illustration.

1.4.4 Other Applications

FEM is one of the most active areas of research, design and development because of its well-developed standard procedures for analyzing discrete systems as an accurate means to predict the behavior of the corresponding continuous system by analysis and synthesis. New application areas are being added to the already existing applications by appropriate modelling, innovation, AI tools, pre- and post-processing techniques including visualization and animation etc. There is a vast scope for advanced research in FEM in terms of new applications, improved computational techniques, convergence, stability, innovative design of products, AI applications, alternative formulations etc. The future is full of possibilities and expectations besides being open ended.

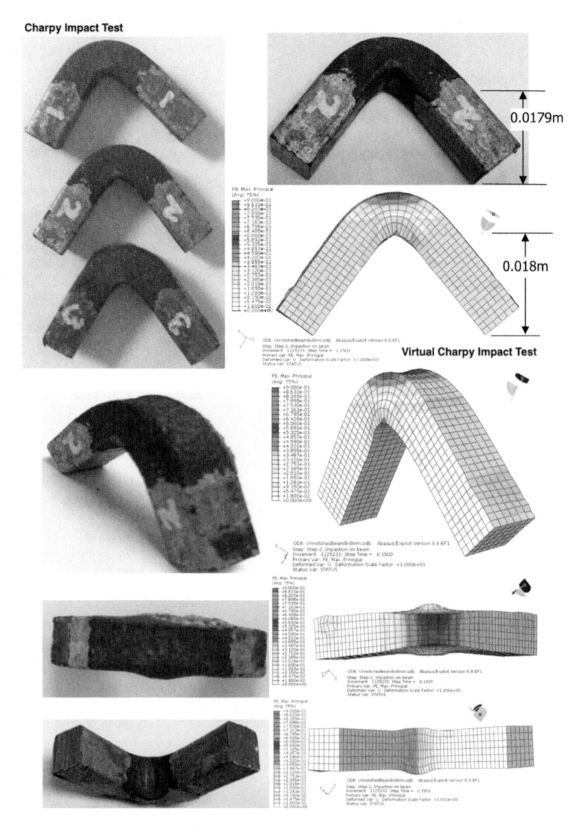

Figure 1.1 Deformation of Unnotched steel beams with 8x8x55mm tested using Charpy impact test and Virtual Charpy impact test

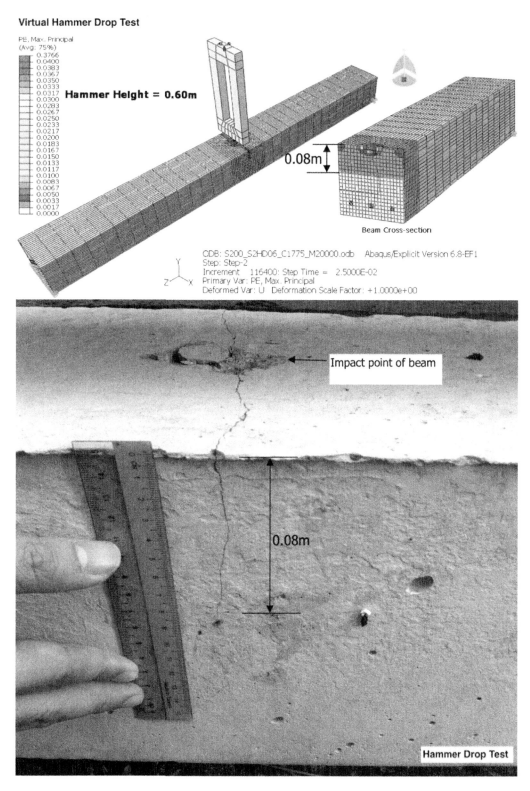

Figure 1.2 Deflection of Simply supported reinforced concrete beam with 200mmx200mm cross-section and 2m span due to impaction of 0.60m height tested using Hammer Drop test and Virtual Hammer Drop test

Chapter 2 Finite Element Method- General Concepts

2.1   General Philosophy

The finite element method was originally developed in aircraft industry to facilitate a refined (approximate) analysis of complex airframe structures. Though the procedure was developed as a concept of structural analysis, the wider basis of the method makes it applicable to a variety of field problems such as soil structure interaction, elasticity, structural analysis, heat conduction, fluid flow, etc. In general, this method is applicable to almost all problems where a variational formulation of the physical phenomenon is feasible. The important characteristics of the finite element procedure are (i) that the method is a general one based on an approximate solution of an extremum problem, (which makes it applicable to many problems) and (ii) unlike the Ritz process, physical quantities which have an obvious meaning are chosen as the parameters (Zienkiewicz, 1971; Zienkiewicz and Taylor, 1989).

Because of its origin and development as a tool for structural analysis, let us first examine the basic philosophy as applied to problems in structural mechanics. The method in its popular form is essentially a generalization of standard procedure in structural analysis known as direct stiffness analysis. The basic concept involved is that every structure may be considered to be an assemblage of finite number of individual structural components or elements. In many engineering problems, analysis of stress and strain in elastic continua is required. In all such problems the number of interconnections between any finite element isolated by some imaginary boundaries and the neighboring elements is infinite. However to make the analysis feasible by this method, the continuum is idealized as an assemblage of one, two or three dimensional elements of proper shape and size, with finite number of inter connections. It is to be noted that this approximation of discretization is purely of a physical nature and that there need be no approximation in the mathematical analysis of the substitute system. This feature distinguishes the finite element technique from finite difference methods in which the exact equations of the actual physical system are solved by approximate mathematical procedures. Another important attribute of the finite element method is its capacity for treating arbitrary material properties such as non-homogeneity, anisotropy, nonlinearity etc., as all of the material properties of the original system can be retained in the discrete elements used in the substitute system. Before we actually go into the finite element procedure let us briefly recapitulate some of the structural analysis concepts involved.

Let Figure 2.1 represent a two dimensional structure assembled from individual components and interconnected at the nodes designated (1) to (n). The joints at the nodes are pinned. Considering the element (a) in Figure 2.1, knowing its characteristics, the forces at the associated nodes 1, 2 and 3 can be uniquely determined using the displacement of these nodes, the distributed load 'p' and initial strain 'ε_0' if any. The initial strain may be due to temperature, shrinkage or simply an initial lack of fit and any other causes. Let the forces and displacements be defined by components (U, V and u, v) in a common coordinate system.

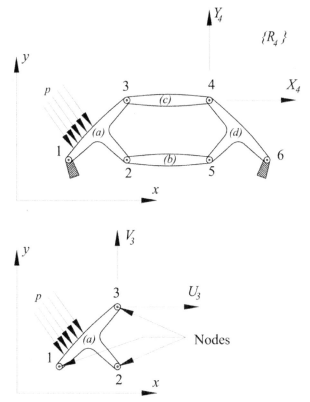

Figure 2.1 A typical structural assemblage using interconnected finite elements

The forces at the nodes of element (a) can be written in matrix form as

$$\{F\}^a = \begin{Bmatrix} F_1 \\ F_2 \\ F_3 \end{Bmatrix}^a = \begin{Bmatrix} U_1 \\ V_1 \\ U_2 \\ V_2 \\ U_3 \\ V_3 \end{Bmatrix}^a \tag{2.1}$$

and the corresponding nodal displacements as

$$\{\delta\}^a = \begin{Bmatrix} \delta_1 \\ \delta_2 \\ \delta_3 \end{Bmatrix}^a = \begin{Bmatrix} u_1 \\ v_1 \\ u_2 \\ v_2 \\ u_3 \\ v_3 \end{Bmatrix}^a \tag{2.2}$$

Assuming elastic behavior, the characteristic relationship between the forces (internal) and the displacements of the element will be of the form

8

$$\{F\}^{a} = [K]^{a}\{\delta\}^{a} + \{F\}^{a}_{p} + \{F\}^{a}_{\varepsilon_{0}} + \{F\}^{a}_{others} \tag{2.3a}$$

To ensure equilibrium of the element a, these equivalent internal nodal forces should be equated to the equivalent nodal forces due to externally applied loads on the element, $\{R\}^{a}$ i.e.

$$\{R\}^{a} = \{F\}^{a} = [K]^{a}\{\delta\}^{a} + \{F\}^{a}_{p} + \{F\}^{a}_{\varepsilon_{0}} + \{F\}^{a}_{others} \tag{2.3b}$$

where $\{R\}^{a}$ = equivalent nodal forces due to externally applied forces on the element.

The first term on the RHS (right hand side) of Eq. (2.7a) i.e. $[K]^{a}\{\delta\}^{a}$ represents the forces due to displacements at the nodes in which $[K]^{a}$ is referred to as element stiffness matrix and $\{F\}^{a}_{p}$ represents the nodal forces required to balance any distributed loads acting on the element and $\{F\}^{a}_{\varepsilon_{0}}$ represent the nodal forces required to balance any initial strains. The term $\{F\}^{a}_{others}$ is included to include forces due to any other internal causes that the element may be subjected to. In subsequent expressions this will not be included with the understanding that forces due to any other causes are implicitly included in $\{F\}^{a}_{p}$, if present.

Similarly stresses, $\{\sigma\}$ at any specified point or points of the element can be written in terms of nodal displacements as

$$\{\sigma\}^{a} = [S]^{a}\{\delta\}^{a} + \{\sigma\}^{a}_{p} + \{\sigma\}^{a}_{\varepsilon_{0}} + \{\sigma\}^{a}_{others} \tag{2.4}$$

The matrix $[K]^{a}$ is known as the element stiffness matrix and $[S]^{a}$ as element stress matrix. To get the complete solution of the structural, assembly, just as the one shown in Figure 2.1, the two conditions to be satisfied are

  (a) Displacement compatibility
  (b) Equilibrium

By listing the nodal displacements for all the elements of the structural assembly, the first condition is taken care of. As can be seen, the overall equilibrium has already been satisfied within each element as expressed by Eqns. (2.3) and (2.4). Hence by writing down the equilibrium conditions at the nodes, the resulting equations will contain the displacements as unknowns. If once these are solved, the rest of the analysis for forces, stresses, etc. can be calculated using the element characteristics. Let us now proceed to the study of the finite element procedure.

2.2   Finite Element Procedure

This can be divided into three phases

  (a) Structural idealization
  (b) Evaluation of the element properties
  (c) Structural analysis of element assemblage

The structural idealization is the process of subdivision of the original system into an assemblage of discrete segments of proper sizes and shapes. Judgement is required as results can be valid only to the extent that the behavior of the substitute structure simulates the actual structure. In general better results can be achieved by finer subdivision.

The objective of the second phase is to find the stiffness or the flexibility of the element, which is an important step in the analysis and will be discussed later.

The third phase is a standard structural problem which will be discussed in subsequent sections. The individual element configurations are of no concern. The same techniques are applicable to systems of one, two or three dimensional elements or any combinations of these. The essential problem is to satisfy the three conditions of equilibrium, compatibility and force-deflections relationships. Either of the basic approaches of structural analysis known as "force method" and "displacement method" or "hybrid methods" can be used. However, the displacement method is preferred (Timoshenko and Goodier, 1951) as it is simpler for formulation and computer programming. The fundamental steps involved are

1. Evaluation of the stiffness properties of the individual structural elements, expressed in any convenient local (element) coordinate system.
2. Transformation of the element stiffness matrices from the local coordinates to a form relating to global coordinates system of the complete structural assemblage. This can be done as explained in Sec 2.2.2.
3. Superposition of the individual element stiffnesses contributing to each nodal point to obtain the total assemblage nodal stiffness matrix $[K]$.
4. Formulation of the equilibrium equations of the assemblage / system by equating the externally applied nodal forces $\{R\}$ and the internal forces from the assemblage of elements as

$$\{R\} = [K]\{\delta\} + \{F\}_p + \{F\}_{\varepsilon_0} \qquad (2.5)$$

and solution of unknown displacements.
5. Evaluation of the element deformations from the computed nodal displacements by kinematic relationships and determination of element forces from the element deformations by means of the element stiffness matrices.

2.2.1 Finite Element Deformation Patterns:

In the analysis of framed structures, the evaluation of the element stiffnesses can be done by simple procedure. The stiffness characteristics of two and three dimensional elements cannot be obtained by equivalent methods because of several obvious reasons. To quote the more important one, the results obtained using element stiffness properties defined in this way would differ greatly from the stresses and deflections of the actual continuum. Hence, in order that the finite element idealization may represent the behavior of the continuum closely, the deformation patterns which may develop in the element have to be prescribed. The choice of the proper deformation pattern is the critical step in the evaluation of element stiffness.

10

Another important point to be considered while prescribing the deformation patterns is the compatibility. It is not always easy to ensure displacement compatibility between adjacent elements. The compatibility condition on such common boundary lines may be violated, though within each element it is obviously satisfied due to uniqueness of displacements implied in its functional representation. However for the simple triangular plane stress elements it is easy to prescribe fully compatible deformation patterns.

Also, by concentrating the equivalent forces at the nodes, equilibrium conditions are satisfied in the overall sense only, local violation of equilibrium conditions within each element and on its boundaries will usually arise. But these artificial boundary forces are local, self-equilibrating effects which have little influence on the general behavior of the structure (Zienkiewicz and Taylor, 1989).

To ensure convergence of the solution by this method the displacement function chosen should satisfy the following criteria.

1. The displacement function chosen should be such that it does not permit straining of an element to occur when the nodal displacements are caused by a rigid body displacement.
2. The displacement function has to be of such a form that if nodal displacements are compatible with a constant strain condition such constant strain will in fact be obtained.
3. The displacement function should be so chosen that the strains at the interface between elements are finite though they may be indeterminate.

Some details of the commonly used element shapes are summarized in Appendix 2A.

Guidelines for choosing distribution functions within any element are discussed in Appendix 2B. Also, the procedure is illustrated using a Constant Strain Triangular (CST) element for easy understanding in the Appendix 2B.

2.2.2 Transformation of Co-ordinates

It is generally convenient to compute the characteristics of an individual element in a local coordinate system best suited to the geometry of the element. Then these can be easily transformed into any convenient global coordinate system chosen for the whole systems for obtaining the global characteristics. Then the total assemblage to represent the system can be obtained by superposing the individual element characteristics and assembling them. A different coordinate system may conveniently be used for every element to facilitate the computation. It is easy to transform the coordinates of the displacement and force components of Equations (2.1 to 2.3) to any other global coordinate system as explained below. It is necessary to do so before an assembly of the structure can be made.

Let the local coordinate system in which the element properties have been evaluated be denoted by prime superscript and the global coordinate system necessary for assembly be non-primed. The displacement components can be transformed by a suitable matrix of direction cosines $[L]$ as

$$\{\delta'\}^a = [L]\{\delta\}^a \tag{2.6a}$$

Equating the work done by the forces in both coordinate systems as force components must perform the same amount of work, we have

$$\left(\{F\}^a\right)^T\{\delta\}^a = \left(\{F'\}^a\right)^T\{\delta'\}^a$$

Inserting Equation (2.6a) in the above equation we get

$$\left(\{F\}^a\right)^T\{\delta\}^a = \left(\{F'\}^a\right)^T[L]\{\delta\}^a$$

i.e. $$\{F\}^a = [L]^T\{F'\}^a \tag{2.6b}$$

The stiffnesses which may be available in local coordinates can be transformed to global ones as follows. The force displacement relationship in local coordinates can be written from Equation (2.3) as:

$$\{F'\}^a = [k']^a\{\delta'\}^a \tag{2.6c}$$

Substituting Eqns. (2.6a) and (2.6b) in Eq. (2.6c), we have

$$\{F\}^a = [L]^T[k']^a[L]\{\delta\}^a = [k]^a\{\delta\}^a$$

Hence, from the above equation it can be noted that

$$[k]^a = [L]^T[k']^a[L] \tag{2.6d}$$

where k' and k are element stiffness matrices in local and global coordinates.

Similarly, the reverse transformations from global to local coordinates can also be done as follows. Noting that $[L]^{-1} = [L]^T$ (since direction cosine matrix L is an orthonormal matrix), from Eqns. (2.6a)

$$\{\delta\}^a = [L]^{-1}\{\delta'\}^a = [L]^T\{\delta'\} \tag{2.6e}$$

$$\{F'\}^a = [L]\{F\}^a \tag{2.6f}$$

$$[k']^a = [L][k]^a[L]^T \tag{2.6g}$$

2.2.3 System / Global Stiffness Matrix, Assembly and System Equilibrium Equations

Once the element stiffness matrices of all elements of the system are evaluated and transformed to the global coordinate system, the global stiffness matrix can be assembled by adding the contributions of individual elements connected at any node as explained below. Accordingly, after assembling the system, the system equations can be written by equating the internal equivalent nodal forces to the externally applied forces and moments as given in Equation (2.5). Equivalent nodal forces due to distributed forces, temperature, and initial lack of fitness and any other internal causes of individual element can all be included in the equilibrium equations, which are explained in this section. Thus the system equilibrium equations can be expressed in the standard form as in Equation (2.5) as follows.

$$\{R\} = [K]\{\delta\} + \{F\}_p + \{F\}_{\varepsilon_0} \tag{2.7}$$

This can be written in the expanded form after assembling and superposing the stiffness characteristics, external and equivalent internal nodal forces contributed by each element connected at each node of the total assembly as:

$$\{R\}_{system} = [K]_{system}\{\delta\}_{system} + \{F\}_p + \{F\}_{\varepsilon_o} = [K]_{system}\begin{Bmatrix} v_1 \\ \theta_1 \\ v_2 \\ \theta_2 \\ \vdots \\ v_n \\ \theta_n \end{Bmatrix} + \sum\begin{Bmatrix} F_1 \\ F_2 \\ F_3 \\ \vdots \\ F_n \end{Bmatrix}_p + \sum\begin{Bmatrix} F_1 \\ F_2 \\ F_3 \\ \vdots \\ F_n \end{Bmatrix}_{\varepsilon_o} \tag{2.8a}$$

where

$$\{R\}_{system} = \begin{Bmatrix} R_1 \\ R_2 \\ \vdots \\ R_n \end{Bmatrix} = [K]_{system}\begin{Bmatrix} v_1 \\ \theta_1 \\ v_2 \\ \theta_2 \\ \vdots \\ v_n \\ \theta_n \end{Bmatrix} + \sum\begin{Bmatrix} F_1 \\ F_2 \\ F_3 \\ \vdots \\ F_n \end{Bmatrix}_p + \sum\begin{Bmatrix} F_1 \\ F_2 \\ F_3 \\ \vdots \\ F_n \end{Bmatrix}_{\varepsilon_o} = \sum_{i=1,2...,n}[k]^i\{\delta\}_{system} + \sum\{F_i\}_p + \sum\{F_i\}_{\varepsilon_o}$$

$$\tag{2.8b}$$

$\{R\}_{system} = \sum \{R\}_i$ = total nodal force matrix of externally applied forces on the system.

This is obtained by adding contributions of all the externally applied forces on the elements connected at each node of the system including applied concentrated forces at each of the node.

$$[K]_{system} = \sum_{i=1,2----n} [k]^i = \text{system / global stiffness matrix of the assemblage}$$

and elements of $[K]_{system}$ can be expressed as:

$$[K_{ij}] = \sum[k_{ij}]^a \tag{2.9}$$

with summation being carried out over all the elements of the system which have with i and j as common nodes, and

$$[K_{ii}] = \sum [k_{ii}]^a$$ where the summation is carried out on all the elements connected at any node i of the assemblage; it may further be noted that $K_{ij} = K_{ji}$ in structural analysis problems indicating that the system stiffness matrix $[K]_{system}$ is symmetric.

Further,

$$\{\delta\}_{system} = \begin{Bmatrix} \delta_1 \\ \delta_2 \\ \vdots \\ \delta_n \end{Bmatrix} = \begin{Bmatrix} v_1 \\ \theta_1 \\ v_2 \\ \theta_2 \\ \vdots \\ v_n \\ \theta_n \end{Bmatrix} = \text{nodal displacement matrix of the system / assemblage} \qquad (2.10)$$

$$\{F\}_p = \sum \{F_i\}_p^a = \sum \begin{Bmatrix} F_1 \\ F_2 \\ F_3 \\ \vdots \\ F_n \end{Bmatrix}_p \quad = \text{total nodal force matrix due to internal forces } (p) \text{ in the elements}$$

$$(2.11)$$

$$\{F\}_{\varepsilon_o} = \sum \{F_i\}_{\varepsilon_o}^a = \sum \begin{Bmatrix} F_1 \\ F_2 \\ F_3 \\ \vdots \\ F_n \end{Bmatrix}_{\varepsilon_o} \quad = \text{total nodal force matrix due to internal causes } (\varepsilon_0) \text{ in the elements}$$

$$(2.12)$$

The above equations can be assembled from the stiffness characteristics of individual elements which can be evaluated as explained in Chapter 3.

In particular, if element stiffness matrix, $[k]^a$ is given, equivalent nodal forces due to internal forces, $\{F\}_p^a$ and equivalent nodal forces due to other internal causes in the element, $\{F\}_{\varepsilon_0}^a$ can be calculated from the stiffness analysis expressions as derived in the next Chapter 3. Alternatively, they can also be calculated using the simple lumped method or consistent method (Section 4.3). In practice, the lumped method is preferred wherein equivalent nodal forces are distributed between the nodes based on the proximity of the resultant load to the corresponding node and other common sense approaches wherever possible.

If different types of elements are used for discretization of the system, then the above element and system matrices and submatrices have to be modified / augmented to comply with the rules of matrix algebra. This can be done by inserting zero coefficients at appropriate locations of the matrices and submatrices involved in the analysis.

Next step is, to apply the boundary conditions specified along the boundary (boundaries) of the system by incorporating them in the above system equilibrium equations. Solving these resulting equations, unknown nodal displacements at all the nodes can be obtained. Once the nodal displacements are solved as above, the displacements, stresses and any other required parameters can be evaluated using the element stiffness characteristics as discussed in Chapter 3.

While simplifying the system of equations for applying boundary conditions, it may be a good practice to use Payne and Iron's Technique (Zienkiewicz, 1971) which allows to retain the assembled system of equations as it is and then modifying only the coefficients in the matrix suitably to obtain the specified boundary values as a part of the solution of the assembled system of simultaneous equations without changing the order of the system. According to this, instead of eliminating the equilibrium equation at which a particular variable / displacement is specified (and the corresponding external force component is unknown) and proceeding with the substitution of that variable / displacement in the remaining equations, the diagonal component the matrix $[K]$ at that point concerned is multiplied by a very large number. Simultaneously, the term on the RHS of the (same) equation is replaced by the same large number multiplied by the prescribed variable / displacement value. This has the same effect of replacing the particular equation by one stating that the displacement being specified is incorporated in the total system of equations. Hence there is no need to replace the equation and the total set of equations can be retained for solutions without changing the order of equations.

The rest of the analysis and design procedure is the same as in any other solid mechanics / structural analysis problem.

This makes the analysis very modular since forces due various external and internal causes can be included in the analysis with the stiffness characteristics of the elements and the system remaining the same while analysis can be carried out with total flexibility for including the forces due to various causes affecting the elements and the system.

Appendix 2A

Commonly Used Finite Element Shapes

2A.1 Some details of Commonly Used Finite Elements

Elements of several geometric shapes are commonly used in finite element analysis for specific applications. General-purpose FEM software codes offer a large choice of element library which include line elements, surface elements, solid elements etc. and also special-purpose elements needed for the analysis as per requirement. Few details of elements commonly used in finite element analysis are given in Table 2A.1 (Budynas, 1999).

Table 2A.1 Common Shapes of Finite Elements

Element reference	Name	Shape	Number of nodes	Applications
Line	Truss		2	Bars in tension or compression (pin-ended).
	Beam		2	Bending–beam theory.
	Frame		2	Combined axial, torsional, and bending response (with and without load stiffening).
Surface	4-noded quadri-lateral		4	Problems of Plane stress or strain, axisymmetry, shear panels, thin flat plates in bending.
	8-noded quadri-lateral		8	Plane stress or strain, thin plates or shells in bending.
	3-noded Triangular		3	Problems of Plane stress or strain, axisymmetry, shear panels, thin flat plates in bending. (Quadrilaterals are preferred wherever possible. Triangular elements are also used for transitions of shapes from quadrilaterals).

16

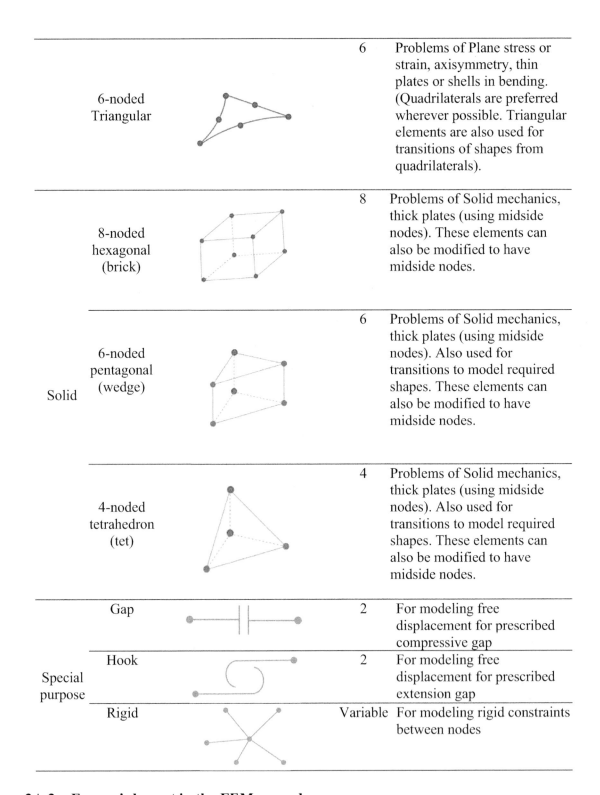

		6	Problems of Plane stress or strain, axisymmetry, thin plates or shells in bending. (Quadrilaterals are preferred wherever possible. Triangular elements are also used for transitions of shapes from quadrilaterals).
	6-noded Triangular		
Solid	8-noded hexagonal (brick)	8	Problems of Solid mechanics, thick plates (using midside nodes). These elements can also be modified to have midside nodes.
	6-noded pentagonal (wedge)	6	Problems of Solid mechanics, thick plates (using midside nodes). Also used for transitions to model required shapes. These elements can also be modified to have midside nodes.
	4-noded tetrahedron (tet)	4	Problems of Solid mechanics, thick plates (using midside nodes). Also used for transitions to model required shapes. These elements can also be modified to have midside nodes.
Special purpose	Gap	2	For modeling free displacement for prescribed compressive gap
	Hook	2	For modeling free displacement for prescribed extension gap
	Rigid	Variable	For modeling rigid constraints between nodes

2A.2   Errors inherent in the FEM procedure:

The finite element method being a numerical technique which discretizes the continuous domain into a discrete system, and carries out intensive computations,   following numerical errors are inherent in the analysis which need to be examined for ensuring the desired accuracy.

1. Computational errors:

These are round-off errors caused due to floating-point calculations, and, the numerical integration and differentiation techniques that are used. Most commercial finite element codes try to reduce these errors and consequently the analyst generally may have to be concerned with errors due to discretization.

2. Discretization errors:

The geometry and the displacement distribution of a real solid / structure / domain / region is not the same as the substitute discretized system using finite elements of limited regular shapes and sizes for obvious reasons. Using a finite number of elements to model the structure introduces errors in matching geometry and the displacement distribution due to the inherent limitations of the elements. The errors caused due to these aspects are called discretization errors.

Appendix 2B

Choice of Displacement Distribution Functions – Example of Constant Strain Triangular (CST) Element

2B.1 Choosing the Polynomials for Displacement / Variable Distribution Functions

Choosing a proper distribution function of the variable within the individual elements of the discretized system / assembly / region / domain is one of the most important steps for an accurate analysis using FEM. These should also be simple enough to carry out algebraic / differential / integral / matrix and other mathematical operations easily for computation / programming and pre- and post-processing steps besides the main analysis. While many options have been explored, use of polynomials is preferred due to their suitability in view of the above requirements.

Further, the completeness and compatibility requirements are important for selecting the order of a polynomial. In addition to the requirement that the pattern should be independent of the orientation of the local coordinate system. This property of the model is known as **geometric isotropy, spatial isotropy, or geometric invariance.** For linear polynomials, the isotropy requirement is usually equivalent to the inclusion of constant strain states of the element. For higher order patterns, it is undesirable to have a preferential coordinate direction, in other words to have displacement shapes which change with a change in local coordinate system. For example, one way to achieve isotropy is to use the Pascal triangle shown in Figure 2B.1, for the choosing variable terms of the two-dimensional, equations (Desai and Abel, 1972).

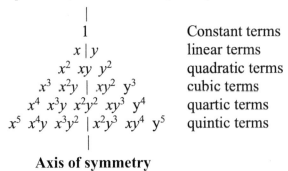

1	Constant terms
$x \mid y$	linear terms
$x^2 \ xy \ y^2$	quadratic terms
$x^3 \ x^2y \mid xy^2 \ y^3$	cubic terms
$x^4 \ x^3y \ x^2y^2 \ xy^3 \ y^4$	quartic terms
$x^5 \ x^4y \ x^3y^2 \mid x^2y^3 \ xy^4 \ y^5$	quintic terms

Axis of symmetry

Figure 2B.1 Pascal Triangle

For the two-dimensional problems, any term from one side of the axis of symmetry of the triangle should not be included without including its counterpart from the other side. For example, to construct a cubic model with eight terms, the following two choices are geometrically isotropic: (1) all the constant, linear and quadratic terms plus the x^3 and y^3 terms; or (2) all the constant, linear, and quadratic terms plus the x^2y and xy^2 terms.

In the following discussion, element degrees of freedom and their selection are discussed. The final consideration in the selection of the displacement polynomial is that the total number of generalized coordinates for an element must be equal to or greater than the number of joint or external degrees of freedom of the element. The usual procedure adopts the same number of generalized coordinates as the degrees of freedom. It is possible to utilize an excess of generalized coordinates to improve the element stiffness matrix, that is, to make it less stiff or more flexible. These excess coordinates are generally associated with internal nodes and improve the approximation of equilibrium within the element. However, they do not improve inter element equilibrium. Therefore, more than a few of such extra coordinates are rarely justifiable; in fact, their inclusion may be detrimental in some cases.

2B.2   Nodal Degrees of Freedom

The nodal displacements, rotations, and/or strains necessary to specify completely the deformation / variable of the finite element are defined as the **degrees of freedom** of the element. The degrees of freedom differ from the generalized coordinates in that each is specifically identified with a single nodal point and represents a displacement (or rotation or strain or other variables) having a clear physical interpretation. We shall distinguish between degrees of freedom occurring at external and internal nodes by referring to them as joint or nodal degrees of freedom and internal degrees of freedom, respectively. In the formulation of individual element properties, however, we need not make this distinction; it is only during the assembly process that the difference becomes important.

The minimum number of degrees of freedom (or generalized coordinates) necessary for a given element is determined by the completeness requirements for convergence, the requirements of geometric isotropy, and the necessity of an adequate representation of the terms in the potential energy functional. Additional degrees of freedom beyond the minimum number may be included for any element by adding secondary external nodes or by specifying as degrees of freedom higher order derivatives of displacements at the primary nodes. The latter approach is preferred because it leads to a more compact numerical formulation in the assembly process. Elements with additional degrees of freedom are called higher order elements. The more such additional degrees of freedom are added, the more will be the flexibility of the element stiffness for an element of given size. However, this improvement in the upper bound to the stiffness will make the formulation of the individual element properties increasingly complex. The nature of this tradeoff may have to be considered while choosing the model.

We may now relate the degrees of freedom and the generalized coordinates by employing the displacement model for example. We can evaluate the displacements at the nodes by substituting the nodal coordinates into the model. For example, using a one dimensional model of the form given by:

$$u = \alpha_1 + \alpha_2 x + \alpha_3 x^2 \ldots \alpha_{n+1} x^n \tag{2B.1}$$

We may express Eq. (2B.1) as:

$$\{u\} = [\phi]\{\alpha\} \tag{2B.2}$$

i.e.,

20

$$\{q\} = \begin{Bmatrix} \{u \;\;(node\;\;1)\} \\ \{u \;\;(node\;\;2)\} \\ ... \\ \{u \;\;(node\;\;n)\} \end{Bmatrix} = \begin{Bmatrix} \{\varphi \;\;(node\;\;1)\} \\ \{\varphi \;\;(node\;\;2)\} \\ ... \\ \{\varphi \;\;(node\;\;n)\} \end{Bmatrix} \{\alpha\} = [A]\{\alpha\} \qquad (2B.3)$$

Here n is the total number of nodes for the element being considered, $\{q\}$ is the vector of nodal displacements, and the notation in parentheses indicates that the dependent variables are assigned their values at the particular node. We may invert equation (2B.3) to give

$$\{\alpha\} = [A]^{-1}\{q\} \qquad (2B.4)$$

where $[A]^{-1}$ is a displacement transformation matrix. Note that $[A]$ is a square matrix; hence, the total number of generalized coordinates equals the total number of joint and internal degrees of freedom. When equation (2B.4) is substituted into equation (2B.2), we can eliminate the generalized coordinates and obtain

$$\{u\} = [\phi][A]^{-1}\{q\} = [N]\{q\} \qquad (2B.5)$$

Equation (2B.5) expresses the displacements $\{u\}$ at any point within the element in terms of the displacements of the nodes $\{q\}$.

One limitation of generalized coordinate displacement models is that it is not always possible to obtain the inverse of matrix $[A]$. The interpolation function representation of displacement models avoids this difficulty (Desai and Abel, 1972).

The above procedure is also explained by applying it to the Constant Strain Triangular (CST) elements discussed in Chapter 3 and Sec. 4.5. However, some details of these CST elements are explained below for more clarity.

2B.3 Example of CST Elements - For Applications to plane problems (plain stress, plane strain, two-dimensional seepage etc.)

Now let us apply this procedure to the problem of plane stress: Consider a typical triangular element 'e' as shown in Figure 2B.2, associated with the nodes i, j, m.

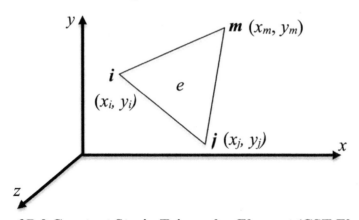

Figure 2B.2 Constant Strain Triangular Element (CST Element)

Then displacement components at any node i can be expressed as:

$$(\delta_i) = \left\{ \begin{array}{c} u_i \\ v_i \end{array} \right\}$$

(2B.6)

where u and v are the displacement components along x and y axes respectively.
Nodal displacement matrix of the element can be written as:

$$\{\delta\}^e = \left\{ \begin{array}{c} \delta_i \\ \delta_j \\ \delta_m \end{array} \right\} = \left\{ \begin{array}{c} u_i \\ v_i \\ u_j \\ v_j \\ u_m \\ v_m \end{array} \right\} \qquad :$$

(2B.7)

The displacements within an element have to be uniquely defined by these six values. Representing displacements by two linear polynomials as follows.

$$u = \alpha_1 + \alpha_2 x + \alpha_3 y$$

(2B.8)

$$v = \alpha_4 + \alpha_5 x + \alpha_6 y$$

The six constants α_1 to α_6 can be solved in terms of the six nodal displacements by substituting the nodal coordinater in the above equation. By solving for these, and substituting back in Eq. (2B.8) we have

$$u = \frac{1}{2\Delta}\left\{ \left(a_i + b_i x + c_i y\right)u_i + \left(a_j + b_j x + c_j y\right)u_j + \left(a_m + b_m x + c_m y\right)u_m \right\}$$

(2B.9)

$$v = \frac{1}{2\Delta}\left\{ \left(a_i + b_i x + c_i y\right)v_i + \left(a_j + b_j x + c_j y\right)v_j + \left(a_m + b_m x + c_m y\right)v_m \right\}$$

(2B.10)

where

$$a_j = x_m y_i - x_i y_m \qquad b_j = y_m - y_i \qquad c_j = x_i - x_m$$

$$a_i = x_j y_m - x_m y_j \qquad b_i = y_j - y_m \qquad c_i = x_m - x_j$$

$$a_m = x_i y_j - x_j y_i \qquad b_m = y_i - y_j \qquad c_m = x_j - x_i$$

(2B.11)

It can be noted that the coefficients in Equation (2B.11) can be written by cyclic permutation of the subscripts i, j and m. And,

$$2\Delta = Det \left| \begin{array}{ccc} 1 & x_i & y_i \\ 1 & x_j & y_j \\ 1 & x_m & y_m \end{array} \right| = 2 \text{ (Area of triangle } ijm) = a_i + b_i x_i + c_i y_i$$

(2B.12)

22

where a_i, b_i, c_i are given by Eq. (2B.11). It may be noted that the above equations assume that the nodes of the CST element $i \rightarrow j \rightarrow m$ are ordered in a counter clockwise manner. *Det* refers to the value of the determinant shown in the above equation.

At this stage we can represent the above relations in the standard form as:

$$(f) = \begin{pmatrix} u \\ v \end{pmatrix} = \begin{pmatrix} IN'_i & IN'_j & IN'_m \end{pmatrix} (\delta)^e \tag{2B.13}$$

where I, is a 2 x 2 Identity matrix and,

$$N'_i = (a_i + b_i x + c_i y)/2\Delta$$
$$N'_j = (a_j + b_j x + c_j y)/2\Delta$$
$$N'_m = (a_m + b_m x + c_m y)/2\Delta \tag{2B.14}$$

The chosen displacement functions ensure displacement compatibility with adjacent elements.

Now the strains at any point can be written as

$$\{\varepsilon\} = \begin{Bmatrix} \varepsilon_x \\ \varepsilon_y \\ \gamma_{xy} \end{Bmatrix} = \begin{Bmatrix} \dfrac{\partial u}{\partial x} \\ \dfrac{\partial v}{\partial y} \\ \dfrac{\partial u}{\partial y} + \dfrac{\partial v}{\partial x} \end{Bmatrix} \tag{2B.15}$$

using Eqns. (2B.9) and (2B.10) the same (at any point in the element) can be written as

$$\{\varepsilon\}^e = \frac{1}{2\Delta} \begin{bmatrix} b_i & 0 & b_j & 0 & b_m & 0 \\ 0 & c_i & 0 & c_j & 0 & c_m \\ c_i & b_i & c_j & b_j & c_m & b_m \end{bmatrix} \{\delta\}^e = [B]^e \{\delta\}^e \tag{2B.16}$$

which defines matrix $[B]^e$, referred to as the strain matrix. It may be noted from Eq. (2B.16) that the strains in the element $\{\varepsilon\}^e$ are constant. Hence the element is called Constant Strain Triangular (CST) element. The CST elements are used for FEM analysis in Chapters 3, 4 and 5 including worked out examples in these chapters.

2B.4 Adopting CST elements for problems of flow through porous media

Figure 2B.3 shows a typical triangular element subjected to fluid flow. It may be noted that the nodes $i \rightarrow j \rightarrow m$ are ordered in a counter clockwise direction to use the element characteristic equations as obtained in Section 2B.3 above. Noting that the fluid potential / total head at any point / node is the same in all directions (unlike different displacements along different directions in solid mechanics problems), the total potential / head distribution function can be simplified from Eq. (2B.2) as:

$$\{\delta\}^e = \{\phi\}^e \qquad (2B.17)$$

i.e. $\qquad \{\delta\}^e = \left\{ \begin{array}{c} \phi_i \\ \phi_j \\ \phi_m \end{array} \right\}$ where i, j, m are the nodes (Fig. 2B.3) $\qquad (2B.18)$

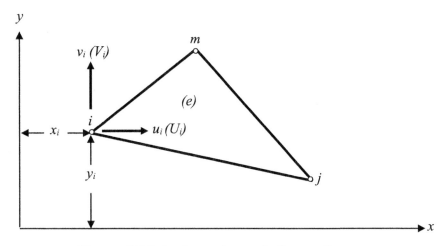

Figure 2B.3 A plane element of a continuum

The distribution function for fluid flow through the triangular element (Figure 2B.3) can be written from Eqns. (2B.9 to 2B.10) as

$$\phi = \frac{1}{2\Delta}\left\{\left(a_i + b_i x + c_i y\right)\phi_i + \left(a_j + b_j x + c_j y\right)\phi_j + \left(a_m + b_m x + c_m y\right)\phi_m\right\} \qquad (2B.19)$$

where

$$\begin{aligned} a_i &= x_j y_m - x_m y_j \\ b_i &= y_j - y_m \\ c_i &= x_j - x_m \end{aligned} \qquad (2B.20)$$

and Δ is the area of the triangular element given by Eq. (2B.12) with other coefficient a_j, b_j, c_j, a_m, b_m, c_m as given in Eq. (2B.11).

Eq. (2B.19) can be expressed as in Eq. (2B.9 and 2B.10) as

$$\phi = \left\{\begin{array}{ccc} N_i & N_j & N_m \end{array}\right\} \left\{ \begin{array}{c} \phi_i \\ \phi_j \\ \phi_m \end{array} \right\} = \left\{\begin{array}{ccc} N_i & N_j & N_m \end{array}\right\}^e \{\phi\}^e \qquad (2B.21)$$

where,

$$\begin{aligned} N_i &= \frac{1}{2\Delta}\left(a_i + b_i x + c_i y\right) \\ N_j &= \frac{1}{2\Delta}\left(a_j + b_j x + c_j y\right) \\ N_m &= \frac{1}{2\Delta}\left(a_m + b_m x + c_m y\right) \end{aligned} \qquad (2B.22)$$

24

Then the gradients of flow can be expressed from Eq. (2B.21 and 2B.22) (similarly as the strain matrix given by Eq. (2B.16) as:

$$
\left\{ \begin{array}{c} Grad \quad \phi \end{array} \right\} = \left\{ \begin{array}{c} \dfrac{\partial \phi}{\partial x} \\[2mm] \dfrac{\partial \phi}{\partial y} \end{array} \right\} = \dfrac{1}{2\Delta} \begin{bmatrix} b_i & b_j & b_m \\ c_i & c_j & c_m \end{bmatrix} \left\{ \begin{array}{c} \phi_i \\ \phi_j \\ \phi_m \end{array} \right\} = [B]\{\phi\}^e
\qquad (2B.23)
$$

Further details of seepage analysis are given in Chapter 6.

2B.5 Isoparametric Elements

Finite Elements that use the same set of shape functions to represent both the element geometry and displacement interpolations (u, v) are called isoparametric elements. The shape functions are defined by natural coordinates, for example, triangle coordinates for triangles and square coordinates for any quadrilateral etc. Prior to the development of isoparametric elements, the element geometry and displacements were not treated equally (this was called superparametric representation). In isoparametric representation, the same coordinates define both the geometry and shape functions, which also represent the displacement interpolations. Consequently the displacement equations (u, v) would become more refined as the order of the element increases, but the geometry would remain the same (simple geometries with straight sides).

The isoparametric elements have the advantages of the ability to map more complex shapes and have compatible geometries. Also, there is no need to distinguish between straight and curved boundaries of isoparametric elements. The disadvantages of isoparametric elements are the possibility of poor (overstiff) performance of low order isoparametric elements and the limited use of isoparametric elements for solving plate bending and shell problems. The completeness condition for plate bending and shell problems (which have a higher variational index, greater than one) is not satisfied using isoparametric representation. They are computationally and mathematically more intensive. The details are given in Zienkiewcz (1972), Zienkiewcz and Taylor (1989), Desai and Abel (1972) and several other books.

Chapter 3

Finite Element Characteristics –

Direct Stiffness Analysis

3.1   Formulation of Finite Element Characteristics using Direct Stiffness Analysis

Evaluating the characteristics of the object as a whole or its equivalent assemblage of finite elements is an essential process for analysing the object. This in turn can be achieved by analyzing the finite elements into which the object is discretized and then analysing the substitute discrete system (representing the continuous system / object) for assessing its performance as a whole. Thus evaluating the characteristics of these finite elements is an essential aspect of the FEM. As explained in Chapter 2, the objects to be analysed may be of general nature i.e. solids or continuua or regions over which a general phenomenon needs to be analysed using FEM. Thus evaluation of the element characteristics constituting the region is a key step in FEM analysis. This is explained in the following sections for elastic solid elements (for example) using direct stiffness analysis while similar procedures can be adopted for other FEM applications in several other areas of research and design.

Stiffness is the measure of rigidity / strength of an object / element i.e. the extent to which it resists deformation in response to an applied force. The complementary concept is flexibility or pliability: the more flexible an object is, the less stiff it is.

The stiffness of any arbitrary element of the assemblage can be evaluated using the assumed deformation patterns as mentioned in the previous Chapter 2. Besides providing boundary compatibility, the number of displacement functions chosen must agree with the number of degrees of nodal displacement freedom of the element as explained in Appendix 2B. The procedure of stiffness analysis will be illustrated with a simple plane stress analysis of a thin slice, but the extension to other problems and areas is similar in procedure.

3.2   Element Stiffness Characteristics – Plane stress and Plane Strain Solid Elements

Consider the region shown in Figure 3.1 subdivided into triangular elements. The following are basic operations.

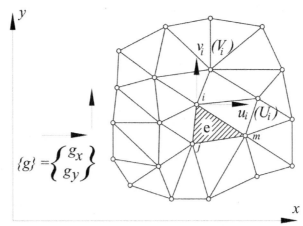

Figure 3.1 Finite element discretization of a plane stress solid

1. Consider a typical element (*e*) of the assembly, associated with nodes *i*, *j*, *m* and straight-line boundaries.

    Let the displacements of any point within the element be defined by $f(x,y)$, as

    $$\{f\} = [N]\{\delta\}^e = [N_i \quad N_j \quad N_m \quad \cdots] \begin{Bmatrix} \delta_i \\ \delta_j \\ \delta_m \\ \vdots \end{Bmatrix}^e \qquad (3.1)$$

    where components of [*N*] are functions of position and $\{\delta\}^e$ represents a listing of nodal displacements for a particular element *e*.

    In the case of plane stress, for example

    $\{f\} = \begin{Bmatrix} u(x,y) \\ v(x,y) \end{Bmatrix}$ which represents horizontal and vertical displacements of a point

    within the element and $\{\delta_i\} = \begin{Bmatrix} u_i \\ v_i \end{Bmatrix}$, the displacements of node '*i*'. The displacement

    functions should satisfy internal compatibility and also maintain boundary compatibility as far as possible, as cited earlier in Appendix 2B.

2. The strains can be evaluated as

    $\{\varepsilon\} = [B]\{\delta\}^e$

    For the plane stress case, for instance

    $$\{\varepsilon\} = \begin{Bmatrix} \varepsilon_{xx} \\ \varepsilon_{yy} \\ \gamma_{xy} \end{Bmatrix} = \begin{Bmatrix} \dfrac{\partial u}{\partial x} \\[2mm] \dfrac{\partial v}{\partial y} \\[2mm] \dfrac{\partial u}{\partial y} + \dfrac{\partial v}{\partial x} \end{Bmatrix} = [B]\{\delta\}^e \qquad (3.2)$$

Knowing N_i, N_j, N_m, matrix $[B]$ can be obtained as explained in Appendix 2B and also in Chapter 4.

3. Assuming linear elastic behavior, the stress-strain relations for the particular case of plane stress, can be expressed as

$$\{\sigma\} = \begin{Bmatrix} \sigma_{xx} \\ \sigma_{yy} \\ \tau_{xy} \end{Bmatrix} = [D]\{\{\varepsilon\} - \{\varepsilon_0\}\} + \{\sigma_0\} \tag{3.3}$$

where $\{\varepsilon_0\}$ and $\{\sigma_0\}$ are the initial strains and stresses if any.

The matrix $[D]$ will be the usual isotropic stress-strain matrix relationship and is written as:

$$[D] = \frac{E}{1-v^2} \begin{Bmatrix} 1 & v & 0 \\ v & 1 & 0 \\ 0 & 0 & \dfrac{1-v}{2} \end{Bmatrix} \tag{3.4}$$

where E and v are the elastic properties of the material i.e., Young's modulus of elasticity and Poisson's ratio respectively.

It may be noted that $[D]$ matrix for the cases of plane strain, 3D-problems, orthotropic materials, axisymmetric problems will be different and are available in standard books / manuals in FEM (Zienkiewicz, 1971). However, for plane strain element $[D]$ can be expressed as (Sec. 4.5):

$$[D] = \frac{E}{(1+v)(1-2v)} \begin{Bmatrix} (1-v) & v & 0 \\ v & (1-v) & 0 \\ 0 & 0 & (1-2v)/2 \end{Bmatrix} \tag{3.5}$$

4. Equivalent nodal forces: the nodal forces which are statically equivalent to the boundary stresses and distributed loads on the element are represented as

$$\{F\}^e = \begin{Bmatrix} F_i \\ F_j \\ F_m \\ \vdots \end{Bmatrix}^e \tag{3.6}$$

The number of components of $\{F_i\}$ should be the same as the components of $\{\delta_i\}$ and be designated in the corresponding directions.

The distributed loads $\{p\}$ are defined as those acting on a unit volume of material within the element with directions corresponding to those of the displacements $[\delta]$ at that point.

For the plane stress case,

$$\{F_i\} = \begin{Bmatrix} U_i \\ V_i \end{Bmatrix} \quad and \quad \{p\} = \begin{Bmatrix} X \\ Y \end{Bmatrix} \tag{3.7}$$

where U, V correspond to the directions of displacements u and v and X and Y are the body force components along x and y directions.

3.3 Evaluation of Equivalent Nodal Forces using Principle of Virtual Displacements

To make the nodal forces statically equivalent to the actual boundary stresses and distributed loads, the simplest procedure is to impose a virtual nodal displacement and to equate the external and internal work done by the various forces and stresses during that displacement.

Taking the virtual displacement matrix at the nodes as $\{\overline{\delta}\}^e$, we have from equations (3.1) and (3.2), the virtual displacement function and virtual strain matrices as:

$$\{\overline{f}\} = [N]\{\overline{\delta}\}^e \quad and \quad \{\overline{\varepsilon}\} = [B]\{\overline{\delta}\}^e \tag{3.8}$$

Then work done by the nodal forces over the corresponding virtual displacements can be written as:

$$\left\{\{\overline{\delta}\}^e\right\}^T \{F\}^e \tag{3.9}$$

The internal work done per unit volume by the stresses and distributed forces is:

$$\{\overline{\varepsilon}\}^T \{\sigma\} \quad - \quad \{\overline{f}\}^T \{p\} \tag{3.10}$$

From equation (3.8), the same can be written as:

$$\left\{\{\overline{\delta}\}^e\right\}^T \left\{[B]^T \{\sigma\} - [N]^T \{p\}\right\} \tag{3.11}$$

Equating external work done to the total internal work obtained by integrating over the volume of the element we have:

$$\left\{\{\overline{\delta}\}^e\right\}^T \{F\} = \left\{\{\overline{\delta}\}^e\right\}^T \left[\int^{Vol} [B]^T \{\sigma\} d(Vol) \quad - \quad \int [N]^T \{p\} d(Vol) \right] \tag{3.12}$$

As this is identically true for any virtual displacement, we have from above

$$\{F\}^e = \int [B]^T \{\sigma\} d(Vol) \quad - \quad \int [N]^T \{p\} d(Vol)$$

$$= \int [B]^T [D][B] d(Vol) \{\delta\}^e - \int [B]^T [D]\{\varepsilon_0\} d(Vol) - \int [N]^T \{p\} d(Vol) \tag{3.13}$$

(from equations (3.1), (3.2) and (3.3)).

This relation can be identified to be the typical characteristic of any structural element as seen earlier in Chapter 2 and Sec. 4.5, which can be rewritten as:

$$\{F\}^e = [K]^e \{\delta\}^e + \{F\}^e_p + \{F\}^e_{\varepsilon_0} + \{F\}^e_{others} \qquad (3.14)$$

The first term on the RHS (right hand side) of Eq. (3.14) i.e. $[K]^e\{\delta\}^e$ represents the forces due to displacements at the nodes in which $[K]^e$ is referred to as the stiffness matrix of the element and $\{F\}^e_p$ represents the nodal forces required to balance any distributed loads acting on the element and $\{F\}^e_{\varepsilon_0}$ represents the nodal forces required to balance any initial strains. The term $\{F\}^e_{others}$ is included to include forces due to any other internal causes that the element may be subjected to. In subsequent expressions this will not be included with the understanding that forces due to any other causes are implicitly included if present in $\{F\}^e_p$.

By comparison of Eq. (3.14) and Eq. (3.13), the stiffness matrix and other parameters can be written as,

$$[K]^e = \int [B]^T [D][B] \, d(Vol) \qquad (3.15)$$

$$\{F\}^e_p = -\int [N]^T \{p\} \, d(Vol) \qquad (3.16)$$

$$\{F\}^e_{\varepsilon_0} = -\int [B]^T [D](\varepsilon_0) \, d(Vol) \qquad (3.17)$$

$$\{F\}^e_{\sigma_0} = \int [B]^T \{\sigma_0\} \, d(Vol) \quad \text{(in case there is an initial stress } \sigma_0\text{)} \qquad (3.18)$$

If initially the system is in equilibrium with residual stresses $\{\sigma_0\}$, then the forces given by Eq. (3.18) will be equal to zero after assembling the elements to form the total system. In case the boundary conditions are specified in terms of displacements, they can be incorporated in the analysis directly while solving the equilibrium equations of the system / assemblage. However, if the boundary conditions are in terms of distributed external forces, say $\{g\}$ per unit area, equivalent nodal forces due to this equal to

$$\{F\}^e_g = -\int [N]^T \{g\} \, d(area) \qquad (3.19)$$

needs to be added on the RHS of Eq. (3.14) for all those elements on the boundary of the system. Alternatively, it can be added as an externally applied force on the element while writing the equilibrium equation of the element with a +ve sign. This is also called the consistent method.

Alternatively, it is also a common practice to evaluate equivalent nodal forces due to distributed / concentrated loads acting on the element / boundary by lumped method instead of the consistent method as expressed in Eq. (3.19). In this method, the boundary loads are distributed among the nodes of the element on a prorate basis depending on the proximity of the resultant load to the nodes concerned, This can be treated either as an internal force with a –ve sign and added to the RHS of the element equilibrium equation or an externally applied load with a +ve sign on the LHS of the element equilibrium equation. These approaches are explained in Chapters 4 and 5. Usually lumped method is preferred due to its simplicity and common sense approach. The element equilibrium equation can be written by equating internal equivalent nodal forces to externally applied forces of the element as:

$$\{R\}^e = \{F\}^e = [K]^e\{\delta\}^e + \{F\}^e_p + \{F\}^e_{\varepsilon_0} + \{F\}^e_{others} \tag{3.20}$$

where $\{R\}^e$ represents the externally applied equivalent nodal forces of the element.

3.4   System / Assemblage Equilibrium Equations

The total solution of the assembly of the finite elements, now follows the standard structural procedures. Once the element characteristics are evaluated, the detailed steps for analyzing the total system have been presented in Chapter 2 and also in Sec.4.5.

In general, external concentrated forces may exist at the nodes in addition to the above internally applied forces on the element. These may be represented as

$$\{R\} = \begin{Bmatrix} R_1 \\ R_2 \\ \vdots \\ R_n \end{Bmatrix} = \sum\{R\}_i \tag{3.21}$$

which have to be added on the LHS of equilibrium equations of the system for ensuring the total system equilibrium at all the nodes. After solving for the unknown nodal displacements from these system equations as explained in Chapter 2, the stresses at any point can be obtained from Eqns. (3.2) and (3.3), i.e.

$$\{\sigma\} = [D][B]\{\delta\}^e - [D]\{\varepsilon_0\} + \{\sigma_0\} \tag{3.22}$$

This equation gives the stresses in the element and the stress matrix and can be identified as

$$\{S\}^e = [D][B] \tag{3.23}$$

The initial stress due to initial strain is:

$$\{\sigma\}^e_{\varepsilon_0} = -[D]\{\varepsilon_0\} \tag{3.24}$$

All the above equations are very general and can be interpreted depending on the problem. For example, if it is of an elastic continuum problem, then the displacements, strains and stresses and forces have the same meaning mentioned in the above sections. However, if it is a beam or a plate or solid of other shapes, the generalized displacements and responses can be interpreted as vertical deflections, slopes, curvatures and generalized forces could be interpreted as bending moments, shear forces etc.

Accordingly, FEM can also be used for analyzing several field problems such as heat conduction, seepage, steady and unsteady fluid flow, current and magnetic field problems etc (Zienkiewicz, 1971, Bowles, 1996, Canale and Chapra, 1989, Desai and Abel, 1972). These are in addition to solid mechanics problems of general shapes, constitutive relationships (linear and nonlinear) external loads, etc. and areas wherever the problems involve minimization of an extremum problem such as minimization of energy (as in the case of elasticity problems).

However, for problems where such a minimization process is not either easily identifiable or difficult to formulate, an alternate formulation of FEM can be carried out using Galerkin's method of weighted residuals wherein the residues are minimized (Zienkiewicz, 1971, Zienkiewicz and Taylor, 1989, Crandall, 1956). This procedure makes FEM very versatile and adaptable to all problems where governing equations are known with proper boundary conditions and (or) initial conditions. All these details are very well explained in standard books on FEM as well as in the large number of general purpose software packages such as SAP, NASTRAN, ANSYS, ABAQUS etc.

Using the general equations discussed in Sections 3.2 to 3.4, we can derive the relevant element characteristics / expressions for a wide variety of structural and solid mechanics problems such as analysis of rods, trusses, the bending of beams and plates, frames, solids etc. which are discussed in detail in Chapter 4. Foundation analysis is essentially an extension of structural / solid mechanics problems and presented in Chapter 5. FEM analysis of field problems i.e. seepage / flow through porous media is discussed in Chapter 6.

3.5   Displacement Method – Equivalent to Minimization of Potential Energy

The displacement method and principle of virtual displacements used in Sec. 3.4 for evaluating stiffness characteristics of the element ensures element equilibrium conditions to be satisfied within the limitations imposed by the assumed displacement distribution functions / patterns (Zienkiewicz and Taylor, 1989). It amounts to restate that if the element equilibrium is ensured then the total potential energy is a minimum for the element for all possible variations of admissible displacements. Thus, the direct stiffness analysis discussed above is an important step in FEM analysis and seeks the solution ensuring minimum potential energy of the element and the system consisting of these artificially discretized physical systems within the limitations of assumed displacement patterns.

However, if true equilibrium of the system / continuum requires an absolute minimum of total potential energy, the FEM being an approximate method, will always provide an approximate potential energy greater than the absolute minimum mentioned above. Thus, a bound on the value of total potential energy of the system is always ensured. Hence FEM can also be perceived as a solution seeking approach using the extremum principles which makes its obvious applicability for a wide range of problems in Engineering, Science and other general areas. Accordingly, application of FEM to seepage and flow through porous media is presented in Chapter 6 as an example.

In the light of the above minimization process implicitly present in the FEM, it is identically the same as the famous Rayleigh-Ritz procedure. The difference is only in the manner in which the displacements are prescribed. In the traditional Rayleigh-Ritz method the displacement patterns are prescribed as expressions valid over the whole region / continuum / system, while in FEM these are prescribed piece / element wise, each nodal parameter influencing only the neighboring elements connected at that node. Also, Rayleigh-Ritz method by its nature is applicable to relatively simple geometric shapes, while FEM can handle complex shapes within the limitations of the chosen element shapes. Accordingly, with FEM, complex and realistic configurations involving regions / phenomena in multi-dimensional framework and in combinations can be assembled and analysed using simple element shapes.

3.6 Alternate Formulations of FEM

3.6.1 Formulation of FEM as an Extremum Problem

As observed in the above section, direct stiffness analysis using displacement approach presented in this chapter can be viewed as identical to the process of minimization of total potential energy of the finite element and the assemblage / total system (may not be the absolute minimum for the total system since FEM is an approximate method). Thus the solution to an extremum problem using Euler's theory of calculus of variations (Crandall, 1956) could be used to extend the FEM concepts presented in Chapters 2 and 3 to a large number of field problems in engineering and science such as seepage and flow through porous media (presented in Chapter 6), heat conduction and thermal sciences problems, fluid mechanics, electrical or magnetic potential studies, torsion and bending of structures, lubrication studies, physical and chemical sciences etc. This approach is a direct extension of the extremum problem implicitly complied in the displacement method presented above in this chapter. One such extension is to analyze seepage and flow through porous media problems - presented in Chapter 6.

Thus, these class of problems in applied physics and mathematics, a variational, extremum principle valid over the whole region / continuum is postulated and the correct solution is the one that minimizes some quantity called a functional, say, χ, which is defined by appropriate integration of the unknown quantities over the whole domain. In most of these phenomena the functional χ generally has a physical meaning such as potential, potential energy etc. In such problems an exact solution of the one (say extremum problem) will automatically be the exact solution of the other (the governing equation of the phenomenon).

There are many other classes of problems in applied sciences and mathematics, where a differential / partial differential / mathematical equation governing the behavior of a variable / variables / phenomenon over a typical infinitesimal region are given. The solution of phenomena may not fall in the category of extremum problems and there may not be any functional χ which can be physically interpreted, like in the extremum problems. The general FEM analysis can also be extended to these classes of problems using the method of weighted residuals (Canale and Chapra, 1989; Crandall, 1956; Zienkiewicz, 1971; Zienkiewicz and Taylor, 1989; Strang and Fix, 2018).

In this approach, there is no need to check whether the problem can be formulated as an extremum problem and hence no need to search for 'the functional' that fits the postulated governing equation of the phenomenon. The method can be extended to a wide range of problems for which the functional may not even exist. Some details of the method are discussed in the following section.

3.6.2 FEM Formulation using Weighted Residuals Method

Consider a set of governing equations to be solved over a domain / region / continuum, V of the form:

$$L\left(\{\phi\}\right) = 0 \tag{3.25}$$

where $\{\phi\}$ is the unknown function / variables describing the phenomenon being analyzed.

Let the boundary conditions be:

$$C\left(\{\phi\}\right) = 0 \tag{3.26}$$

which need to be satisfied on the boundary, S of the domain, V.

As in FEM, let the domain be discretized into an assemblage of small elements of simple and suitable / varying sizes, shapes and dimensions which can be as closely represent the domain as possible. The trial solution satisfying the boundary conditions can be expressed in a similar form as in FEM for any element, e as:

$$\{\phi\} = [N]\{\phi\}^e \tag{3.27}$$

where $[N]$ are functions that can be obtained from the assumed distribution functions as explained in Chapters 2, 3 and 6.

Substituting Eq. (3.27) for $\{\phi\}$ in Eq. (3.25) there will be a residue $\{R\}^e$ for each element since Eq. (3.27) is only a trial solution and not an exact solution of Eq. (3.25). Accordingly,

$$L\left(\{\phi\}\right)^e = L\left([N]\{\phi\}^e\right) = \{R\}^e \neq 0 \tag{3.28}$$

The next best solution is when $\{R\}^e$ is least at all points of the element e i.e., sub domain V^e which is a sub domain of V. An obvious and simple approach to realize this is to at least make the weighted residual of $\{R\}^e$ over the entire element to be zero though $\{R\}^e$ should be zero ideally which is not so. The same can be expressed as

$$\int_{V^e} W\{R\}^e \, dV^e = 0 \tag{3.29}$$

where W is any function of the coordinates, referred to as weighting function. If the number of unknown parameters of $\{\phi\}^e$ are n, then n linearly independent functions, W_i are chosen to satisfy the above Eq. (3.29) which will result in appropriate number of simultaneous equations *as*:

$$\int_{V^e} W\{R\}^e \, dV^e = \int_{V^e} W_i L\left([N]\{\phi\}^e\right) dV^e = 0 \quad , (i = 1, 2, ..., n) \qquad (3.30)$$

The above processes are referred to as weighted residual methods. The above equation will have $\{\phi\}^e$ as unknowns and forcing function at the nodes and boundary conditions if these are elements near the boundary of the domain, V. Adding up the Eqns. (3.30) for all the elements of the domain, V will result in the global / system weighted residual equations which include unknown nodal values of ϕ at all the nodes, and forcing functions at the nodes similar to the system equilibrium equations assembled in direct stiffness analysis discussed in Chapters 2 and 3. These global weighted residual equations will include contributions of weighted residuals from all those elements connected at each node of the domain as well as forcing functions contributed by all elements connected at each node. Then by applying the boundary conditions and solving the resulting simplified equations, we can obtain the unknown values of the functions / variables, ϕ at all the nodes in the domain.

Depending on the choice of the weighting functions, several classical approaches can be adopted for the solution as described below.

1. Collocation Method:

The residual Eq. (3.30) is set equal to zero at n points in the domain V^e. This provides n simultaneous algebraic equations involving the unknown $\{\phi\}^e$. The location of the points is arbitrary but is usually chosen such that V^e is covered more or less uniformly by a simple pattern.

2. Subdomain Method:

The domain V^e is subdivided into n sub domains, usually according to a simple pattern. The integral of the residual Eq. (3.30) over each subdomain is then set equal to zero to provide n simultaneous equations involving $\{\phi\}^e$.

3. Galerkin's Method:

In this method, weighted integrals are taken over the entire domain using each of the known function $[N]$ and are equated to zero i.e.

$$\int_{V^e} N_i R^e \, dV^e = \int_{V^e} N_i L\left([N]\{\phi\}^e\right) dV^e = 0, \quad (i = 1, ..., n) \qquad (3.31)$$

These provide n equations involving $\{\phi\}^e$. The weighing functions here are the same functions $[N]$ used in expressing $\{\phi\}$ as given in Eq. (3.27). In general this method leads to the best approximation (Zienkiewicz, 1971).

4. Method of Least Squares:

In this method, the integral of the square of the residual is minimized with respect to the undetermined parameters to provide n simultaneous equations as follows. If L is a linear operator, then

$$\frac{\partial}{\partial N_i} \int_{V^e} \left\{ R^e \right\}^2 dV^e = 0 \quad , \quad (i = 1, ..., n) \tag{3.32}$$

The resulting n simultaneous equations are used to formulate the element residue equations and subsequently the system residue equations are obtained by assembling all the element characteristics as mentioned in the above paragraphs.

5. Other Methods:

There are many possible approaches to minimize the objective functions such as total potential energy, equations which can be formulated using extremum principle, weighted residue using trial family expressions, minimizing residue using extremum principle such as stationery functional method etc. in literature (Canale and Chapra, 1989; Crandall, 1956; Zienkiewicz, 1971; Zienkiewicz and Taylor, 1989; Strang and Fix, 2018). Also new approaches are evolving continuously using mainly the above principles and are blended with FEM as the FEM procedure is very general as explained in Chapter 2.

3.6.3 General Comments

It should be emphasized that the most important (and most difficult) step in all these methods is the selection of the trial family, Eq. (3.27). The purpose of the above criteria is merely to pick the "best" approximation out of a given family. Good results cannot be obtained if good approximations are not included within the trial family. Theoretically, if enough independent components are included in $\{\varphi\}^e$ (Eq. 3.27), good approximations are possible. However, the principal attraction of these methods lies in the possibility of obtaining good approximations with a limited number of adjustable parameters.

Consideration should be given to symmetry or any other special characteristics of the solution which may be known while choosing the trial family of functions.

When the system of Eq. (3.25) is linear, the equations for the $[N]$ used by any of the weighted residual methods will also be linear.

For a particular trial family like the one in Eq. (3.27) which satisfies all the boundary conditions of a linear-equilibrium problem the four weighted residual criteria will in general produce slightly different approximations. If there is an equivalent extremum problem and the Ritz method is applied to the same trial family, the approximation obtained turns out to be identical with that obtained by the Galerkin's criterion. Thus, Galerkin's method provides the optimum weighted residual criterion in the sense that the approximation so obtained is one which renders ϕ stationary for all variations within the given trial family (Canale and Chapra, 1989; Crandall, 1956).

Several applications using direct stiffness analysis are presented in Chapters 4 and 5. In Chapter 6, applications of flow through porous media are presented using extremum principles.

Chapter 4 FEM for the Analysis of Rods, Trusses, Beams, Frames and Elastic Solids

FEM is one of the most popular and preferred methods used for structural analysis / solid mechanics problems and analysis of structural components. The most commonly used structural components in engineering practice are rods, trusses, beams, frames and elastic solids. Analysis of these components using FEM is discussed in detail in the following sections.

4.1 Rod element / one dimensional truss element:

The axially loaded one dimensional rod element is shown in Figure 4.1.1 with two nodes i (x_i, 0) and j (x_j, 0) with its axis along x direction.

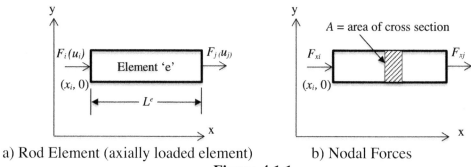

a) Rod Element (axially loaded element)   b) Nodal Forces

Figure 4.1.1

This is also referred to as one dimensional truss element which forms the basis for development of general 2D or 3D (space) truss elements in subsequent sections.

Following the FEM stiffness analysis as explained in Chapter 3, the displacement distribution function $u(x)$ can be expressed as:

$$f = u(x) = a_1 + a_2 x = \begin{bmatrix} N_i & N_j \end{bmatrix} \begin{Bmatrix} u_i \\ u_j \end{Bmatrix} = [N]\{\delta\}^e \qquad \text{4.1 (1)}$$

where a_1 and a_2 are arbitrary constants, N_i and N_j are functions of position of the nodes and $\{\delta\}^e$ is the listing of nodal displacements of the particular element to ensure unique displacement within the element. The order of polynomial can be chosen depending on the degree of freedom of the element as described in Appendix 2B.

Accordingly,

$$f(x = x_i) = u_i = a_1 + a_2 x_i$$
$$f(x = x_j) = u_j = a_1 + a_2 x_j \qquad \text{4.1 (2)}$$

Solving Equations 4.1 (2), the arbitrary constants a_1 and a_2 can be obtained as:

$$a_1 = \frac{u_i x_j - u_j x_i}{x_j - x_i} \qquad\qquad a_2 = \frac{u_j - u_i}{x_j - x_i} \qquad \text{4.1 (3)}$$

Substituting a_1 and a_2 from Equations 4.1 (2) in Equation 4.1 (1),

$$u(x) = \frac{x_j - x}{L^e} u_i + \frac{x - x_i}{L^e} u_j = \begin{bmatrix} N_i & N_j \end{bmatrix} \begin{Bmatrix} u_i \\ u_j \end{Bmatrix} = [N]\{\delta\}^e \qquad \text{4.1 (4)}$$

where $L^e = x_j - x_i = L =$ Length of the element / rod   and

$$N_i = \frac{x_j - x}{L^e} \qquad\qquad N_j = \frac{x - x_i}{L^e} \qquad \text{4.1 (5)}$$

N_i, N_j are referred to as interpolations functions.
The axial strain of the element can now be expressed as:

$$\varepsilon_{xx} = \frac{\partial u}{\partial x} = \frac{1}{L^e}(u_j - u_i) = \frac{1}{L^e}\begin{bmatrix} -1 & 1 \end{bmatrix} \begin{Bmatrix} u_i \\ u_j \end{Bmatrix} = [B]^e\{\delta\}^e \qquad \text{4.1 (6)}$$

where $[B]^e = \dfrac{1}{L^e}\{-1 \quad 1\}$ \qquad\qquad\qquad 4.1 (7)

The axial stress in the elastic element can be expressed as:

$$(\sigma_{xx})^e = E^e\varepsilon_{xx} = \left(\frac{E}{L}\right)^e \begin{bmatrix} -1 & 1 \end{bmatrix} \begin{Bmatrix} u_i \\ u_j \end{Bmatrix} = E^e[B]^e\{\delta\}^e = [S]^e\{\delta\}^e \qquad \text{4.1 (8)}$$

where

$E^e =$ modulus of elasticity of the element

$[S]^e = E^e[B]^e =$ stress matrix

The nodal forces (Figure 4.1.1b) can be expressed as:
(noting σ_x is positive at i while the force is negative.)

$$F_{xi} = -(\sigma_{xx}A)^e \qquad\qquad F_{xj} = (\sigma_{xx}A)^e \qquad \text{4.1 (9)}$$

where A_e is the area of cross section of the rod. These can be expressed in matrix form using Eq. 4.1 (8) as:

$$\begin{Bmatrix} v_i \\ v_j \end{Bmatrix}^e = \begin{Bmatrix} F_{xi} \\ F_{xj} \end{Bmatrix}^e = A^e \begin{bmatrix} -1 \\ 1 \end{bmatrix} \{\sigma_x\}^e = \left(\frac{AE}{L}\right)^e \begin{bmatrix} -1 \\ 1 \end{bmatrix}\begin{bmatrix} -1 & 1 \end{bmatrix} \begin{Bmatrix} u_i \\ u_j \end{Bmatrix}$$

$$= \left(\frac{AE}{L}\right)^e \begin{bmatrix} 1 & -1 \\ -1 & 1 \end{bmatrix} \begin{Bmatrix} u_i \\ u_j \end{Bmatrix}^e = [k]^e\{\delta\}^e \qquad \text{4.1 (10)}$$

Equation 4.1 (10) relates nodal forces to nodal displacements and $[k]^e$ may be noted to be the stiffness matrix of element 'e'. Thus, for the rod element:

$$[k]^e = \left(\frac{AE}{L}\right)_e \begin{bmatrix} 1 & -1 \\ -1 & 1 \end{bmatrix} = k_e \begin{bmatrix} 1 & -1 \\ -1 & 1 \end{bmatrix} \qquad \text{4.1 (11)}$$

where

$$k_e = \left(\frac{AE}{L}\right)_e \qquad \text{4.1 (12)}$$

is referred to as the spring constant of the axial rod element.

4.1.1 Stiffness matrix from general direct stiffness analysis (FEM)

The stiffness matrix can also be obtained using the FEM formulation (Chapter 3) as follows.

From Equation 4.1(8), noting that:

$$\sigma_{xx} = E\{\varepsilon_{xx}\} = [D]\{\varepsilon_{xx}\} \qquad \text{(from Chapter 3)} \qquad\qquad 4.1(13)$$

$[D] = E$ for axially loaded rod element

$$[k]^e = \int [B]^T [D][B] d \text{ vol} \qquad \text{(from Chapter 3)} \qquad\qquad 4.1(14)$$

From Equation 4.1 (7):

$$[B]^e = \frac{1}{L^e} \left\{ \begin{array}{cc} -1 & 1 \end{array} \right\} \qquad\qquad 4.1 (15)$$

Then from Equation 4.1(14)

$$[k]^e = \int \frac{1}{L^e} \left\{ \begin{array}{cc} -1 & 1 \end{array} \right\}^T E \frac{1}{L^e} \left\{ \begin{array}{cc} -1 & 1 \end{array} \right\} A^e dx$$

$$= \frac{1}{L^e} \left\{ \begin{array}{c} -1 \\ 1 \end{array} \right\} E \cdot \frac{1}{L^e} \left\{ \begin{array}{cc} -1 & 1 \end{array} \right\} A^e L^e = \left(\frac{AE}{L} \right)^e \left[\begin{array}{cc} 1 & -1 \\ -1 & 1 \end{array} \right] \qquad\qquad 4.1 (16)$$

<u>Summary:</u>

The stiffness matrix can also be derived using Rayleigh-Ritz method (Budynas, 1999, Zienkiewicz and Taylor, 1989) for obtaining the stiffness characteristics of any finite element in general (and rod element for example as discussed in this section), the steps (Chapter 3) for direct stiffness analysis can be summarized as follows.

1. Define the geometric and, material properties of the element. Identify the nodes, displacements and loads associated with corresponding degree of freedom of the nodes / element.
2. Assume proper displacement distribution (*f*) functions (preferably choosing polynomial of appropriate order as discussed in Appendix 2B) with as many arbitrary constants as there are degrees of freedom, solve for these unknown constants in terms of nodal displacements and substitute them in the displacement distribution functions. Thus, the displacement functions are now in the form {*f*}= [*N*]{δ } as discussed in Chapter 3.
3. Express the strains in terms of nodal displacements from the equations {*f*}= [*N*]{δ }. Identify the [*B*] matrix from the resulting expressions.
4. a) For elements of simple shape and size:
    Obtain the stresses from the stress-strain relations of the material. Evaluate the nodal forces using the stresses thus obtained by multiplying stresses by corresponding geometric properties (areas) for simple elements.
    From the resulting expression relating nodal forces to the nodal displacement of the element, the stiffness matrix [*k*] can be identified as the coefficient matrix multiplying the nodal displacements.

b) For finite elements of general shape and size:

Having obtained the matrices $[D]$ (step 2) and $[B]$ (step 3), the stiffness matrix can be obtained from Chapter 3 as $[k] = \int_{vol} [B]^T [D][B](d\text{volume})$.

4.1.2 System consisting of several rod elements – system stiffness matrix:

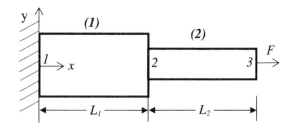

Figure 4.1.2 Axially loaded structure with two elements

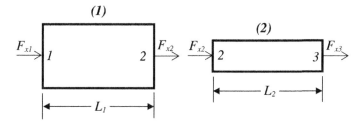

Figure 4.1.3 Free-body diagram of the two elements

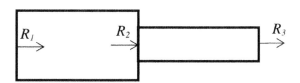

Figure 4.1.4 System with applied forces at the nodes

An axially loaded structure with two rod elements as illustrated in Figure 4.1.2 and the free body diagram of the individual elements are shown in Figures 4.1.3 and 4.1.4. The elements stiffness matrices (from Section 4.1.1) are:

$$\begin{Bmatrix} F_{x1} \\ F_{x2} \end{Bmatrix} = \left(\frac{AE}{L}\right)_1 \begin{pmatrix} 1 & -1 \\ -1 & 1 \end{pmatrix} \begin{Bmatrix} u_1 \\ u_2 \end{Bmatrix} = \begin{pmatrix} k_1 & -k_1 \\ -k_1 & k_1 \end{pmatrix} \begin{Bmatrix} u_1 \\ u_2 \end{Bmatrix}$$
4.1 (17)

where $k_1 = \left(\dfrac{AE}{L}\right)_1$

$$\begin{Bmatrix} F_{x2} \\ F_{x3} \end{Bmatrix} = \left(\frac{AE}{L}\right)_2 \begin{pmatrix} 1 & -1 \\ -1 & 1 \end{pmatrix} \begin{Bmatrix} u_2 \\ u_3 \end{Bmatrix} = \begin{pmatrix} k_1 & -k_1 \\ -k_1 & k_1 \end{pmatrix} \begin{Bmatrix} u_2 \\ u_3 \end{Bmatrix}$$
4.1 (18)

where $k_2 = \left(\dfrac{AE}{L}\right)_2$

If the forces R_1, R_2, R_3 are applied at the nodes 1, 2 and 3 respectively, (Figure 4.1.4), then equilibrium at each node is satisfied when

$$\left(F_{x1}\right)_1 = R_1 \qquad\qquad \left(F_{x2}\right)_1 + \left(F_{x2}\right)_2 = R_2 \qquad\qquad \left(F_{x3}\right)_2 = R_3 \qquad\qquad 4.1\ (19)$$

Using Equations 4.1 (17) and 4.1 (18), Equation 4.1 (19) can be expressed in matrix form as:

$$\left\{ \begin{array}{c} (F_{x1})_1 \\ (F_{x2})_1 + (F_{x2})_2 \\ (F_{x3})_2 \end{array} \right\} = \left[\begin{array}{ccc} k_1 & -k_1 & 0 \\ -k_1 & k_1+k_2 & -k_2 \\ 0 & -k_2 & k_2 \end{array} \right] \left\{ \begin{array}{c} u_1 \\ u_2 \\ u_3 \end{array} \right\} = \left\{ \begin{array}{c} R_1 \\ R_2 \\ R_3 \end{array} \right\} \qquad 4.1\ (20)$$

or

$$[k]_{system} \{\delta\}_{system} = \{R\}_{system} \qquad\qquad 4.1(21)$$

where

$$[k]_{system} = \left[\begin{array}{ccc} k_1 & -k_1 & 0 \\ -k_1 & (k_1+k_2) & -k_2 \\ 0 & -k_2 & k_2 \end{array} \right] = \text{system / global stiffness matrix} \qquad 4.1\ (22a)$$

$$\{\delta\}_{system} = \left\{ \begin{array}{c} u_1 \\ u_2 \\ u_3 \end{array} \right\} = \text{system displacement matrix} \qquad 4.1\ (22b)$$

$$\{R\}_{system} = \left\{ \begin{array}{c} R_1 \\ R_2 \\ R_3 \end{array} \right\} = \text{system applied forces matrix} \qquad 4.1\ (22c)$$

The Equation 4.1 (21) can be noticed to be the system equilibrium equation. It may also be noticed from Equation 4.1 (22a) that the stiffnesses of elements (1) and (2), k_1 and k_2 of elements get added up at the common node 2 (joining node) where elements (1) and (2) are connected to form the total assembly i.e. the total system.

Once the system stiffness matrix ($[k]_{system}$, given Equation 4.1 (22)) is assembled, analysis of the system can be carried using the system equilibrium equation 4.1(20). The system equilibrium equations can be simplified after applying the specified boundary conditions (i.e. u_1, u_2, u_3, or $\{\delta\}_{system}$) and applied nodal forces (i.e. R_1, R_2, R_3, or $\{R\}_{system}$). From these simplified equations for the specific problem, the unknown displacements and (or) forces can be solved using matrix algebra. Once all the parameters of the system are solved, any of the required displacements, strains, stresses and nodal forces in any of the individual elements can be computed using Equations 4.1 (5), 4.1 (6), 4.1 (8) and 4.1 (10) respectively. These can further be used to derive other parameters needed in the analysis. Thus, once $[k]_{system}$ is assembled, the system can be analyzed for different boundary conditions, different nodal forces, variations in material and geometric properties of individual elements and other imposed conditions on the elements using the system equilibrium equation – Equation 4.1 (20). The procedure can be further extended to systems comprising of any number of such rod elements or also can be integrated with other elements with different degrees of freedom and characteristics. These aspects are illustrated in the examples of Section 4.6.

4.1.3 Application of distributed loads:

If the rod is subjected to distributed loads, such as those due to body forces per unit volume, $\{p\} = \{p_x \ p_y \ p_z\}$ or distributed forces on the boundary surface, $\{q\} = \{q_x \ q_y \ q_z\}$ or concentrated forces applied intermittently, these can be converted as equivalent external nodal forces $\{P\} = \{f_{x1} \ f_{x2} \ ...\}$ using either equivalent lumped force approach or consistent matrix approach.

A. Equivalent lumped force approach:

This is a simple approach usually followed in most of the analysis and design problems and is based on the judicious lumping of the forces at the nodes of the element. Since it is a one dimensional rod element with axis along x-axis, only forces along x direction are considered i.e. $p_x = p$ and (or) $q_x = q$ and (or) $P_x = P$.

i. Distributed body forces p /unit volume (Figure 4.1.5a)
In this case, the total body force applied:
$R_b = p$. Volume of the Rod $= p. LA$ 4.1 (23)
where L and A are the length and area of cross-section of the rod.
Now R_b can be lumped equally at both the nodes (if p is constant) as

$$R_{bi} = R_{bj} = \frac{p.LA}{2}$$ 4.1 (24)

In case p is not constant, the total load can be lumped at the nodes judiciously.

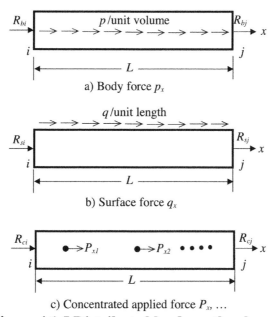

a) Body force p_x

b) Surface force q_x

c) Concentrated applied force $P_x, ...$

Figure 4.1.5 Distributed loads on the element

42

ii. Distributed surface loads, q /unit length (Figure 4.1.5b)
In this case, the total applied load on the surface:

$$R_s = q.L \qquad\qquad\qquad 4.1\ (25)$$

R_s can now be lumped equally (if q is constant) as:

$$R_{si} = R_{sj} = \frac{q.L}{2} \qquad\qquad\qquad 4.1\ (26)$$

If q is not constant, the total load can be lumped judiciously after locating the point of action of the resultant load.

iii. Concentrated intermittent loads (Figure 4.1.5c)
In this case, the total load can be lumped ($R_c = P_1 + P_2 + ...$) at the nodes judiciously after locating the point of applications of the resultant load. However, it is preferable to refine the elements such that the nodes of the refined element coincide with the point of application of these intermittent concentrated loads.

These equivalent externally applied nodal forces can be added at the corresponding nodes of the system i.e. $\{R\}$ matrix of the system (right hand side of Equation 4.1 (22) for example) for further analysis. The rest of the analysis can be carried out further in the usual manner.

B. Consistent force approach:
The equivalent applied nodal forces due to distributed loads and moments (if applicable) can be obtained using the expressions derived in direct stiffness analysis i.e. Chapter 3.

Accordingly

$$R_b = -\{F\}_p^e = \int [N]^T \{p\} dvol \qquad\qquad\qquad 4.1\ (27)$$

and can be expressed as below.

i. Distributed body forces, p /unit volume (Figure 4.1.5a)

$$R_b = \int [N]^T \{p\} dvol = \int_0^L \begin{Bmatrix} N_i(x)pAdx \\ N_j(x)pAdx \end{Bmatrix} \qquad\qquad 4.1\ (28)$$

If p is constant, Equation 4.1 (28) can be simplified (after substituting for $N_i(x)$ and $N_j(x)$ from Equation 4.1 (5) and carrying out the integration along x) as:

$$R_b = \begin{Bmatrix} R_{bi} \\ R_{bj} \end{Bmatrix} = \begin{Bmatrix} pLA/2 \\ pLA/2 \end{Bmatrix} \qquad\qquad\qquad 4.1\ (29)$$

It can be noted that the forces given by Equation 4.1 (29) is identically same as in Equation 4.1 (24), which were obtained using lumped parameter approach. This may not be the same if p is not constant or p is acting on a part of the rod.

ii. Distributed surface forces, q /unit length (Figure 4.1.5b)
In this case, since the q is applied as a force / unit length of the rod, Equation 4.1 (27) needs to be modified as (Zienkiewicz and Taylor, 1989):

$$R_s = \left\{ \begin{array}{c} R_{si} \\ R_{sj} \end{array} \right\} = -\{F\}_q^e = \int [N]^T \{q\} dx = \int_0^L \left\{ \begin{array}{c} N_i(x)\, q\, dx \\ N_j(x)\, q\, dx \end{array} \right\} \qquad \text{4.1 (30a)}$$

$N_i(x), N_j(x)$ are given in Equation 4.1 (5).

If q is constant, Equation 4.1 (30) simplifies as:

$$\left\{ \begin{array}{c} R_{si} \\ R_{sj} \end{array} \right\} = \left\{ \begin{array}{c} qL/2 \\ qL/2 \end{array} \right\} \qquad \text{4.1 (30b)}$$

These can be noted to be the same as those given by lumped parameters approach given in Equation 4.1 (26). If q is not constant and acting on part of the surface, Equation 4.1 (30) has to be integrated along x in the usual manner.

iii. Concentrated intermittent loads (Figure 4.1.5c)
In this case, the integration of Equation 4.1 (30) is to be carried out using the N values at the points of application of the concentrated intermittent loads replacing $q\, dx$ as the concentrated load. However, it is preferable to refine the elements such that the nodes of the refined elements coincide with the points of application of the concentrated loads.
The rest of the procedure can be carried out with the modified system equilibrium equations as summarized in the previous section i.e. lumped force approach. Some examples are illustrated in Sec. 4.6.

4.1.4 Applications to Thermal Strain, Initial Strain / Initial Lack of Fitness / Residual Stress

The modifications needed in the equilibrium equations of the element (Chapter 3) and the system (such as Equation 4.1 (22)) can be obtained from Chapter 3.

i. Effect of Temperature:
The strain $\{\varepsilon_o\}$ due to temperature in the rod element can be expressed as:
$$\varepsilon_0 = \alpha T \qquad \text{4.1 (31)}$$

where α is the coefficient of thermal expansion of the material of the element and T is the change in temperature in °C.

The equivalent nodal forces and stresses due to strain $\{\varepsilon_o\}$ can be expressed using the general equation (obtained using FEM stiffness analysis as explained in Chapter 3) as:
$$\{F\}_{\varepsilon_o}^e = -\int [B]^T [D]\{\varepsilon_o\} d\text{vol} = -\int [B]^T [D]\{\varepsilon_o\} A\, dx \qquad \text{4.1 (32)}$$

where A = area of cross-section of the rod, and
$$\{\sigma\}^e = -[D]\{\varepsilon_o\} \qquad \text{4.1 (33)}$$

where $[B]$ and $[D]$ for the one dimensional rod element are given by Equations 4.1 (16) and (14). Substituting for $[B]$ and $[D]$ in Equations 4.1 (32) and (33), we get

$$\{F\}^e_{\varepsilon_o} = \left\{ \begin{array}{c} F_i \\ F_j \end{array} \right\}_{\varepsilon_o} = -\frac{1}{L}\begin{bmatrix} -1 \\ 1 \end{bmatrix} E\alpha TA.L = \left\{ \begin{array}{c} EA\alpha T \\ -EA\alpha T \end{array} \right\} \qquad \text{4.1 (34)}$$

and the initial stress due to temperature is

$$\{\sigma\}_{\varepsilon_o} = -E\varepsilon_o \qquad \text{4.1 (35)}$$

These additional internal equivalent nodal forces can be added at the corresponding nodes of the element which can then be incorporated in the system equilibrium equations as described in Chapter 3.

Accordingly, for the element subjected to temperature, the equivalent nodal forces can be expressed as:

$$\{F\}^e = \underset{\text{(Stiffness)}}{[k]^e\{\delta\}^e} + \underset{\left(\substack{\text{Temperature/ Initial strain}\\ \text{/Initial lack of fitness}}\right)}{\{F\}^e_{\varepsilon_o}} + \underset{\text{(Body forces)}}{\{F\}^e_p} + \underset{\text{(Surface forces)}}{\{F\}^e_S} + \underset{\text{(Initial stress)}}{\{F\}^e_{\sigma_o}}$$

where: 4.1 (36)

$[k]^e$ = The element stiffness matrix

$\{\delta\}^e$ = Element displacement matrix

$\{F\}^e_{\varepsilon_o}$ = Equivalent nodal forces due to temperature or initial strain / lack of fitness

$\{F\}^e_p$ = Equivalent nodal forces due to distributed body forces

$\{F\}^e_s$ = Equivalent nodal forces due to distributed surface forces

$\{F\}^e_{\sigma_o}$ = Equivalent nodal forces due to initial stresses in the element

However, forces due to initial stress get nullified at the nodes while assembling the system as stresses are equal and opposite at every point and hence do not contribute to the system equilibrium equations (Zienkiewicz and Taylor, 1989).

The total stress at any point in the element can also be expressed in a similar manner (Chapter 3) as:

$$\{\sigma\} = \{\sigma\}_\delta + \{\sigma\}_{\varepsilon_o} + \{\sigma\}_p + \{\sigma\}_s + \{\sigma\}_{\sigma_o}$$

$$\{\sigma\} = \underset{\text{(due to } \delta)}{[D][B]\{\delta\}} + \underset{\left(\substack{\text{Temperature}\\ \text{/Strain}}\right)}{\{\sigma\}_{\varepsilon_o}} + \underset{\text{(Body forces)}}{\{\sigma\}_p} + \underset{\text{(Surface forces)}}{\{\sigma\}_s} + \underset{\text{(Initial stress)}}{\{\sigma\}_{\sigma_o}}$$

4.1 (37)

These values are expressed for the rod element as discussed in the above sections as well as in Chapter 3.

Equation 4.1 (36) can now be used while assembling the system with several elements to formulate the system equilibrium equations as discussed in Section 4.1.2 (and in Chapter 3).

ii. Effect of Initial Strain / lack of Fitness:

If the rod element is subjected to initial strain or lack of fitness, then the internal equivalent nodal forces can be expressed as:

$$\{F\}_{\varepsilon_o}^e = \left\{ \begin{array}{c} F_i \\ F_j \end{array} \right\}_{\varepsilon_o} = -\int [B]^T [D]\{\varepsilon_o\} A\,dx = -\frac{1}{L}\left\{ \begin{array}{c} -1 \\ 1 \end{array} \right\} E\varepsilon_o . A.L = \left\{ \begin{array}{c} EA\varepsilon_o \\ -EA\varepsilon_o \end{array} \right\} \qquad 4.1\ (38)$$

i.e. replace $\alpha\,T$ by ε_0 in Equation 4.1 (34).

The stress in the element is

$$\{\sigma\}_{\varepsilon_o} = -\{E\,\varepsilon_o\} \qquad\qquad 4.1\ (39)$$

Equivalent nodal forces due to initial lack of fitness:

If the element is longer by ΔL, then the initial strain for fitting the rod element in the assembly is given by:

$$\varepsilon_o = \frac{\Delta L}{L} \qquad\qquad 4.1\ (40)$$

The corresponding forces and stresses can be obtained from Equations 4.1 (38) and (39) by replacing ε_o as given in Equation 4.1 (40).

If the element is shorter by ΔL, then the initial strain in the element becomes:

$$\varepsilon_o = -\frac{\Delta L}{L} \qquad\qquad 4.1\ (41)$$

The equivalent nodal forces and stresses can be obtained using Equations 4.1 (38) and (39) by replacing ε_o as given in Equation 4.1 (41).

iii. Initial / Residual Stresses:

If there is an initial / residual stress σ_o in the element, the equivalent nodal forces can be expressed as illustrated in Figure 4.1.6 and Equation 4.1 (42).

Figure 4.1.6 Effect of Initial Stress

$$\{F\}_{\sigma_o}^e = \left\{ \begin{array}{c} F_i \\ F_j \end{array} \right\}_{\sigma_o} = \left\{ \begin{array}{c} -\sigma_o A \\ \sigma_o A \end{array} \right\} \qquad\qquad 4.1\ (42)$$

where A is the area of cross-section of the element.

However, these forces due to internal stresses are self-equilibrating at every point for equilibrium. Hence, they get nullified and do not appear in the system equilibrium equations (Zienkiewicz and Taylor, 1989).

4.1.5 Examples

Examples of rods are presented in Section 4.6.1.

4.2 The two and three dimensional truss elements

The stiffness characteristics of two and three dimensional truss elements can now be obtained using the characteristics of one dimensional bar / rod element (Section 4.1) with appropriate coordinate transformation as discussed in Chapter 2. The same is illustrated for a two dimensional truss element in the following section.

4.2.1 The two dimensional truss element

Consider a typical 2-D truss element, e shown in Figure 4.2.1.

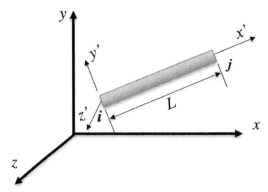

Figure 4.2.1 2D- truss element

x, y, z are global coordinates with the coordinates of i and j as (x_i, y_i) and (x_j, y_j) omitting the z coordinate as the element is in the xy plane (two dimensional element). To use the expressions for the stiffness characteristics of the one dimensional truss / bar / rod element, x', y' can be chosen as the local coordinates by choosing x' axis along the axial direction of the element $i\,j$.

The element characteristics of the above element e in the local coordinate system are [Section 4.1.1 and Chapter 2] related to global coordinate system as follows.

Stiffness matrix: $[k'] = \dfrac{AE}{L}\begin{bmatrix} 1 & -1 \\ -1 & 1 \end{bmatrix}$

Strain- displacement matrix: $[B'] = \dfrac{1}{L}\begin{bmatrix} -1 & 1 \end{bmatrix}$　　　　　　　4.2 (1)

Nodal Displacement Matrix	$: \{\delta'\} = [T]\{\delta\}$
Nodal Force Matrix	$: \{F'\} = [k']\{\delta'\}$
Nodal Force in Global Coordinates	$: \{F\} = [k]\{\delta\} = [T]^{T}\{F'\}$
Nodal Displacement in Global Coordinates	$: \{\delta\} = [T]^{T}\{\delta'\}$
Element Stiffness Matrix in Global Coordinates	$: [k] = [T]^{T}[k'][T]$　　　4.2 (2)

where parameters with prime " $'$ " refer to local coordinates and parameters without prime refer to global coordinates and

L = Length of the element and

$[\,T\,]$ is called the global transformation matrix (or the direction cosine matrix) of the local coordinates with reference to the global coordinates. i.e.

$$[T] = \begin{bmatrix} l & m & 0 & 0 \\ 0 & 0 & l & m \end{bmatrix} \qquad\qquad 4.2\ (3)$$

where

$$l = \cos\phi_x = \frac{x_j - x_i}{L}$$

$$m = \cos\phi_y = \frac{y_j - y_i}{L} \qquad\qquad 4.2\ (4)$$

$$\phi_x = \tan^{-1}\left(\frac{y_i - y_j}{x_j - x_i}\right)$$

$$\phi_y = \tan^{-1}\left(\frac{x_j - x_i}{y_j - y_i}\right) \qquad\qquad 4.2\ (5)$$

Thus the stiffness in global coordinates can be obtained as:

$$[k]_{global} = [T]^T [k'][T] = \begin{bmatrix} l & 0 \\ m & 0 \\ 0 & l \\ 0 & m \end{bmatrix} \frac{AE}{L}\begin{bmatrix} 1 & -1 \\ -1 & 1 \end{bmatrix}\begin{bmatrix} l & m & 0 & 0 \\ 0 & 0 & l & m \end{bmatrix} = \frac{AE}{L}\begin{bmatrix} l^2 & lm & -l^2 & -lm \\ lm & m^2 & -lm & -m^2 \\ -l^2 & -lm & l^2 & lm \\ -lm & -m^2 & lm & m^2 \end{bmatrix} \quad 4.2\ (6)$$

in which $[k']$ is the stiffness in the local coordinates (stiffness of one dimensional bar / rod element) as given in Equations 4.2(1) and

A = area of cross-section of the element
E = modulus of elasticity of the element
L = length of the element
l, m = components of the global transformation matrix / direction cosine matrix (Equation 4.2 (4))

  In truss systems, the node numbers do not follow each other as in the case of one dimensional bar / rod element. Similarly assembling the system / global stiffness matrix knowing the individual element characteristics is slightly more strenuous though the procedure is the same as in the 1D – bar element system discussed in Section 4.1. To simplify the process, we can mark the displacement vector according to the node number and the degree of freedom. Thus, for a system with n nodes, the global / system displacement vector can be written as:

$$\{\delta\}_{system} = \begin{Bmatrix} u_1 \\ v_1 \\ u_2 \\ v_2 \\ \vdots \\ u_n \\ v_n \end{Bmatrix} \qquad\qquad 4.2\ (7)$$

To simplify the assembly process further, the following procedures can be used:

1. For a given element, always choose the lower node number as i and the higher node number as j.

2. After forming the element equilibrium equations in terms of element degrees of freedom, expand them gradually to the system equilibrium equations, adding the element characteristics to the stiffnesses and forces as per the contributions of the connecting elements at these nodes with respect to the system degrees of freedom. This will simplify into a summation process in a systematic manner. The same is illustrated in the examples of trusses given in Section 4.6.

4.2.2 The three dimensional truss element (Space Truss Element)

The characteristics of the 3D - truss element can also be derived in the same manner as the 2D - truss element (Section 4.2.1).

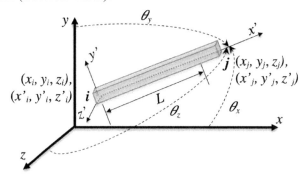

Figure 4.2.2

Consider a 3D – truss element as shown in Figure 4.2.2. The nodes i and j have (x_i, y_i, z_i) and (x_j, y_j, z_j) as the global coordinates and (x'_i, y'_i, z'_i) and (x'_j, y'_j, z'_j) as the local coordinates. θ_x, θ_y, θ_z are the angles the axis of element makes with the x, y, z axes respectively.

L is the length of the 3D–truss element $= \sqrt{(x_j - x_i)^2 + (y_j - y_i)^2 + (z_j - z_i)^2}$.   4.2 (8)

The directional cosines of the 3D - element can be written as:

$$l = \cos\theta_x = \frac{x_j - x_i}{L} \qquad m = \cos\theta_y = \frac{y_j - y_i}{L} \qquad n = \cos\theta_z = \frac{z_j - z_i}{L} \qquad 4.2\ (9)$$

The expressions for transforming the 1D- characteristics into 3D- characteristics are the same as given Equations 4.2 (1) and 4.2 (2) (Section 4.2). Accordingly,

$$\{\delta'\}_{(local)} = \begin{Bmatrix} u_i' \\ u_j' \end{Bmatrix} = [T]\{\delta\} = \begin{bmatrix} l & m & n & 0 & 0 & 0 \\ 0 & 0 & 0 & l & m & n \end{bmatrix} \begin{Bmatrix} u_i \\ v_i \\ w_i \\ u_j \\ v_j \\ w_j \end{Bmatrix} = [T]\{\delta\}_{global} \qquad 4.2\ (10)$$

where u, v, w are the displacement components in three dimensions for the nodes i and j (denoted by the suffices of the displacements u, v and w). The forces, displacements and stiffness characteristics can now be transformed from local to global and vice versa using Equation 4.2 (2) with the coordinates transformation matrix in 3 dimensions as given in Equation 4.2 (11).
i.e.

$$[T] = \begin{bmatrix} l & m & n & 0 & 0 & 0 \\ 0 & 0 & 0 & l & m & n \end{bmatrix}$$ 4.2 (11)

The global stiffness matrix $[k]_{global}$ / $[k]_{system}$ can be obtained using the Equation 4.2 (2) as:

$$[k]_{system} = [k]_{global} = [T]^T [k'][T] = \underset{(local)}{\begin{bmatrix} l & 0 \\ m & 0 \\ n & 0 \\ 0 & l \\ 0 & m \\ 0 & n \end{bmatrix}} \frac{AE}{L} \begin{bmatrix} 1 & -1 \\ -1 & 1 \end{bmatrix} \begin{bmatrix} l & m & n & 0 & 0 & 0 \\ 0 & 0 & 0 & l & m & n \end{bmatrix}$$

$$[k]_{system} = [k]_{global} = \frac{AE}{L} \begin{bmatrix} l^2 & lm & ln & -l^2 & -lm & -ln \\ ml & m^2 & mn & -ml & -m^2 & -mn \\ nl & nm & n^2 & -nl & -nm & -n^2 \\ -l^2 & -lm & -ln & l^2 & lm & ln \\ -ml & -m^2 & -mn & ml & m^2 & mn \\ -nl & -nm & -n^2 & nl & nm & n^2 \end{bmatrix}$$ 4.2 (12)

where $[k']$ is the local stiffness matrix of the 1D- element (in local coordinates) as given in Equation 4.2 (1) i.e.

$$[k']_{local} = \frac{AE}{L} \begin{bmatrix} 1 & -1 \\ -1 & 1 \end{bmatrix}$$ 4.2 (13)

It can be seen that $[k]_{system / global}$ and $[k']_{local}$ are symmetric and positive definite.

The rest of the procedure is the same as in the FEM steps i.e. assembling the system, formulating the equilibrium equations, applying boundary conditions and solving the unknown parameters from the resulting consistent system of equations. The method is illustrated in the examples of 3D-truss elements presented in Section 4.6.

4.2.3 Distributed loads, Thermal effects, Lack of fitness

These effects can be first calculated by treating it as 1D- truss element in local coordinates as discussed in Section 4.1 (Section 4.1.3 and Section 4.1.4) and then the forces (and displacements if any) can be transformed to global coordinates (using the general equations for coordinate transformations either in 2D or 3D problems) i.e. Equation 4.2 (2). Using these values in global coordinates, the assembly of the system can be carried out to formulate the system equilibrium equations. Some examples are presented in Section 4.6.

4.2.4 Examples

Examples of trusses are presented in Section 4.6.2.

4.3 Beam Elements

4.3.1 Planar Beam Elements

Beams are most commonly used structural elements and carry transverse loads by bending following Euler-Bernoulli bending theory (Crandall and Dahl, 1972). A typical planar beam element is shown in Figure 4.3.1a and b with two nodes i, j and positive forces V_i, V_j and moments M_i, M_j and vertical displacement v_i, v_j and slopes θ_i, θ_j shown at the two nodes respectively. The nodal coordinates are i $(x = 0)$, j $(x = L)$ where L is the length of the beam.

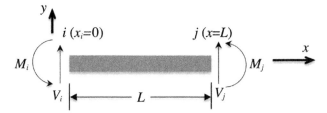

a) Nodal loads (positive as per solid mechanics)

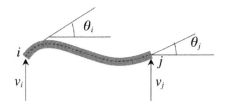

b) Nodal displacements and slopes

Figure 4.3.1

It may be noted that shear forces and bending moments are considered positive as per bending theory as shown below in Figure 4.3.2 (not as per mechanics of solids convention) as shown in Figure 4.3.1.

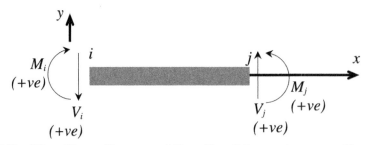

Figure 4.3.2 Positive Shear Force and Bending Moments as per Bending Theory

As usual x, y, z coordinate axes (without prime " $'$ ") refer to the global / system coordinates and if they are mentioned with prime, they refer to the local coordinates.

Following the standard FEM, direct stiffness analysis (Chapter 3), the beam element is associated with generalized deflections, i.e. vertical displacement v and slope θ

$= dv/dx$ at any point along the axis of the beam. Accordingly, the nodal displacement matrix can be expressed as:

$$\{\delta\} = \begin{Bmatrix} v_i \\ \theta_i \\ v_j \\ \theta_j \end{Bmatrix}$$

4.3 (1)

It may be noted that the displacement distribution function $[N(x)]$ has to be chosen only for vertical deflection 'v'. Since the slope θ (dv/dx) is the derivative of v with respect to x (coordinate direction along the axis of the beam). Hence, v can be expressed as a polynomial in 'x' compatible with the degrees of freedom of the nodal displacements i.e. v_i, θ_i, v_j, θ_j. Therefore,

$$v(x) = a_1 + a_2x + a_3x^2 + a_4x^3$$

4.3 (2)

where a_1 to a_4 are the four arbitrary constants to be solved in terms of the nodal coordinates v_i, θ_i, v_j, θ_j. Thus, the displacement distribution function is consistent with the four nodal displacements of the two nodal beam elements. Then, this represents fully the third order polynomial in x (fully conforming):

From Equation 4.3 (2),

$$\theta = \frac{dv}{dx} = a_2 + 2a_3x + 3a_4x^2$$

4.3 (3a)

Now substituting the nodal displacements and their x-coordinate in Equations 4.3 (1) and 4.3 (2), results in:

$$v_i = a_1$$
$$\theta_i = a_2$$
$$v_j = a_1 + a_2L + a_3L^2 + a_4L^3$$
$$\theta_j = a_2 + 2a_3L + 3a_4L^2$$

4.3 (3b)

Solving for a_1 to a_4 from the four Equations 4.3 (3b) and substituting them in Equation 4.3 (1) and rearranging the equation, we get

$$v(x) = \left[1 - 3\left(\frac{x}{L}\right)^2 + 2\left(\frac{x}{L}\right)^3\right]v_i + \left[x - \frac{2x^2}{L} + \frac{x^3}{L^2}\right]\theta_i + \left[3\left(\frac{x}{L}\right)^2 - 2\left(\frac{x}{L}\right)^3\right]v_j + \left[-\frac{x^2}{L} + \frac{x^3}{L^2}\right]\theta_j$$

4.3 (4)

$$v(x) = (N_v)_i v_i + (N_\theta)_i \theta_i + (N_v)_j v_j + (N_\theta)_j \theta_j = \left\{(N_v)_i \quad (N_\theta)_i \quad (N_v)_j \quad (N_\theta)_j\right\} \begin{Bmatrix} v_i \\ \theta_i \\ v_j \\ \theta_j \end{Bmatrix}$$

$$v(x) = [N]\{\delta\}$$

4.3 (5)

Thus, the displacement distribution function is expressed in standard form as is the usual procedure in FEM analysis. The next step is to write strain displacement relation. The second derivative is the generalized strain for beam bending (i.e. the curvature). Accordingly,

$$\frac{d^2v}{dx^2} = \frac{d^2}{dx^2}[N] = \begin{bmatrix} \frac{d^2(N_v)_i}{dx^2} & \frac{d^2(N_\theta)_i}{dx^2} & \frac{d^2(N_v)_j}{dx^2} & \frac{d^2(N_\theta)_j}{dx^2} \end{bmatrix} \begin{Bmatrix} v_i \\ \theta_i \\ v_j \\ \theta_j \end{Bmatrix} = \begin{bmatrix} -\frac{6}{L^2}+\frac{12x}{L^3} & -\frac{4}{L}+\frac{6x}{L^2} & \frac{6}{L^2}-\frac{12x}{L^3} & -\frac{2}{L}+\frac{6x}{L^2} \end{bmatrix} \begin{Bmatrix} v_i \\ \theta_i \\ v_j \\ \theta_j \end{Bmatrix}$$

$$\frac{d^2v}{dx^2} = \begin{bmatrix} B_1 & B_2 & B_3 & B_4 \end{bmatrix} \{\delta\} = [B]\{\delta\}$$

4.3 (6)

Next the stress-strain relationship can be identified for beam bending. This is the bending moment-curvature relationship for beams as per classical bending theory i.e.

$$M = EI\left(\frac{d^2v}{dx^2}\right)$$

4.3 (7)

This is similar to $(\sigma) = [D]\{\varepsilon\}$ in FEM procedure.

Accordingly, [D] can be identified from Equation 4.3 (7) as:

D = EI (Flexure rigidity of the prismatic beams), where:
E = Modulus of elasticity of beam material
I = Area moment of inertia of beam cross section about the axis of bending

Now the stiffness matrix $\begin{bmatrix} k \end{bmatrix}$ for the beam element can be written as per the standard expression from FEM (Chapter 3) as:

$$[k] = \int_L B^T DB \, dx = \int_L \begin{bmatrix} B_1 \\ B_2 \\ B_3 \\ B_4 \end{bmatrix} (EI) \begin{bmatrix} B_1 & B_2 & B_3 & B_4 \end{bmatrix} dx$$

4.3 (8)

$$[k] = (EI)\int \begin{bmatrix} B_1^2 & B_1B_2 & B_1B_3 & B_1B_4 \\ B_1B_2 & B_2^2 & B_2B_3 & B_2B_4 \\ B_1B_3 & B_2B_3 & B_3^2 & B_3B_4 \\ B_1B_4 & B_2B_4 & B_3B_4 & B_4^2 \end{bmatrix} dx$$

4.3 (9)

After integrating with respect to x (0 to L), the above equation can be simplified as:

$$[k] = \left(\frac{EI}{L^3}\right) \begin{bmatrix} 12 & 6L & -12 & 6L \\ 6L & 4L^2 & -6L & 2L^2 \\ -12 & -6L & 12 & -6L \\ 6L & 2L^2 & -6L & 4L^2 \end{bmatrix}$$

4.3 (10)

It can be noted that [k] is a symmetric and positive definite matrix.

To obtain the corresponding equilibrium equation of the element, consider the general loading on the beam element as shown in Figure 4.3.3.

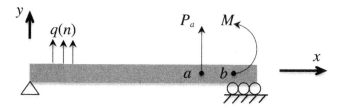

Figure 4.3.3 Beam with General Loadings

The loads and moments on the beams may include distributed forces $q(x)$, concentrated loads P_a, concentrated moments M_b at several points along the beam axis. The total potential energy which is the sum of the internal strain energy and the work potential can be written as (Budynas, 1999):

$$PE = \Pi_e = \frac{1}{2}\{\delta\}^T[k]\{\delta\} - V_i v_i - V_j v_j - M_i \theta_i - M_j \theta_j \qquad 4.3\,(11)$$

Minimisation of potential energy with respect to nodal displacements $\{\delta\}=\{v_i\ \theta_i\ v_j\ \theta_j\}^T$ results in:

$$[k]\{\delta\} = \left(\frac{EI}{L^3}\right)\begin{bmatrix} 12 & 6L & -12 & 6L \\ 6L & 4L^2 & -6L & 2L^2 \\ -12 & -6L & 12 & -6L \\ 6L & 2L^2 & -6L & 4L^2 \end{bmatrix}\begin{Bmatrix} v_i \\ \theta_i \\ v_j \\ \theta_j \end{Bmatrix} = \begin{Bmatrix} V_i \\ M_i \\ V_j \\ M_j \end{Bmatrix} \qquad 4.3\,(12)$$

where V_i, M_i, V_j, M_j are the applied forces and moments at the two nodes, i and j.

It may be noted that θ, M and V can also be interpreted as slope, bending moments and shear forces at any cross section of the beam as per bending theory and are given by (Kameswara Rao, 2011),

$$\theta = \frac{dv}{dx} \qquad M = EI\frac{d^2v}{dx^2} \qquad V = -EI\frac{d^3v}{dx^3} \qquad 4.3\,(13)$$

where v is the vertical displacement at that cross section.
These parameters can be calculated at any cross section of the beam from Equations 4.3(4) and 4.3 (6) and subsequent derivatives for calculating slope θ (involving first derivative of v), bending moment M (involving second derivative of v), shear force V (involving third derivative of v)as expressed in the above Equation 4.3 (13) (Sec.4.3), once the deflections and the forces at the nodes are obtained from the FEM analysis. It may be observed that FEM analysis gives forces V, and moments M, as per mechanics convention which can be interpreted as shear forces and bending moments as per bending theory convention as may be required for design.

4.3.2 System / global stiffness matrix, assembly and system equilibrium equation

Once the element stiffness matrix is obtained, the global stiffness matrix can be assembled by adding the contributions of individual elements connected at any node as explained in Sections 4.1 and 4.2. After assembling the system, the system equations can be written by equating the internal equivalent nodal forces to the externally applied forces and moments at each node contributed by the elements connected at that node. The individual element contributions of forces and moments are given in Equation 4.3 (12). Effects due to distributed forces, temperature, initial lack of fitness and any peculiarities of individual element(s) (if they are present) can all be included in the equilibrium equations, which are explained in the subsequent section. However, it may be noted that temperature will not affect the transverse forces though it produces axial forces. Thus, the system equilibrium equations can be expressed in the standard form (as discussed in Chapters 2 and 3) as follows.

$$\{F\}=[K]_{system}\{\delta\}+\{F\}_p+\{F\}_{\varepsilon_o}=[K]_{system}\begin{Bmatrix} u_1 \\ v_1 \\ u_2 \\ v_2 \\ \vdots \\ u_n \\ v_n \end{Bmatrix}+\Sigma\begin{Bmatrix} F_{1p} \\ F_{2p} \\ \vdots \\ F_{np} \end{Bmatrix}+\Sigma\begin{Bmatrix} F_{1\varepsilon_o} \\ F_{2\varepsilon_o} \\ \vdots \\ F_{n\varepsilon_o} \end{Bmatrix}$$

4.3 (14)

The next step is to apply the boundary conditions of the system i.e. substituting the values of deflection, slope or other prescribed conditions in the system displacement matrix. The resulting equations will result in a consistent system of simultaneous equations in unknown displacements, slopes and some forces at the nodes other than the boundary nodes whose values are known from the boundary conditions. Once the unknown displacements and slopes and forces are known, then the bending moments, shear forces etc. can be computed using Equation 4.3 (12) for each element. Then bending moment diagram (BMD) and shear force diagrams (SFD) can be drawn for the system. The stresses etc. can then be computed using the classical bending theory by equations:

$$-\frac{f}{y}=-\frac{M}{I}=\frac{E}{R}$$

4.3 (15)

where

f = stress at any point (depth) of the beam

y = the y coordinate of the point measured in the y direction from the neutral axis of the beam

M = bending moment at that cross-section of the beam

I = area moment of inertial of the cross-section of the beam about the neutral axis of the beam

E = modulus of elasticity of the beam material

R = radius of curvature of the beam due to bending

Similarly, the stresses due to shear force at the cross section can be computed using the bending theory (Crandall and Dahl, 1972). Some examples of beams subjected to loads and moments are presented in Section 4.6.

4.3.3 Distributed loads

The equivalent nodal forces due to distributed loads on beam element shown in Figure 4.3.4 can be computed as explained in the Section 4.1 for rod element.

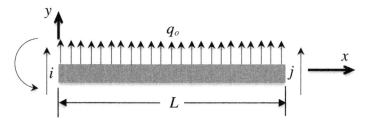

Figure 4.3.4 Distributed loads on the beam

For applications in FEM all loads have to be reduced to equivalent nodal loads as follows. Following the same procedure as used in the case of rod / bar element (Sec. 4.1), the distributed load can be lumped equally as equivalent nodal load at both the nodes i.e.

$$\begin{Bmatrix} V_i \\ M_i \\ V_j \\ M_j \end{Bmatrix}_{q_o} = \begin{Bmatrix} \dfrac{q_o L}{2} \\ 0 \\ \dfrac{q_o L}{2} \\ 0 \end{Bmatrix}$$

4.3 (16)

This is called the lumped load method which is most commonly used. If the distributed load is only on a part of the beam or non-uniform, the load can be lumped at both the nodes as a static equivalent load depending on the static reaction of the load at both the nodes.

A better equivalent method is referred to as "consistent method" in which the equivalent nodal loads and moments are computed using equivalent work done by the distributed loads and the corresponding static nodal loads and moments.

Accordingly, the work done by the distributed load can be written as:

$$w_p = \int_L q(x) v(x) dx = -q_0 \int_L v(x) dx$$

4.3 (17)

where $v(x)$ is the vertical displacement as a function of x along the axis of the beam. Substituting for $v(x)$ from Equation 4.3 (4) in Equation 4.3 (17) and integrating it, we get,

$$w_p = -q_o \left(\frac{L}{2} v_i + \frac{L^2}{12} \theta_i + \frac{L}{2} v_j - \frac{L^2}{12} \theta_j \right) = -\{ v_i \quad \theta_i \quad v_j \quad \theta_j \} \begin{Bmatrix} \dfrac{q_o L}{2} \\ \dfrac{q_o L^2}{12} \\ \dfrac{q_o L}{2} \\ -\dfrac{q_o L^2}{12} \end{Bmatrix}$$

4.3 (18)

From the minimization principle (Rayleigh-Ritz method), the equivalent nodal forces can be obtained by equating the first derivatives of w_p with respect to corresponding nodal displacements, $v_i, \theta_i, v_j, \theta_j$ to zero as given below

56

Thus, the equivalent nodal loads for the uniform distributed loads (u.d.l.) on beam can be expressed using the consistent method as:

$$\begin{Bmatrix} V_i \\ M_i \\ V_j \\ M_j \end{Bmatrix} = \begin{Bmatrix} \dfrac{q_o L}{2} \\ \dfrac{q_o L^2}{12} \\ \dfrac{q_o L}{2} \\ -\dfrac{q_o L^2}{12} \end{Bmatrix}$$

4.3 (19)

These are shown in Figure 4.3.5 for easy understanding and application.

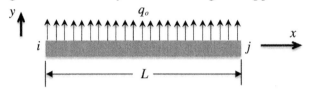

a) Uniformly distributed load applied on the beam

b) Equivalent nodal loads using Lumped method

c) Equivalent nodal loads using Consistent method

Figure 4.3.5 Equivalent nodal loads due to uniformly distributed load (u.d.l.)

It is important to note that the consistent method gives exact results for displacements (irrespective of the number of elements) while lumped method gives slightly higher displacements than the exact values (Budynas, 1999).

Once the system equilibrium equations are formed, the effect of several variations in boundary conditions, non-uniformity of cross sections, loading, pin connected beam elements, uneven support conditions including spring supports etc. can be analysed with suitable manipulations in these system equations. Some examples of distributed loading on beams are presented in Section 4.6.

4.3.4 Beams in two-plane bending

In the previous sections (Section 4.3.1 to 4.3.3) the stiffness analysis is presented for bending of beams in the $x - y$ plane (bending about the z axis). The same can be extended to two plane bending of beams (bending in the *xy* plane about *z* axis and bending in the *xz* plane about *y* axis) by adding the corresponding degrees of freedom for bending with associated notations of displacements, forces and moments as explained below.

The element equilibrium equations for beams bending in *xy* plane can be re-written from Equation 4.3 (12) by adding the appropriate – subscripts to the forces, displacement / slope and moment of inertia flexural rigidity as:

$$
\begin{Bmatrix} V_{yi} \\ M_{zi} \\ V_{yj} \\ M_{zj} \end{Bmatrix} = \left(\frac{EI_z}{L^3}\right) \begin{bmatrix} 12 & -6L & -12 & 6L \\ 6L & 4L^2 & -6L & 2L^2 \\ -12 & -6L & 12 & -6L \\ 6L & 2L^2 & -6L & 4L^2 \end{bmatrix} \begin{Bmatrix} v_i \\ \theta_{zi} \\ v_j \\ \theta_{zj} \end{Bmatrix}
$$

4.3 (20)

Similarly, the above equilibrium equation for bending of the beam in *xz* plane can be written with appropriate modifications in the subscripts as:

$$
\begin{Bmatrix} V_{zi} \\ M_{yi} \\ V_{zj} \\ M_{yj} \end{Bmatrix} = \left(\frac{EI_y}{L^3}\right) \begin{bmatrix} 12 & 6L & -12 & -6L \\ -6L & 4L^2 & 6L & 2L^2 \\ -12 & 6L & 12 & 6L \\ -6L & 2L^2 & 6L & 4L^2 \end{bmatrix} \begin{Bmatrix} w_i \\ \theta_{yi} \\ w_j \\ \theta_{yj} \end{Bmatrix}
$$

4.3 (21)

where *w* denotes the deflection in the *z* direction.

Now the equilibrium equation for two plane bending of the beam can be written by combining the Equations 4.3 (20) and 4.3 (21) and adding the degrees of freedom. Thus the nodal displacement vector becomes $\{v_i \; w_i \; (\theta_y)_i \; (\theta_z)_i \; v_j \; w_j \; (\theta_y)_j \; (\theta_z)_j\}$. The rearranged element equilibrium matrix for two plane bending can be written by expanding the matrix to accommodate the 8 degrees of freedom corresponding to the 8 components of the displacement matrix as:

$$
\begin{Bmatrix} (V_y)_i \\ (V_z)_i \\ (M_y)_i \\ (M_z)_i \\ (V_y)_j \\ (V_z)_j \\ (M_y)_j \\ (M_z)_j \end{Bmatrix} = \left(\frac{E}{L^3}\right) \begin{bmatrix} 12I_z & 0 & 0 & 6I_zL & -12I_z & 0 & 0 & 6I_zL \\ 0 & 12I_y & -6I_yL & 0 & 0 & -12I_y & -6I_yL & 0 \\ 0 & -6I_yL & 4I_yL^2 & 0 & 0 & 6I_yL & 2I_yL^2 & 0 \\ 6I_zL & 0 & 0 & 4I_zL^2 & -6I_zL & 0 & 0 & 2I_zL^2 \\ -12I_z & 0 & 0 & -6I_zL & 12I_z & 0 & 0 & -6I_zL \\ 0 & -12I_y & 6I_yL & 0 & 0 & 12I_y & 6I_yL & 0 \\ 0 & -6I_yL & 2I_yL^2 & 0 & 0 & 6I_yL & 4I_yL^2 & 0 \\ 6I_zL & 0 & 0 & 2I_zL^2 & -6I_zL & 0 & 0 & 4I_zL^2 \end{bmatrix} \begin{Bmatrix} v_i \\ w_i \\ (\theta_y)_i \\ (\theta_z)_i \\ v_j \\ w_j \\ (\theta_y)_j \\ (\theta_z)_j \end{Bmatrix}
$$

4.3 (22)

All other procedures for assembling, forming system equilibrium equations, applying boundary conditions, treating the distributed loads and other specific attributes of the beam, solving the resulting system of equations will remain the same as for beams with single plane bending or for that matter any FEM analysis of structural elements. However, manual computations will become laborious as the size of the simultaneous equations to be solved increases.

4.3.5 Examples
Examples of beams are presented in Section 4.6.3.

4.4 The Frame Element

Most of the professional / commercial FEM software programs for structural analysis use this element as it is a versatile element. The frame element can be assembled by adding the axial stiffness (of the bar / rod element) and a similar torsional element stiffness to the stiffness matrix of the two plane bending element. The same is explained below.

Rewriting the stiffness equations of the axial rod / bar element from Section 4.4.11 as:

$$\begin{Bmatrix} N_{xi} \\ N_{xj} \end{Bmatrix} = \left(\frac{AE}{L} \right) \begin{bmatrix} 1 & -1 \\ -1 & 1 \end{bmatrix} \begin{Bmatrix} u_i \\ u_j \end{Bmatrix}$$

4.4 (1)

where N_{xi}, N_{xj} are the nodal axial forces (instead of V_i, V_j) of the rod element with the rest of the parameters being the same as defined in Section 4.1.

The load displacement relationship for torsional element is very similar to the axially loaded bar element and can be written as (Budynas, 1999):

$$\begin{Bmatrix} T_{xi} \\ T_{xj} \end{Bmatrix} = \left(\frac{EJ}{2(1+v)L} \right) \begin{bmatrix} 1 & -1 \\ -1 & 1 \end{bmatrix} \begin{Bmatrix} \theta_{zi} \\ \theta_{zj} \end{Bmatrix}$$

4.4 (2)

where
T_{xi}, T_{xj} = torsional moments at nodes i and j
J = polar moment of inertial of the frame element. Second moment of the area of cross section about the axial direction.
E = modulus of elasticity of frame material
v = Poisson's ration
θ_{zi}, θ_{zj} = torsional rotations at i and j
L = length of the frame element

Adding these four additional degrees of freedom i.e. u_i, u_j, θ_{zi}, θ_{zj} to the two plane bending element (8 degrees of freedom) and expanding the matrix to accommodate these $8 + 4 = 12$ degrees of freedom, the element equilibrium equation of the frame element can be written as:

$$
\begin{Bmatrix}
(N_x)_i \\ (V_y)_i \\ (V_z)_i \\ (T_x)_i \\ (M_y)_i \\ (M_z)_i \\ (N_x)_j \\ (V_y)_j \\ (V_z)_j \\ (T_x)_j \\ (M_y)_j \\ (M_z)_j
\end{Bmatrix}
= \left(\frac{E}{L} \right)_e
\begin{bmatrix}
A & 0 & 0 & 0 & 0 & 0 & -A & 0 & 0 & 0 & 0 & 0 \\
0 & 12I_z/L^2 & 0 & 0 & 0 & 6I_z/L & 0 & -12I_z/L^2 & 0 & 0 & 0 & 6I_z/L \\
0 & 0 & 12I_y/L^2 & 0 & -6I_y/L & 0 & 0 & 0 & -12I_y/L^2 & 0 & -6I_y/L & 0 \\
0 & 0 & 0 & J/[2(1+v)] & 0 & 0 & 0 & 0 & 0 & -J/[2(1+v)] & 0 & 0 \\
0 & 0 & -6I_y/L & 0 & 4I_y & 0 & 0 & 0 & 6I_y/L & 0 & 2I_y & 0 \\
0 & 6I_z/L & 0 & 0 & 0 & 4I_z & 0 & -6I_z/L & 0 & 0 & 0 & 2I_z \\
-A & 0 & 0 & 0 & 0 & 0 & A & 0 & 0 & 0 & 0 & 0 \\
0 & -12I_z/L & 0 & 0 & 0 & -6I_z/L & 0 & 12I_z/L^2 & 0 & 0 & 0 & -6I_z/L \\
0 & 0 & -12I_y/L^2 & 0 & 6I_y/L & 0 & 0 & 0 & 12I_y/L^2 & 0 & 6I_y/L & 0 \\
0 & 0 & 0 & -J/[2(1+v)] & 0 & 0 & 0 & 0 & 0 & J/[2(1+v)] & 0 & 0 \\
0 & 0 & -6I_y/L & 0 & 2I_y & 0 & 0 & 0 & 6I_y/L & 0 & 4I_y & 0 \\
0 & 6I_z/L & 0 & 0 & 0 & 2I_z & 0 & -6I_z/L & 0 & 0 & 0 & 4I_z
\end{bmatrix}
\begin{Bmatrix}
u_i \\ v_i \\ w_i \\ (\theta_x)_i \\ (\theta_y)_i \\ (\theta_z)_i \\ u_j \\ v_j \\ w_j \\ (\theta_x)_j \\ (\theta_y)_j \\ (\theta_z)_j
\end{Bmatrix}
$$

4.4 (3)

4.4.1 Three dimensional frame element – space frame element

The characteristics of frame element discussed in the above sections for elements using x y z as local coordinates. For applications in 3-D-space, the frame element characteristics can be transformed using coordinate transformation matrices in a similar manner as was done for 1-D-axial bar element for applications in 2-D and 3-D truss problems (space truss). The resulting matrices become very large to handle manually (Budynas, 1999) and even simple applications need computers and software packages and (or) mathematical software programs like MATLAB etc. For more details, users can refer to Budynas, 1999.

4.4.2 Buckling of beams and other problems

Using the characteristics of basic elements discussed in this chapter, problems of buckling and other complex situations can be analyzed using FEM (Budynas, 1999, Zienkiewicz and Taylor, 1989, Rajasekharan, 1993).

4.5 Elastic Solids

The problems involving 2D and 3D elastic solids can be analysed using FEM as discussed in Chapter 3, which are otherwise very difficult to solve using theory of elasticity equations. For easy understanding of the FEM analysis of elastic solids, two dimensional elastic solid elements are discussed below.

4.5.1 Two dimensional elastic solids – plane stress and plane strain problems

Two dimensional elasticity problems were some of the earliest successful applications of FEM analysis for solving complex problems. The formulations and general equations for deriving the Finite Element characteristics of elastic solid elements are discussed in detail in Chapter 3. The details for obtaining these characteristics are discussed below. However, the rest of the solution of FEM steps such as assembling the system, formulating the equilibrium equations, applying the boundary conditions and solving for unknown parameters such as displacements, forces, etc. from the resulting system of equations (which are consistent), computation of stresses and strains and other required parameters etc. are the same as in FEM analysis of solids as discussed in the above sections of this chapter and in Chapter 2 and 3.

4.5.2 Element Characteristics

A typical triangular element of plane elastic solid / continuum is shown in Figure 4.5.1 with i, j, m as nodes numbered in an anticlockwise direction.

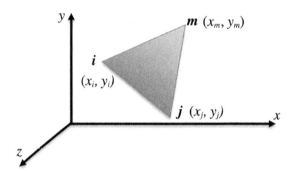

Figure 4.5.1 Plane Elastic Triangular Element

As usual, the first step in deriving the FE characteristics of the element is to express the displacement functions consistent with the nodal degrees of freedom of the element. Following the notation used in Chapter 3, the nodal displacement matrix is:

$$\{\delta\}^e = \begin{Bmatrix} \delta_i \\ \delta_j \\ \delta_m \end{Bmatrix} = \begin{Bmatrix} u_i \\ v_i \\ u_j \\ v_j \\ u_m \\ v_m \end{Bmatrix}$$

4.5 (1)

where the displacement vectors of the nodes are δ_i, δ_j, δ_m and each of these has two components (u_i, v_i), (u_j, v_j) and (u_m, v_m) being a two dimensional plane element. For a 3D element, the displacement degrees of freedom at each node would be three i.e. u, v, w along x, y, z direction.

Thus, for a plane element the displacements within the element have to be uniquely defined as polynomials which can have six constants. Then, they can be expressed as two linear polynomials as:

$$u(x) = \alpha_1 + \alpha_2 x + \alpha_3 y$$
$$v(x) = \alpha_4 + \alpha_5 x + \alpha_6 y \qquad\qquad 4.5\,(2)$$

The six arbitrary constants α_1 to α_6 can be expressed in terms of the six nodal displacements by substituting the coordinates of each of the nodes i, j, m in the above equations and solving for these constants from the resulting equations as follows (Chapters 2 and 3). Accordingly,

$$\{\delta\}^e = \begin{Bmatrix} u_i \\ v_i \\ u_j \\ v_j \\ u_m \\ v_m \end{Bmatrix} = \begin{Bmatrix} \alpha_1 + \alpha_2 x_i + \alpha_3 y_i \\ \alpha_4 + \alpha_5 x_i + \alpha_6 y_i \\ \alpha_1 + \alpha_2 x_j + \alpha_3 y_j \\ \alpha_4 + \alpha_5 x_j + \alpha_6 y_j \\ \alpha_1 + \alpha_2 x_m + \alpha_3 y_m \\ \alpha_4 + \alpha_5 x_m + \alpha_6 y_m \end{Bmatrix} \qquad\qquad 4.5\,(3)$$

Solving for the arbitrary constants α_1 to α_6 in terms of the six nodal displacements by substituting their values in Equation 4.5 (3), results in expressing the displacement functions of the element in terms of nodal displacements as:

$$\{f\} = \begin{Bmatrix} u(x) \\ v(x) \end{Bmatrix} = [N]\{\delta\}^e \qquad\qquad 4.5\,(4)$$

where

$$u(x) = \frac{1}{2\Delta}\left\{ (a_i + b_i x + c_i y)u_i + (a_j + b_j x + c_j y)u_j + (a_m + b_m x + c_m y)u_m \right\}$$
$$v(x) = \frac{1}{2\Delta}\left\{ (a_i + b_i x + c_i y)v_i + (a_j + b_j x + c_j y)v_j + (a_m + b_m x + c_m y)v_m \right\} \qquad 4.5\,(5)$$

$$2\Delta = a_i + a_j + a_m = \begin{vmatrix} 1 & x_i & y_i \\ 1 & x_j & y_j \\ 1 & x_m & y_m \end{vmatrix} = 2 \times (\text{area of the triangular element } i,j,m) \qquad 4.5\,(6)$$

$$a_i = x_j y_m - x_m y_j \qquad b_i = y_j - y_m \qquad c_i = x_m - x_j$$
$$a_j = x_m y_i - x_i y_m \qquad b_j = y_m - y_i \qquad c_j = x_i - x_m$$
$$a_m = x_i y_j - x_j y_i \qquad b_m = y_i - y_j \qquad c_m = x_j - x_i \qquad 4.5\,(7)$$

It can be noted that the coefficients in Equation 4.5 (7) can be written by cyclic permutation of the subscripts i, j, m.

Thus, the displacement distribution functions have been uniquely expressed as in Equation 4.5 (4) and can also be expressed as:

$$\{f\} = \begin{Bmatrix} u(x) \\ v(x) \end{Bmatrix} = \begin{bmatrix} IN'_i & IN'_j & IN'_m \end{bmatrix} \begin{Bmatrix} \delta_i \\ \delta_j \\ \delta_m \end{Bmatrix} = [N]\{\delta\}^e \qquad 4.5\,(8)$$

where I is a 2×2 identity matrix $= \begin{bmatrix} 1 & 0 \\ 0 & 1 \end{bmatrix}$ and

$$N'_i = (a_i + b_i x + c_i y)/(2\Delta)$$
$$N'_j = (a_j + b_j x + c_j y)/(2\Delta)$$
$$N'_m = (a_m + b_m x + c_m y)/(2\Delta) \qquad 4.5\ (9)$$

It can be noted that the displacement functions being linear polynomials, they satisfy complete compatibility of displacements with adjacent elements (including the boundary between the elements).

Then the next step would be to obtain the strains of the element. For the plane elastic solid element, the elastic strains can be written as in Chapter 3 (Zienkiewicz and Taylor, 1989):

$$\{\varepsilon\} = \left\{ \begin{array}{c} \varepsilon_{xx} \\ \varepsilon_{yy} \\ \gamma_{xy} \end{array} \right\} = \left\{ \begin{array}{c} \partial u / \partial x \\ \partial v / \partial y \\ \partial u / \partial y + \partial v / \partial x \end{array} \right\} \qquad 4.5\ (10)$$

Substituting for $u(x)$ and $v(x)$ from Equation 4.5 (5) in Equation 4.5 (10), we get:

$$\{\varepsilon\} = \frac{1}{2\Delta} \begin{bmatrix} b_i & 0 & b_j & 0 & b_m & 0 \\ 0 & c_i & 0 & c_j & 0 & c_m \\ c_i & b_i & c_j & b_j & c_m & b_m \end{bmatrix} \left\{ \begin{array}{c} u_i \\ v_i \\ u_j \\ v_j \\ u_m \\ v_m \end{array} \right\} = [B]\{\delta\}^e \qquad 4.5\ (11)$$

where

$$[B] = \frac{1}{2\Delta} \begin{bmatrix} b_i & 0 & b_j & 0 & b_m & 0 \\ 0 & c_i & 0 & c_j & 0 & c_m \\ c_i & b_i & c_j & b_j & c_m & b_m \end{bmatrix} \qquad 4.5\ (12)$$

It may be noted that Equation 4.5 (11) involves only constants which implies that the strains are constant over the entire triangular element. Hence these are also called Constant Strain Triangular (CST) elements.

4.5.3 Stress-strain relations (Constitutive equations)

The next step in FEM is to write the stress-strain relations (constitutive equations) of the material of the element (Chapter 3). For a linear elastic material, these can be written using modulus of elasticity, E and Poisson's ratio, v, as (using expression given in Chapter 3):

$$\{\sigma\} = \begin{bmatrix} \sigma_{xx} & \tau_{zy} \\ \tau_{yx} & \sigma_{yy} \end{bmatrix} = [D]\{\varepsilon\} = [D] \left\{ \begin{array}{c} \varepsilon_{xx} \\ \varepsilon_{yy} \\ \gamma_{xy} \end{array} \right\} = [D] \left\{ \begin{array}{c} \dfrac{\partial u}{\partial x} \\ \dfrac{\partial v}{\partial y} \\ \dfrac{\partial u}{\partial y} + \dfrac{\partial v}{\partial x} \end{array} \right\} \qquad 4.5\ (13)$$

using Equations 4.4, Equation 4.5 (13) can be written as:

$$\{\sigma\}=[D][B]\{\delta\}^e=[S]\{\delta\}^e \qquad\qquad 4.5\ (14)$$

where $[B]$ is given by Equation 4.5 (12) and $[S]$ is referred to as stress matrix given by:

$$[S]=[D][B] \qquad\qquad 4.5\ (15)$$

However $[D]$ matrix involving element material properties E and v are different for plane stress and plane strain elements (Zienkiewicz and Taylor, 1989) and can be expressed as (Chapter 3):

For plane stress element:

$$[D]=\frac{E}{\left(1-v^2\right)}\begin{bmatrix} 1 & v & 0 \\ v & 1 & 0 \\ 0 & 0 & (1-v)/2 \end{bmatrix} \qquad\qquad 4.5\ (16)$$

For plane strain element:

$$[D]=\frac{E}{(1+v)(1-2v)}\begin{bmatrix} (1-v) & v & 0 \\ v & (1-v) & 0 \\ 0 & 0 & (1-2v)/2 \end{bmatrix} \qquad\qquad 4.5\ (17)$$

4.5.4 Stiffness matrix of the plane elastic solid element (k matrix)

Using the standard expressions derived to express the stiffness matrix of the finite element, we can obtain the same as follows (Chapter 3):

$$[k]=\int_{Vol}[B]^T[D][B](dvol)=\int_{Vol}[B]^T[D][B]\,dxdydz=[B]^T[D][B]\int_{Vol}dx\,dy\,dz \qquad 4.5\ (18)$$

Substituting for $[B]$ and $[D]$ from Equations 4.5 (12) and 4.5 (16) for plane stress case and 4.5 (17) for plane strain case, and noting that thicknesses of the elements are constant in the z direction, Equation 4.5 (18) can be simplified as:

For plane stress element:

$$[k]=\int_{Vol}[B]^T[D][B]\,dxdydz=[B]^T[D][B]\int dx\,dy\,dz=[B]^T[D][B]\cdot t\int dx\,dy=[B]^T[D][B]t\Delta \qquad 4.5\ (19)$$

where t is the thickness (along the z direction) of the plane stress element and Δ as the area $i\ j\ m$, of the triangular element. It may be noted that in the integral (Equation 4.5 (18)) matrices $[B]$ and $[D]$ are constants (not functions of x, y, z) than the integral gives the volume of the element $= t\ \Delta$. The matrix $[D]$ for plane stress case is given by Equation 4.5 (16).

For plane strain element:

$$[k]=[B]^T[D][B]\cdot t\Delta \qquad\qquad 4.5\ (20)$$

where t can be taken as a unit for analysis of plane strain element. For plane strain case, $[D]$ is given by Equation 4.5 (17) and Δ is the area of the triangular element ijm.

The rest of the procedure of FEM analysis such as assembling the system equations etc. are the same as explained earlier for other solids in this chapter. The effects on the nodal forces due to initial strain / temperature, distributed body forces, boundary loads etc. can be computed as follows (Chapters 2 and 3).

4.5.5 Nodal forces due to initial strain

The effect of initial strain / temperature on the equivalent nodal forces can be expressed (as given in Chapter 3) as:

$$\{F\}^{\varepsilon}_{\varepsilon_o} = -\int_{vol} [B]^T [D]\{\varepsilon_o\} dvol = -[B]^T [D]\{\varepsilon_o\} t\Delta \qquad 4.5\ (21)$$

where $[B]$ and $[D]$ are given by Equations 4.5 (12) and 4.5 (16) for plane stress, 4.5 (17) for plane strain respectively, and

t = thickness of the plane elastic element in case of plane stress

= 1 for plane strain

Δ = area of the triangular *ijm* of the element

If the initial strain is due to change in temperature, $\{\varepsilon_o\}$ can be expressed using the thermal properties of the element as follows.

If the change in temperature (increase or decrease) of the element is T and α is the coefficient of thermal expansion, then $\{\varepsilon_o\}$ in Equation 4.5 (21) can be expressed as:

Initial strain due to temperature

$$\{\varepsilon_o\} = \left\{ \begin{array}{c} \alpha T \\ \alpha T \\ 0 \end{array} \right\} \qquad 4.5\ (22)$$

as there will be no shear strain if the temperature change is uniform throughout the element. Then the nodal forces can be computed using Equations 4.5 (21) and 4.5 (22) and included in the FEM analysis (Chapters 2 and 3).

The effect of initial stresses / residual stresses if any can be similarly computed (Zienkiewicz and Taylor, 1989), though they will not contribute to the nodal forces of the element as they are self-equilibrating at every point.

4.5.6 Equivalent nodal forces due to distributed body forces or boundary forces

The equivalent nodal forces due to externally applied load / distributed body forces / boundary loads etc. can be computed using either the lumped method or the consistent method as explained below.

a. Lumped method:
The effect of the externally applied concentrated or distributed loads on the body or boundaries can be computed using equivalent static lumped method using judicious approach for lumping.

For example, if the load is uniformly distributed on the / in the body of the element (such as weight, other component of loads applied within the body or on the surface of the body), the resultant total load may be lumped equally at all the nodes. If the distribution is not uniform, then the total loads may be lumped judiciously (approximately) depending on the location of the point of action of the resultant. Similarly, if the loads (distributed or concentrated) are applied on the boundary, they may be lumped either equally or judiciously at the neighboring nodes depending on the point of action of each / resultant load. This is a simple method which is commonly practiced. This approach is similar to the method used in the case of other solids discussed in this Chapter.

b. Consistent method using FEM expression:

Using the expressions derived for the evaluation of equivalent nodal forces due to distributed body forces $\{p\}$ given in Chapter 3, we can express the nodal forces as:

$$\{F\}_p^e = -\int [N]^T \{p\} dvolume \tag{4.5 (23)}$$

The matrix $[N]$ depends on the element and involves functions x, y, z etc. The distributed forces $\{p\}$ can be prescribed as the matrix of distributed forces applied along each coordinate direction depending on the dimensionality of the element. Then Equation 4.5 (23) can be evaluated after carrying out the integration.

Similarly, if the distributed loads $\{g\}$ are applied on the boundary, then the equivalent nodal forces can be expressed as (Chapter 3):

$$\{F\}_g^e = -\int\limits_{boundary} [N]^T \{g\}(darea) \tag{4.5 (24)}$$

Since the boundary loads are in term of force / unit area, the integral has to be carried out along the boundary which could also involve the coordinate functions in the matrix $[N]$. Thus, the integration in Equations 4.5 (23) and 4.5 (24) could be quite complicated to perform and may have to be carried out numerically.

The consistent method is presented for the plane elastic triangular element (CST). Let the distributed force / body forces, shown in Figure 4.5.1 be:

$$\{p\} = \left\{ \begin{array}{c} X \\ Y \end{array} \right\} \tag{4.5 (25)}$$

where X and Y are components of body forces (forces / unit volume, such as self-weight etc.) along x and y axis. Then:

$$\{F\}_p^e = -t \int [N]^T \left\{ \begin{array}{c} X \\ Y \end{array} \right\} dx dy \tag{4.5 (26)}$$

where t is the thickness of the element. Assuming the X and Y are constant, Equation 4.5 (26) can be rewritten as:

$$\{F\}_p^e = -t \left\{ \begin{array}{c} X \\ Y \end{array} \right\} \int [N]^T dx dy \tag{4.5 (27)}$$

The matrix $[N]$ involves linear term of x and y as given in Equation 4.5 (8). In the special case with the origin of the coordinates is chosen as the center of gravity (C.G.) of the triangle / centroid of the triangle, the integration process gets simplified since integration of the terms involving $\int x\,dy\,dx = \int y\,dx\,dy = 0$,

Accordingly, Equation 4.5 (27) simplifies as:

$$\{F\}_p^e = \left\{ \begin{array}{c} F_i \\ F_j \\ F_m \end{array} \right\}_p - \left\{ \begin{array}{c} X \\ Y \\ X \\ Y \\ X \\ Y \end{array} \right\} \frac{t\Delta}{3} \qquad\qquad 4.5\ (28)$$

Equation 4.5 (28) implies that the total loads are equally distributed at the three nodes $i, j,$ m. This fact corresponds to the judicious distribution of the body forces equally at the three nodes, assumed implicitly in the "lumped method / judicious approach" method above.

Using the above approaches, the effects of various external influences on the elastic solid elements can be computed and included in the FEM analysis in a similar manner as applied to other solids discussed in this chapter.

The rest of the FEM procedure is the same as for other solids explained in this Chapter i.e. assembling the system stiffness matrix, formulating system equilibrium equations, applying boundary conditions, solving the unknown parameters (displacements and forces) from the resulting consistent system of simultaneous equations etc.. Once the system parameters are solved, the derived parameters such as stresses and strains etc. in each element can be obtained using the stiffness characteristics of each element.

4.5.7 Examples

Examples of FEM anlysis of plane elastic solids using triangular element (CST) are presented in Section 4.6.

4.6 Summary of FEM Worked Out Examples for Rods, Trusses, Beams and Solids

4.6.1 Examples of Rods

1. Example 4R.1:

For the axially loaded structure shown in Figure 4.6.1, obtain the deflections at all the nodes, 1 to 4, wall reactions, stresses and nodal forces for each element (1 to 3) using FEM. Compare the values from the conventional method. Take E =100GPa (100 x 10^9 N/m²). A_1 to A_3 are areas of cross section and L_1 to L_3 are the lengths of the elements respectively. What will be the revised forces if element 3 is heated up by 100°C (α = 10^{-5}/°C)?

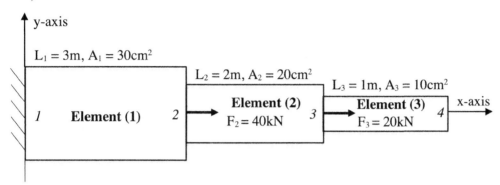

Figure 4.6.1

Solution:

1. The stiffness of each element:

a) Element 1= $k_1 = \dfrac{E_1 A_1}{L_1} = \dfrac{(100 \times 10^9)(30 \times 10^{-4})}{3} = 10^8 N/m$

b) Element 2= $k_2 = \dfrac{E_2 A_2}{L_2} = \dfrac{(100 \times 10^9)(20 \times 10^{-4})}{2} = 10^8 N/m$

c) Element 3= $k_3 = \dfrac{E_3 A_3}{L_3} = \dfrac{(100 \times 10^9)(10 \times 10^{-4})}{1} = 10^8 N/m$

2. The element stiffness matrix:

a) Element 1= $[k_1] = k_1 \begin{matrix} & u_1 & u_2 & \\ \begin{bmatrix} 1 & -1 \\ -1 & 1 \end{bmatrix} & \begin{matrix} u_1 \\ u_2 \end{matrix} \end{matrix}$

b) Element 2= $[k_2] = k_2 \begin{matrix} & u_2 & u_3 & \\ \begin{bmatrix} 1 & -1 \\ -1 & 1 \end{bmatrix} & \begin{matrix} u_2 \\ u_3 \end{matrix} \end{matrix}$

c) Element 3= $[k_3] = k_3 \begin{matrix} & u_3 & u_4 & \\ \begin{bmatrix} 1 & -1 \\ -1 & 1 \end{bmatrix} & \begin{matrix} u_3 \\ u_4 \end{matrix} \end{matrix}$

3. The system stiffness matrix:

$$[k]_{system} = 10^8 \begin{array}{cccc} u_1 & u_2 & u_3 & u_4 \end{array}$$

$$[k]_{system} = 10^8 \begin{bmatrix} 1 & -1 & 0 & 0 \\ -1 & 1+1 & -1 & 0 \\ 0 & -1 & 1+1 & -1 \\ 0 & 0 & -1 & 1 \end{bmatrix} \begin{array}{c} u_1 \\ u_2 \\ u_3 \\ u_4 \end{array} = 10^8 \begin{bmatrix} 1 & -1 & 0 & 0 \\ -1 & 2 & -1 & 0 \\ 0 & -1 & 2 & -1 \\ 0 & 0 & -1 & 1 \end{bmatrix}$$

4. Based on the static equilibrium state, the system nodal force vector, stiffness matrix and displacement vector can be expressed as following:

i.e. $\{F\} = [k]\{\delta\} = \{R\}$

$$\{F\} = 10^8 \begin{bmatrix} 1 & -1 & 0 & 0 \\ -1 & 2 & -1 & 0 \\ 0 & -1 & 2 & -1 \\ 0 & 0 & -1 & 1 \end{bmatrix} \begin{Bmatrix} u_1 = 0 \\ u_2 \\ u_3 \\ u_4 \end{Bmatrix} = \begin{Bmatrix} R_{1x} \\ 40 \times 10^3 \\ 20 \times 10^3 \\ 0 \end{Bmatrix}$$

By solving the equation, the displacement and reaction forces are computed.
$u_2 = 0.6 \times 10^{-3}$m, $u_3 = 0.8 \times 10^{-3}$m, $u_4 = 0.8 \times 10^{-3}$m, $R_{1x} = -60$ kN

5. The element stresses:

a) Element 1 = $\sigma_1 = \dfrac{E_1}{L_1} (-1 \quad 1) \begin{Bmatrix} u_1 \\ u_2 \end{Bmatrix} = \dfrac{100 \times 10^9}{3} (-1 \quad 1) \begin{Bmatrix} 0 \\ 0.6 \times 10^{-3} \end{Bmatrix} = 20 \times 10^6 \, N/m^2$

b) Element 2 = $\sigma_2 = \dfrac{E_2}{L_2} (-1 \quad 1) \begin{Bmatrix} u_2 \\ u_3 \end{Bmatrix} = \dfrac{100 \times 10^9}{2} (-1 \quad 1) \begin{Bmatrix} 0.6 \times 10^{-3} \\ 0.8 \times 10^{-3} \end{Bmatrix} = 10 \times 10^6 \, N/m^2$

c) Element 3 = $\sigma_3 = \dfrac{E_3}{L_3} (-1 \quad 1) \begin{Bmatrix} u_3 \\ u_4 \end{Bmatrix} = \dfrac{100 \times 10^9}{1} (-1 \quad 1) \begin{Bmatrix} 0.8 \times 10^{-3} \\ 0.8 \times 10^{-3} \end{Bmatrix} = 0$

6. The nodal forces for each element:

a) Element 1 = $\{F\}_1 = k_1 \begin{pmatrix} 1 & -1 \\ -1 & 1 \end{pmatrix} \begin{Bmatrix} u_1 \\ u_2 \end{Bmatrix} = 10^8 \begin{pmatrix} 1 & -1 \\ -1 & 1 \end{pmatrix} \begin{Bmatrix} 0 \\ 0.6 \times 10^{-3} \end{Bmatrix} = \begin{Bmatrix} -60 \\ 60 \end{Bmatrix} kN$

b) Element 2 = $\{F\}_2 = k_2 \begin{pmatrix} 1 & -1 \\ -1 & 1 \end{pmatrix} \begin{Bmatrix} u_2 \\ u_3 \end{Bmatrix} = 10^8 \begin{pmatrix} 1 & -1 \\ -1 & 1 \end{pmatrix} \begin{Bmatrix} 0.6 \times 10^{-3} \\ 0.8 \times 10^{-3} \end{Bmatrix} = \begin{Bmatrix} -20 \\ 20 \end{Bmatrix} kN$

c) Element 3 = $\{F\}_3 = k_3 \begin{pmatrix} 1 & -1 \\ -1 & 1 \end{pmatrix} \begin{Bmatrix} u_3 \\ u_4 \end{Bmatrix} = 10^8 \begin{pmatrix} 1 & -1 \\ -1 & 1 \end{pmatrix} \begin{Bmatrix} 0.8 \times 10^{-3} \\ 0.8 \times 10^{-3} \end{Bmatrix} = \begin{Bmatrix} 0 \\ 0 \end{Bmatrix} kN$

7. The thermal nodal force vector due to the heat up:
a) Element 1:

$$\{F\}_1 = \begin{Bmatrix} E_1 A_1 \alpha \Delta T_1 \\ -E_1 A_1 \alpha \Delta T_1 \end{Bmatrix} = \begin{Bmatrix} 100 \times 10^9 \times 30 \times 10^{-4} \times 10^{-5} \times 0 \\ -100 \times 10^9 \times 30 \times 10^{-4} \times 10^{-5} \times 0 \end{Bmatrix} = \begin{Bmatrix} 0 \\ 0 \end{Bmatrix}$$

b) Element 2:

$$\{F\}_2 = \begin{Bmatrix} E_2 A_2 \alpha \Delta T_2 \\ -E_2 A_2 \alpha \Delta T_2 \end{Bmatrix} = \begin{Bmatrix} 100 \times 10^9 \times 20 \times 10^{-4} \times 10^{-5} \times 0 \\ -100 \times 10^9 \times 20 \times 10^{-4} \times 10^{-5} \times 0 \end{Bmatrix} = \begin{Bmatrix} 0 \\ 0 \end{Bmatrix}$$

c) Element 3:

$$\{F\}_3 = \left\{ \begin{array}{c} E_3 A_3 \alpha \Delta T_3 \\ -E_3 A_3 \alpha \Delta T_3 \end{array} \right\} = \left\{ \begin{array}{c} 100 \times 10^9 \times 10 \times 10^{-4} \times 10^{-5} \times 100 \\ -100 \times 10^9 \times 10 \times 10^{-4} \times 10^{-5} \times 100 \end{array} \right\} = \left\{ \begin{array}{c} 10^5 \\ -10^5 \end{array} \right\} N$$

8. After the heated up, the revised forces:

$$\{F\} = [k]\{\delta\} + \{F\}_p + \{F\}_\varepsilon = \{R\}$$

$$\{F\} = 10^8 \begin{bmatrix} 1 & -1 & 0 & 0 \\ -1 & 2 & -1 & 0 \\ 0 & -1 & 2 & -1 \\ 0 & 0 & -1 & 1 \end{bmatrix} \left\{ \begin{array}{c} u_1 = 0 \\ u_2 \\ u_3 \\ u_4 \end{array} \right\} + \{0\} + \left\{ \begin{array}{c} 0 \\ 0+0 \\ 0+10^5 \\ -10^5 \end{array} \right\} = \left\{ \begin{array}{c} R_{1x} \\ 40 \times 10^3 \\ 20 \times 10^3 \\ 0 \end{array} \right\}$$

$$\therefore 10^8 \begin{bmatrix} 1 & -1 & 0 & 0 \\ -1 & 2 & -1 & 0 \\ 0 & -1 & 2 & -1 \\ 0 & 0 & -1 & 1 \end{bmatrix} \left\{ \begin{array}{c} 0 \\ u_2 \\ u_3 \\ u_4 \end{array} \right\} + \left\{ \begin{array}{c} 0 \\ 0 \\ 10^5 \\ -10^5 \end{array} \right\} = \left\{ \begin{array}{c} R_{1x} \\ 40 \times 10^3 \\ 20 \times 10^3 \\ 0 \end{array} \right\}$$

By solving the equation, the displacement and reaction forces are computed.
$u_2 = 0.6 \times 10^{-3}$ m, $u_3 = 0.8 \times 10^{-3}$ m, $u_4 = 1.8 \times 10^{-3}$ m, $R_{1x} = -60$ kN

9. Then calculate the nodal forces for each element:

a) Element 1:

$$\{F\}_1 = k_1 \left(\begin{array}{cc} 1 & -1 \\ -1 & 1 \end{array} \right) \left\{ \begin{array}{c} u_1 \\ u_2 \end{array} \right\} = 10^8 \left(\begin{array}{cc} 1 & -1 \\ -1 & 1 \end{array} \right) \left\{ \begin{array}{c} 0 \\ 0.6 \times 10^{-3} \end{array} \right\} = \left\{ \begin{array}{c} -60 \\ 60 \end{array} \right\} kN$$

b) Element 2:

$$\{F\}_2 = k_2 \left(\begin{array}{cc} 1 & -1 \\ -1 & 1 \end{array} \right) \left\{ \begin{array}{c} u_2 \\ u_3 \end{array} \right\} = 10^8 \left(\begin{array}{cc} 1 & -1 \\ -1 & 1 \end{array} \right) \left\{ \begin{array}{c} 0.6 \times 10^{-3} \\ 0.8 \times 10^{-3} \end{array} \right\} = \left\{ \begin{array}{c} -20 \\ 20 \end{array} \right\} kN$$

c) Element 3:

$$\{F\}_3 = [k]\{\delta\} + \{F\}_{\varepsilon_o} = 10^8 \left(\begin{array}{cc} 1 & -1 \\ -1 & 1 \end{array} \right) \left\{ \begin{array}{c} 0.8 \times 10^{-3} \\ 1.8 \times 10^{-3} \end{array} \right\} + \left\{ \begin{array}{c} 10^5 \\ -10^5 \end{array} \right\} = \left\{ \begin{array}{c} 0 \\ 0 \end{array} \right\} kN$$

2. Example 4R.2:

For the axially loaded structure shown in Figure 4.6.2, obtain the deflections at all the nodes, 1 to 4, wall reactions, stresses and nodal forces for each element (1 to 3) using FEM. Compare the values from the conventional method. Take $E = 100\text{GPa}$ (100×10^9 N/m²). A_1 to A_3 are areas of cross section and L_1 to L_3 are the lengths of the elements respectively.

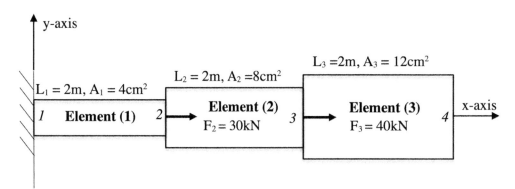

Figure 4.6.2

Solution:

1. The stiffness of each element:

a) Element 1= $k_1 = \dfrac{E_1 A_1}{L_1} = \dfrac{(100 \times 10^9)(4 \times 10^{-4})}{2} = 2 \times 10^7 \, N/m$

b) Element 2= $k_2 = \dfrac{E_2 A_2}{L_2} = \dfrac{(100 \times 10^9)(8 \times 10^{-4})}{2} = 4 \times 10^7 \, N/m$

c) Element 3= $k_3 = \dfrac{E_3 A_3}{L_3} = \dfrac{(100 \times 10^9)(12 \times 10^{-4})}{2} = 6 \times 10^7 \, N/m$

2. The element stiffness matrix:

a) Element 1= $[k_1] = 10^7 \begin{array}{c} \\ \\ \end{array} \begin{matrix} u_1 & u_2 \\ \left[\begin{matrix} 2 & -2 \\ -2 & 2 \end{matrix} \right. & \left. \begin{matrix} u_1 \\ u_2 \end{matrix} \right] \end{matrix}$ b) Element 2= $[k_2] = 10^7 \begin{matrix} u_2 & u_3 \\ \left[\begin{matrix} 4 & -4 \\ -4 & 4 \end{matrix} \right. & \left. \begin{matrix} u_2 \\ u_3 \end{matrix} \right] \end{matrix}$

c) Element 3= $[k_3] = 10^7 \begin{matrix} u_3 & u_4 \\ \left[\begin{matrix} 6 & -6 \\ -6 & 6 \end{matrix} \right. & \left. \begin{matrix} u_3 \\ u_4 \end{matrix} \right] \end{matrix}$

3. The system stiffness matrix:

$$[k]_{system} = 10^7 \begin{matrix} u_1 & u_2 & u_3 & u_4 & \\ \left[\begin{matrix} 2 & -2 & 0 & 0 \\ -2 & 2+4 & -4 & 0 \\ 0 & -4 & 4+6 & -6 \\ 0 & 0 & -6 & 6 \end{matrix} \right. & \left. \begin{matrix} u_1 \\ u_2 \\ u_3 \\ u_4 \end{matrix} \right] \end{matrix} = 10^7 \begin{bmatrix} 2 & -2 & 0 & 0 \\ -2 & 6 & -4 & 0 \\ 0 & -4 & 10 & -6 \\ 0 & 0 & -6 & 6 \end{bmatrix}$$

4. Based on the static equilibrium state, the system nodal force vector, stiffness matrix and displacement vector can be expressed as following:

$$\{F\}=[k]\{\delta\}=\{R\}$$

$$\{F\}=10^7\begin{bmatrix} 2 & -2 & 0 & 0 \\ -2 & 6 & -4 & 0 \\ 0 & -4 & 10 & -6 \\ 0 & 0 & -6 & 6 \end{bmatrix}\begin{Bmatrix} u_1=0 \\ u_2 \\ u_3 \\ u_4 \end{Bmatrix}=\begin{Bmatrix} R_{1x} \\ 30\times10^3 \\ 40\times10^3 \\ 0 \end{Bmatrix}$$

By solving the equation, the displacement and reaction forces are computed.

$u_2= 3.5\times10^{-3}$m, $u_3= 4.5\times10^{-3}$m, $u_4= 4.5\times10^{-3}$m, $R_{1x}= -70$ kN

5. The element stresses:

a) Element 1= $\sigma_1=\dfrac{E_1}{L_1}\begin{pmatrix} -1 & 1 \end{pmatrix}\begin{Bmatrix} u_1 \\ u_2 \end{Bmatrix}=\dfrac{100\times10^9}{2}\begin{pmatrix} -1 & 1 \end{pmatrix}\begin{Bmatrix} 0 \\ 3.5\times10^{-3} \end{Bmatrix}=175\times10^6\,N/m^2$

b) Element 2= $\sigma_2=\dfrac{E_2}{L_2}\begin{pmatrix} -1 & 1 \end{pmatrix}\begin{Bmatrix} u_2 \\ u_3 \end{Bmatrix}=\dfrac{100\times10^9}{2}\begin{pmatrix} -1 & 1 \end{pmatrix}\begin{Bmatrix} 3.5\times10^{-3} \\ 4.5\times10^{-3} \end{Bmatrix}=50\times10^6\,N/m^2$

c) Element 3= $\sigma_3=\dfrac{E_3}{L_3}\begin{pmatrix} -1 & 1 \end{pmatrix}\begin{Bmatrix} u_3 \\ u_4 \end{Bmatrix}=\dfrac{100\times10^9}{2}\begin{pmatrix} -1 & 1 \end{pmatrix}\begin{Bmatrix} 4.5\times10^{-3} \\ 4.5\times10^{-3} \end{Bmatrix}=0$

6. The nodal forces for each element:

a) Element 1= $\{F\}_1=k_1\begin{pmatrix} 1 & -1 \\ -1 & 1 \end{pmatrix}\begin{Bmatrix} u_1 \\ u_2 \end{Bmatrix}=10^7\begin{pmatrix} 2 & -2 \\ -2 & 2 \end{pmatrix}\begin{Bmatrix} 0 \\ 3.5\times10^{-3} \end{Bmatrix}=\begin{Bmatrix} -70 \\ 70 \end{Bmatrix}kN$

b) Element 2= $\{F\}_2=k_2\begin{pmatrix} 1 & -1 \\ -1 & 1 \end{pmatrix}\begin{Bmatrix} u_2 \\ u_3 \end{Bmatrix}=10^7\begin{pmatrix} 4 & -4 \\ -4 & 4 \end{pmatrix}\begin{Bmatrix} 3.5\times10^{-3} \\ 4.5\times10^{-3} \end{Bmatrix}=\begin{Bmatrix} -40 \\ 40 \end{Bmatrix}kN$

c) Element 3= $\{F\}_3=k_3\begin{pmatrix} 1 & -1 \\ -1 & 1 \end{pmatrix}\begin{Bmatrix} u_3 \\ u_4 \end{Bmatrix}=10^7\begin{pmatrix} 6 & -6 \\ -6 & 6 \end{pmatrix}\begin{Bmatrix} 4.5\times10^{-3} \\ 4.5\times10^{-3} \end{Bmatrix}=\begin{Bmatrix} 0 \\ 0 \end{Bmatrix}kN$

7. The conventional method:
   a) For Element 1:

i) $\sum F_x=0 \rightarrow -70+F_{x1}=0 \rightarrow \therefore F_{x1}=70kN$

ii) $\sigma_{x1}=\dfrac{F_{x1}}{A_1}=\dfrac{70kN}{4\times10^{-4}m^2}==175\times10^6\,N/m^2$

iii) $\dfrac{u_2-u_1}{L_1}=\dfrac{\sigma_{x1}}{E}=\dfrac{u_2-0}{2}=\dfrac{175\times10^6}{100\times10^9}N/m^2 \rightarrow \therefore u_2=3.5\times10^{-3}m$

b) Element 2:

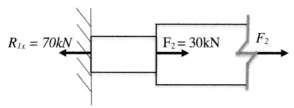

i) $\sum F = 0 \rightarrow -70 + 30 + F_2 = 0 \rightarrow \therefore F_2 = 40kN$

ii) $\sigma_{x2} = \dfrac{F_{x2}}{A_2} = \dfrac{40kN}{8 \times 10^{-4} m^2} = 50 \times 10^6 \, N / m^2$

iii) $\dfrac{u_3 - u_2}{L_2} = \dfrac{\sigma_{x2}}{E} = \dfrac{u_3 - 3.5 \times 10^{-3}}{2} = \dfrac{50 \times 10^6}{100 \times 10^9} \, N / m^2 \rightarrow \therefore u_3 = 4.5 \times 10^{-3} m$

b) Element 3:

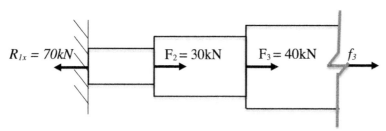

i) $\sum F = 0 \rightarrow -70 + 30 + 40 + F_3 = 0 \rightarrow \therefore F_3 = 0kN$

ii) $\sigma_{x3} = \dfrac{F_{x3}}{A_3} = \dfrac{0kN}{12 \times 10^{-4} m^2} = 0 N / m^2$

iii) $\dfrac{u_4 - u_3}{L_3} = \dfrac{\sigma_{x3}}{E} = \dfrac{u_4 - 4.5 \times 10^{-3}}{2} = \dfrac{0 \times 10^6}{100 \times 10^9} \, N / m^2 \rightarrow \therefore u_4 = 4.5 \times 10^{-3} m$

3. Example 4R.3:

For the axially loaded structure shown in Figure 4.6.3 obtain the deflections at all the nodes, 1 to 4, wall reactions, stresses and nodal forces for each element (1 to 3) using FEM. Compare the values from the conventional method. Take $E = 150$GPa (150×10^9 N/m²). A_1 to A_3 are areas of cross section of the elements.

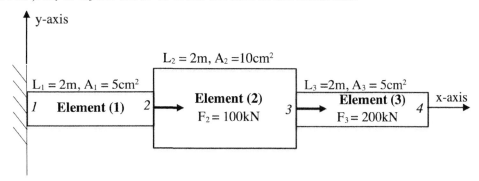

Figure 4.6.3

Solution:

1. The stiffness of each element:

a) Element 1 = $k_1 = \dfrac{E_1 A_1}{L_1} = \dfrac{(150 \times 10^9)(5 \times 10^{-4})}{2} = 37.5 \times 10^6 \, N/m$

b) Element 2 = $k_2 = \dfrac{E_2 A_2}{L_2} = \dfrac{(150 \times 10^9)(10 \times 10^{-4})}{2} = 75 \times 10^6 \, N/m$

c) Element 3 = $k_3 = \dfrac{E_3 A_3}{L_3} = \dfrac{(150 \times 10^9)(5 \times 10^{-4})}{2} = 37.5 \times 10^6 \, N/m$

2. The element stiffness matrix:

a) Element 1 = $[k_1] = 37.5 \times 10^6 \begin{bmatrix} u_1 & u_2 & \\ 1 & -1 & u_1 \\ -1 & 1 & u_2 \end{bmatrix}$

b) Element 2 = $[k_2] = 37.5 \times 10^6 \begin{bmatrix} u_2 & u_3 & \\ 2 & -2 & u_2 \\ -2 & 2 & u_3 \end{bmatrix}$

c) Element 3 = $[k_3] = 37.5 \times 10^6 \begin{bmatrix} u_3 & u_4 & \\ 1 & -1 & u_3 \\ -1 & 1 & u_4 \end{bmatrix}$

3. The system stiffness matrix:

$$[k]_{system} = 37.5 \times 10^6 \begin{bmatrix} u_1 & u_2 & u_3 & u_4 & \\ 1 & -1 & 0 & 0 & u_1 \\ -1 & 1+2 & -2 & 0 & u_2 \\ 0 & -2 & 2+1 & -1 & u_3 \\ 0 & 0 & -1 & 1 & u_4 \end{bmatrix} = 37.5 \times 10^6 \begin{bmatrix} 1 & -1 & 0 & 0 \\ -1 & 3 & -2 & 0 \\ 0 & -2 & 3 & -1 \\ 0 & 0 & -1 & 1 \end{bmatrix}$$

4. Based on the static equilibrium state, the system nodal force vector, stiffness matrix and displacement vector can be expressed as following:

$$\{F\}=[k]\{\delta\}=\{R\}$$

$$\{F\}=37.5\times10^6\begin{bmatrix}1 & -1 & 0 & 0\\ -1 & 3 & -2 & 0\\ 0 & -2 & 3 & -1\\ 0 & 0 & -1 & 1\end{bmatrix}\begin{Bmatrix}u_1=0\\ u_2\\ u_3\\ u_4\end{Bmatrix}=\begin{Bmatrix}R_{1x}\\ 100\times10^3\\ 200\times10^3\\ 0\end{Bmatrix}$$

By solving the equation, the displacement and reaction forces are computed.

$u_2= 8\times10^{-3}$m, $u_3= 10.67\times10^{-3}$m, $u_4= 10.67\times10^{-3}$m, $R_{1x}= -300$ kN

5. The element stresses:

a) Element 1: $\sigma_1=\dfrac{E_1}{L_1}\begin{pmatrix}-1 & 1\end{pmatrix}\begin{Bmatrix}u_1\\ u_2\end{Bmatrix}=\dfrac{150\times10^9}{2}\begin{pmatrix}-1 & 1\end{pmatrix}\begin{Bmatrix}0\\ 8\times10^{-3}\end{Bmatrix}=600\times10^6\,N/m^2$

b) Element 2: $\sigma_2=\dfrac{E_2}{L_2}\begin{pmatrix}-1 & 1\end{pmatrix}\begin{Bmatrix}u_2\\ u_3\end{Bmatrix}=\dfrac{150\times10^9}{2}\begin{pmatrix}-1 & 1\end{pmatrix}\begin{Bmatrix}8\times10^{-3}\\ 10.67\times10^{-3}\end{Bmatrix}=200\times10^6\,N/m^2$

c) Element 3: $\sigma_3=\dfrac{E_3}{L_3}\begin{pmatrix}-1 & 1\end{pmatrix}\begin{Bmatrix}u_3\\ u_4\end{Bmatrix}=\dfrac{150\times10^9}{2}\begin{pmatrix}-1 & 1\end{pmatrix}\begin{Bmatrix}10.67\times10^{-3}\\ 10.67\times10^{-3}\end{Bmatrix}=0$

6. The nodal forces for each element:

a) Element 1:

$$\{F\}_1=k_1\begin{pmatrix}1 & -1\\ -1 & 1\end{pmatrix}\begin{Bmatrix}u_1\\ u_2\end{Bmatrix}=37.5\times10^6\begin{pmatrix}1 & -1\\ -1 & 1\end{pmatrix}\begin{Bmatrix}0\\ 8\times10^{-3}\end{Bmatrix}=\begin{Bmatrix}-300\\ 300\end{Bmatrix}kN$$

b) Element 2:

$$\{F\}_2=k_2\begin{pmatrix}1 & -1\\ -1 & 1\end{pmatrix}\begin{Bmatrix}u_2\\ u_3\end{Bmatrix}=37.5\times10^6\begin{pmatrix}2 & -2\\ -2 & 2\end{pmatrix}\begin{Bmatrix}8\times10^{-3}\\ 10.67\times10^{-3}\end{Bmatrix}=\begin{Bmatrix}-200\\ 200\end{Bmatrix}kN$$

c) Element 3:

$$\{F\}_3=k_3\begin{pmatrix}1 & -1\\ -1 & 1\end{pmatrix}\begin{Bmatrix}u_3\\ u_4\end{Bmatrix}=37.5\times10^6\begin{pmatrix}1 & -1\\ -1 & 1\end{pmatrix}\begin{Bmatrix}10.67\times10^{-3}\\ 10.67\times10^{-3}\end{Bmatrix}=\begin{Bmatrix}0\\ 0\end{Bmatrix}kN$$

7. The conventional method:

a) For Element 1:

i) $\sum F_x=0\rightarrow-300+F_{x1}=0\rightarrow\therefore F_{x1}=300\,kN$

ii) $\sigma_{x1}=\dfrac{F_{x1}}{A_1}=\dfrac{300kN}{5\times10^{-4}m^2}=600\times10^6\,N/m^2$

iii) $\dfrac{u_2 - u_1}{L_1} = \dfrac{\sigma_{x1}}{E} = \dfrac{u_2 - 0}{2} = \dfrac{600 \times 10^6}{150 \times 10^9} \, N/m^2 \rightarrow \therefore u_2 = 8 \times 10^{-3} \, m$

b) Element 2:

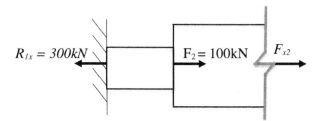

i) $\sum F_x = 0 \rightarrow -300 + 100 + F_{x2} = 0 \rightarrow \therefore F_{x2} = 200 \, kN$

ii) $\sigma_{x2} = \dfrac{F_{x2}}{A_2} = \dfrac{200kN}{10 \times 10^{-4} m^2} = 200 \times 10^6 \, N/m^2$

iii) $\dfrac{u_3 - u_2}{L_2} = \dfrac{\sigma_{x2}}{E} = \dfrac{u_3 - 8 \times 10^{-3}}{2} = \dfrac{200 \times 10^6}{150 \times 10^9} \, N/m^2 \rightarrow \therefore u_3 = 10.67 \times 10^{-3} \, m$

c) Element 3:

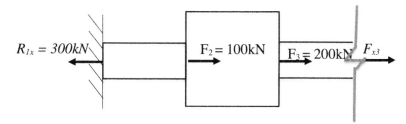

i) $\sum F_x = 0 \rightarrow -300 + 100 + 200 + F_{x3} = 0 \rightarrow \therefore F_{x3} = 0 \, kN$

ii) $\sigma_{x3} = \dfrac{F_{x3}}{A_3} = \dfrac{0kN}{5 \times 10^{-4} m^2} = 0 \, N/m^2$

iii) $\dfrac{u_4 - u_3}{L_3} = \dfrac{\sigma_{x3}}{E} = \dfrac{u_4 - 10.67 \times 10^{-3}}{2} = \dfrac{0 \times 10^6}{150 \times 10^9} \, N/m^2 \rightarrow \therefore u_4 = 10.67 \times 10^{-3} \, m$

4. Example 4R.4:

For the axially loaded structure shown in Figure 4.6.4, obtain the deflections at all the nodes, 1 to 4, wall reactions, stresses and nodal forces for each element (1 to 3) using FEM. Compare the values from the conventional method. Take $E = 150$GPa (150×10^9 N/m²). A_1 to A_3 are areas of cross section and L_1 to L_3 are the lengths of the elements respectively. Calculate the temperature up to which the elements have to be heated such that node 4 touches the boundary CD shown in the Figure iv. Coefficient of thermal expansion, $\alpha = 10^{-6}/^0C$.

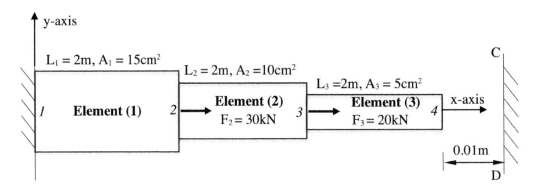

Figure 4.6.4

Solution:

1. The stiffness of each element:

a) Element 1= $k_1 = \dfrac{E_1 A_1}{L_1} = \dfrac{(150 \times 10^9)(15 \times 10^{-4})}{2} = 112.5 \times 10^6 \, N/m$

b) Element 2= $k_2 = \dfrac{E_2 A_2}{L_2} = \dfrac{(150 \times 10^9)(10 \times 10^{-4})}{2} = 75 \times 10^6 \, N/m$

c) Element 3= $k_3 = \dfrac{E_3 A_3}{L_3} = \dfrac{(150 \times 10^9)(5 \times 10^{-4})}{2} = 37.5 \times 10^6 \, N/m$

2. The element stiffness matrix:

a) Element 1: $[k_1] = 112.5 \times 10^6 \begin{bmatrix} u_1 & u_2 \\ 1 & -1 & u_1 \\ -1 & 1 & u_2 \end{bmatrix}$ b) Element 2: $[k_2] = 75 \times 10^6 \begin{bmatrix} u_2 & u_3 \\ 1 & -1 & u_2 \\ -1 & 1 & u_3 \end{bmatrix}$

c) Element 3: $[k_3] = 37.5 \times 10^6 \begin{bmatrix} u_3 & u_4 \\ 1 & -1 & u_3 \\ -1 & 1 & u_4 \end{bmatrix}$

3. The system stiffness matrix:

$$[k]_{system} = 37.5 \times 10^6 \begin{bmatrix} u_1 & u_2 & u_3 & u_4 \\ 3 & -3 & 0 & 0 & u_1 \\ -3 & 3+2 & -2 & 0 & u_2 \\ 0 & -2 & 2+1 & -1 & u_3 \\ 0 & 0 & -1 & 1 & u_4 \end{bmatrix} = 37.5 \times 10^6 \begin{bmatrix} 3 & -3 & 0 & 0 \\ -3 & 5 & -2 & 0 \\ 0 & -2 & 3 & -1 \\ 0 & 0 & -1 & 1 \end{bmatrix}$$

77

4. Based on the static equilibrium state, the system nodal force vector, stiffness matrix and displacement vector can be expressed as following:

$$\{F\} = [k]\{\delta\} = \{R\}$$

$$\{F\} = 37.5 \times 10^6 \begin{bmatrix} 3 & -3 & 0 & 0 \\ -3 & 5 & -2 & 0 \\ 0 & -2 & 3 & -1 \\ 0 & 0 & -1 & 1 \end{bmatrix} \begin{Bmatrix} u_1 = 0 \\ u_2 \\ u_3 \\ u_4 \end{Bmatrix} = \begin{Bmatrix} R_{1x} \\ 30 \times 10^3 \\ 20 \times 10^3 \\ 0 \end{Bmatrix}$$

By solving the equation, the displacement and reaction forces are computed.
$u_2 = 0.444 \times 10^{-3}$m, $u_3 = 0.711 \times 10^{-3}$m, $u_4 = 0.711 \times 10^{-3}$m, $R_{1x} = -50$ kN

5. The element stresses:

a) Element 1: $\sigma_1 = \dfrac{E_1}{L_1}\begin{pmatrix} -1 & 1 \end{pmatrix}\begin{Bmatrix} u_1 \\ u_2 \end{Bmatrix} = \dfrac{150 \times 10^9}{2}\begin{pmatrix} -1 & 1 \end{pmatrix}\begin{Bmatrix} 0 \\ 0.444 \times 10^{-3} \end{Bmatrix} = 33.3 \times 10^6 \, N/m^2$

b) Element 2: $\sigma_2 = \dfrac{E_2}{L_2}\begin{pmatrix} -1 & 1 \end{pmatrix}\begin{Bmatrix} u_2 \\ u_3 \end{Bmatrix} = \dfrac{150 \times 10^9}{2}\begin{pmatrix} -1 & 1 \end{pmatrix}\begin{Bmatrix} 0.444 \times 10^{-3} \\ 0.711 \times 10^{-3} \end{Bmatrix} = 20 \times 10^6 \, N/m^2$

c) Element 3: $\sigma_3 = \dfrac{E_3}{L_3}\begin{pmatrix} -1 & 1 \end{pmatrix}\begin{Bmatrix} u_3 \\ u_4 \end{Bmatrix} = \dfrac{150 \times 10^9}{2}\begin{pmatrix} -1 & 1 \end{pmatrix}\begin{Bmatrix} 0.711 \times 10^{-3} \\ 0.711 \times 10^{-3} \end{Bmatrix} = 0$

6. The nodal forces for each element:
a) Element 1=

$$\{F\}_1 = k_1\begin{pmatrix} 1 & -1 \\ -1 & 1 \end{pmatrix}\begin{Bmatrix} u_1 \\ u_2 \end{Bmatrix} = 37.5 \times 10^6\begin{pmatrix} 3 & -3 \\ -3 & 3 \end{pmatrix}\begin{Bmatrix} 0 \\ 0.444 \times 10^{-3} \end{Bmatrix} = \begin{Bmatrix} -50 \\ 50 \end{Bmatrix} kN$$

b) Element 2=

$$\{F\}_2 = k_2\begin{pmatrix} 1 & -1 \\ -1 & 1 \end{pmatrix}\begin{Bmatrix} u_2 \\ u_3 \end{Bmatrix} = 37.5 \times 10^6\begin{pmatrix} 2 & -2 \\ -2 & 2 \end{pmatrix}\begin{Bmatrix} 0.444 \times 10^{-3} \\ 0.711 \times 10^{-3} \end{Bmatrix} = \begin{Bmatrix} -20 \\ 20 \end{Bmatrix} kN$$

c) Element 3=

$$\{F\}_3 = k_3\begin{pmatrix} 1 & -1 \\ -1 & 1 \end{pmatrix}\begin{Bmatrix} u_3 \\ u_4 \end{Bmatrix} = 37.5 \times 10^6\begin{pmatrix} 1 & -1 \\ -1 & 1 \end{pmatrix}\begin{Bmatrix} 0.711 \times 10^{-3} \\ 0.711 \times 10^{-3} \end{Bmatrix} = \begin{Bmatrix} 0 \\ 0 \end{Bmatrix} kN$$

7. The conventional method:
a) For Element 1:

i) $\sum F_x = 0 \rightarrow -50 + F_{x1} = 0 \rightarrow \therefore F_{x1} = 50 kN$

ii) $\sigma_{x1} = \dfrac{F_{x1}}{A_1} = \dfrac{50kN}{15 \times 10^{-4} m^2} = 33.3 \times 10^6 N/m^2$

iii) $\dfrac{u_2 - u_1}{L_1} = \dfrac{\sigma_{x1}}{E} = \dfrac{u_2 - 0}{2} = \dfrac{33.3 \times 10^6}{150 \times 10^9} N/m^2 \rightarrow \therefore u_2 = 0.444 \times 10^{-3} m$

b) Element 2:

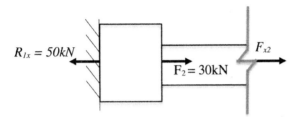

i) $\sum F_x = 0 \rightarrow -50 + 30 + F_{x2} = 0 \rightarrow \therefore F_{x2} = 20kN$

ii) $\sigma_{x2} = \dfrac{F_{x2}}{A_2} = \dfrac{20kN}{10 \times 10^{-4} m^2} = 20 \times 10^6 N/m^2$

iii) $\dfrac{u_3 - u_2}{L_2} = \dfrac{\sigma_{x2}}{E} = \dfrac{u_3 - 0.444 \times 10^{-3}}{2} = \dfrac{20 \times 10^6}{150 \times 10^9} N/m^2 \rightarrow \therefore u_3 = 0.711 \times 10^{-3} m$

c) Element 3:

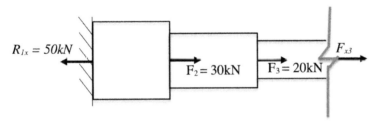

i) $\sum F_x = 0 \rightarrow -50 + 30 + 20 + F_{x3} = 0 \rightarrow \therefore F_{x3} = 0kN$

ii) $\sigma_{x3} = \dfrac{F_{x3}}{A_3} = \dfrac{0kN}{5 \times 10^{-4} m^2} = 0 N/m^2$

iii) $\dfrac{u_4 - u_3}{L_3} = \dfrac{\sigma_{x3}}{E} = \dfrac{u_4 - 0.711 \times 10^{-3}}{2} = \dfrac{0 \times 10^6}{150 \times 10^9} N/m^2 \rightarrow \therefore u_4 = 0.711 \times 10^{-3} m$

8. The temperature for heating the elements such that node 4 touches the boundary CD is:

$$\{F\} = [k]\{\delta\} + \{F\}_\varepsilon = \{R\} \; \rightarrow \qquad \text{where } \{F_{\varepsilon o}\}^e = \begin{Bmatrix} EA\alpha\Delta T \\ -EA\alpha\Delta T \end{Bmatrix} \; \rightarrow$$

i.e.

$$37.5 \times 10^6 \begin{bmatrix} 3 & -3 & 0 & 0 \\ -3 & 5 & -2 & 0 \\ 0 & -2 & 3 & -1 \\ 0 & 0 & -1 & 1 \end{bmatrix} \begin{Bmatrix} 0 \\ u_2 \\ u_3 \\ 0.01 \end{Bmatrix} + 150 \times 10^9 \times 10^{-6} \times \Delta T \begin{Bmatrix} 15 \times 10^{-4} \\ -15 \times 10^{-4} + 10 \times 10^{-4} \\ -10 \times 10^{-4} + 5 \times 10^{-4} \\ -5 \times 10^{-4} \end{Bmatrix} = \begin{Bmatrix} R_{1x} \\ 30 \times 10^3 \\ 20 \times 10^3 \\ 0 \end{Bmatrix}$$

$$37.5 \times 10^6 \begin{Bmatrix} -3u_2 \\ 5u_2 - 2u_3 \\ -2u_2 + 3u_3 - 0.01 \\ -u_3 + 0.01 \end{Bmatrix} + 15 \times \Delta T \begin{Bmatrix} 15 \\ -5 \\ -5 \\ -5 \end{Bmatrix} = \begin{Bmatrix} R_{1x} \\ 30 \times 10^3 \\ 20 \times 10^3 \\ 0 \end{Bmatrix}$$

By solving the equation, the displacement and reaction forces are computed.
$u_2 = 3.54 \times 10^{-3}$m, $u_3 = 6.904 \times 10^{-3}$m, $\Delta T = 1548.15^0$C, $R_{1x} = -50$ kN

5. Example 4R.5:

For the axially loaded structure shown in Figure 4.6.5, obtain the deflections at all the nodes, 1 to 4, wall reactions, stresses and nodal forces for each element (1 to 3) using FEM. Compare the values from the conventional method. Take $E = 50\text{GPa}$ (50×10^9 N/m²). A_1 to A_3 are areas of cross section of the elements. If all the elements are heated up by 30°C, find the displacements and wall reaction. Coefficient of thermal expansion, $\alpha = 10^{-6}$/°C.

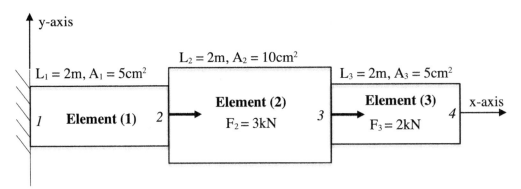

Figure 4.6.5

Solution:

1. The stiffness of each element:

   a) Element 1= $k_1 = \dfrac{E_1 A_1}{L_1} = \dfrac{(50 \times 10^9)(5 \times 10^{-4})}{2} = 12.5 \times 10^6 \, N/m$

   b) Element 2= $k_2 = \dfrac{E_2 A_2}{L_2} = \dfrac{(50 \times 10^9)(10 \times 10^{-4})}{2} = 25 \times 10^6 \, N/m$

   c) Element 3= $k_3 = \dfrac{E_3 A_3}{L_3} = \dfrac{(50 \times 10^9)(5 \times 10^{-4})}{2} = 12.5 \times 10^6 \, N/m$

2. The element stiffness matrix:

   a) Element 1= $[k_1] = 12.5 \times 10^6 \begin{array}{cc} u_1 & u_2 \\ \begin{bmatrix} 1 & -1 \\ -1 & 1 \end{bmatrix} & \begin{array}{c} u_1 \\ u_2 \end{array} \end{array}$

   b) Element 2= $[k_2] = 25 \times 10^6 \begin{array}{cc} u_2 & u_3 \\ \begin{bmatrix} 1 & -1 \\ -1 & 1 \end{bmatrix} & \begin{array}{c} u_2 \\ u_3 \end{array} \end{array}$

   c) Element 3= $[k_3] = 12.5 \times 10^6 \begin{array}{cc} u_3 & u_4 \\ \begin{bmatrix} 1 & -1 \\ -1 & 1 \end{bmatrix} & \begin{array}{c} u_3 \\ u_4 \end{array} \end{array}$

3. The system stiffness matrix:

$$[k]_{system} = 12.5 \times 10^6 \begin{bmatrix} \overset{u_1}{1} & \overset{u_2}{-1} & \overset{u_3}{0} & \overset{u_4}{0} \\ -1 & 1+2 & -2 & 0 \\ 0 & -2 & 2+1 & -1 \\ 0 & 0 & -1 & 1 \end{bmatrix} \begin{matrix} u_1 \\ u_2 \\ u_3 \\ u_4 \end{matrix} = 12.5 \times 10^6 \begin{bmatrix} 1 & -1 & 0 & 0 \\ -1 & 3 & -2 & 0 \\ 0 & -2 & 3 & -1 \\ 0 & 0 & -1 & 1 \end{bmatrix}$$

4. Based on the static equilibrium state, the system nodal force vector, stiffness matrix and displacement vector can be expressed as following:

$$\{F\} = [k]\{\delta\} = \{R\}$$

$$\{F\} = 12.5 \times 10^6 \begin{bmatrix} 1 & -1 & 0 & 0 \\ -1 & 3 & -2 & 0 \\ 0 & -2 & 3 & -1 \\ 0 & 0 & -1 & 1 \end{bmatrix} \begin{Bmatrix} u_1 = 0 \\ u_2 \\ u_3 \\ u_4 \end{Bmatrix} = \begin{Bmatrix} R_{1x} \\ 3 \times 10^3 \\ 2 \times 10^3 \\ 0 \end{Bmatrix}$$

By solving the equation, the displacement and reaction forces are computed.
$u_2 = 0.4 \times 10^{-3}$m, $u_3 = 0.48 \times 10^{-3}$m, $u_4 = 0.48 \times 10^{-3}$m, $R_{1x} = -5$ kN

5. The element stresses:

a) Element 1 = $\sigma_1 = \dfrac{E_1}{L_1}(-1 \quad 1)\begin{Bmatrix} u_1 \\ u_2 \end{Bmatrix} = \dfrac{50 \times 10^9}{2}(-1 \quad 1)\begin{Bmatrix} 0 \\ 0.4 \times 10^{-3} \end{Bmatrix} = 10 \times 10^6 \, N/m^2$

b) Element 2 = $\sigma_2 = \dfrac{E_2}{L_2}(-1 \quad 1)\begin{Bmatrix} u_2 \\ u_3 \end{Bmatrix} = \dfrac{50 \times 10^9}{2}(-1 \quad 1)\begin{Bmatrix} 0.4 \times 10^{-3} \\ 0.48 \times 10^{-3} \end{Bmatrix} = 2 \times 10^6 \, N/m^2$

c) Element 3 = $\sigma_3 = \dfrac{E_3}{L_3}(-1 \quad 1)\begin{Bmatrix} u_3 \\ u_4 \end{Bmatrix} = \dfrac{50 \times 10^9}{2}(-1 \quad 1)\begin{Bmatrix} 0.48 \times 10^{-3} \\ 0.48 \times 10^{-3} \end{Bmatrix} = 0$

6. The nodal forces for each element:

a) Element 1 = $\{F\}_1 = k_1 \begin{pmatrix} 1 & -1 \\ -1 & 1 \end{pmatrix}\begin{Bmatrix} u_1 \\ u_2 \end{Bmatrix} = 12.5 \times 10^6 \begin{pmatrix} 1 & -1 \\ -1 & 1 \end{pmatrix}\begin{Bmatrix} 0 \\ 0.4 \times 10^{-3} \end{Bmatrix} = \begin{Bmatrix} -5 \\ 5 \end{Bmatrix} kN$

b) Element 2 = $\{F\}_2 = k_2 \begin{pmatrix} 1 & -1 \\ -1 & 1 \end{pmatrix}\begin{Bmatrix} u_2 \\ u_3 \end{Bmatrix} = 12.5 \times 10^6 \begin{pmatrix} 2 & -2 \\ -2 & 2 \end{pmatrix}\begin{Bmatrix} 0.4 \times 10^{-3} \\ 0.48 \times 10^{-3} \end{Bmatrix} = \begin{Bmatrix} -2 \\ 2 \end{Bmatrix} kN$

c) Element 3 = $\{F\}_3 = k_3 \begin{pmatrix} 1 & -1 \\ -1 & 1 \end{pmatrix}\begin{Bmatrix} u_3 \\ u_4 \end{Bmatrix} = 12.5 \times 10^6 \begin{pmatrix} 1 & -1 \\ -1 & 1 \end{pmatrix}\begin{Bmatrix} 0.48 \times 10^{-3} \\ 0.48 \times 10^{-3} \end{Bmatrix} = \begin{Bmatrix} 0 \\ 0 \end{Bmatrix} kN$

7. The conventional method:
   a) For Element 1:

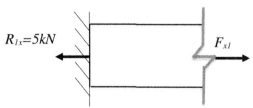

$R_{1x}=5kN$

F_{x1}

i) $\sum F_x = 0 \rightarrow -5 + F_{x1} = 0 \rightarrow \therefore F_{x1} = 5kN$

ii) $\sigma_{x1} = \dfrac{F_{x1}}{A_1} = \dfrac{5kN}{5 \times 10^{-4} m^2} = 10 \times 10^6 \, N/m^2$

iii) $\dfrac{u_2 - u_1}{L_1} = \dfrac{\sigma_{x1}}{E} = \dfrac{u_2 - 0}{2} = \dfrac{10 \times 10^6}{50 \times 10^9} \, N/m^2 \rightarrow \therefore u_2 = 0.4 \times 10^{-3} m$

b) Element 2:

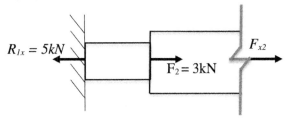

i) $\sum F_x = 0 \rightarrow -5 + 3 + F_{x2} = 0 \rightarrow \therefore F_{x2} = 2kN$

ii) $\sigma_{x2} = \dfrac{F_{x2}}{A_2} = \dfrac{2kN}{10 \times 10^{-4} m^2} = 2 \times 10^6 \, N/m^2$

iii) $\dfrac{u_3 - u_2}{L_2} = \dfrac{\sigma_{x2}}{E} = \dfrac{u_3 - 0.4 \times 10^{-3}}{2} = \dfrac{2 \times 10^6}{50 \times 10^9} \, N/m^2 \rightarrow \therefore u_3 = 0.48 \times 10^{-3} m$

c) Element 3:

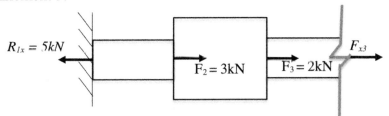

i) $\sum F_x = 0 \rightarrow -5 + 3 + 2 + F_{x3} = 0 \rightarrow \therefore F_{x3} = 0kN$

iii) $\sigma_{x3} = \dfrac{F_{x3}}{A_3} = \dfrac{0kN}{5 \times 10^{-4} m^2} = 0 \, N/m^2$

iii) $\dfrac{u_4 - u_3}{L_3} = \dfrac{\sigma_{x3}}{E} = \dfrac{u_4 - 0.48 \times 10^{-3}}{2} = \dfrac{0 \times 10^6}{50 \times 10^9} \, N/m^2 \rightarrow \therefore u_4 = 0.48 \times 10^{-3} m$

8. The thermal nodal force vector due to the heat up:
   a) Element 1:

   $$\{F\}_1 = \left\{ \begin{array}{c} E_1 A_1 \alpha\, \Delta T_1 \\ -E_1 A_1 \alpha\, \Delta T_1 \end{array} \right\} = \left\{ \begin{array}{c} 50\times10^9\times5\times10^{-4}\times10^{-6}\times30 \\ -50\times10^9\times5\times10^{-4}\times10^{-6}\times30 \end{array} \right\} = \left\{ \begin{array}{c} 750 \\ -750 \end{array} \right\} N$$

   b) Element 2:

   $$\{F\}_2 = \left\{ \begin{array}{c} E_2 A_2 \alpha\, \Delta T_2 \\ -E_2 A_2 \alpha\, \Delta T_2 \end{array} \right\} = \left\{ \begin{array}{c} 50\times10^9\times10\times10^{-4}\times10^{-6}\times30 \\ -50\times10^9\times10\times10^{-4}\times10^{-6}\times30 \end{array} \right\} = \left\{ \begin{array}{c} 1500 \\ -1500 \end{array} \right\} N$$

   c) Element 3:

   $$\{F\}_3 = \left\{ \begin{array}{c} E_3 A_3 \alpha\, \Delta T_3 \\ -E_3 A_3 \alpha\, \Delta T_3 \end{array} \right\} = \left\{ \begin{array}{c} 50\times10^9\times5\times10^{-4}\times10^{-6}\times30 \\ -50\times10^9\times5\times10^{-4}\times10^{-6}\times30 \end{array} \right\} = \left\{ \begin{array}{c} 750 \\ -750 \end{array} \right\} N$$

9. After the system is heated up, the revised equilibrium equation can be expressed as:

   $$\{F\} = [k]\{\delta\} + \{F\}_p + \{F\}_\varepsilon = \{R\}$$

   $$\{F\} = 12.5\times10^6 \begin{bmatrix} 1 & -1 & 0 & 0 \\ -1 & 3 & -2 & 0 \\ 0 & -2 & 3 & -1 \\ 0 & 0 & -1 & 1 \end{bmatrix} \left\{ \begin{array}{c} u_1 = 0 \\ u_2 \\ u_3 \\ u_4 \end{array} \right\} + \{0\} + \left\{ \begin{array}{c} 750 \\ -750+1500 \\ -1500+750 \\ -750 \end{array} \right\} = \left\{ \begin{array}{c} R_{1x} \\ 3\times10^3 \\ 2\times10^3 \\ 0 \end{array} \right\}$$

   $$12.5\times10^6 \begin{bmatrix} 1 & -1 & 0 & 0 \\ -1 & 3 & -2 & 0 \\ 0 & -2 & 3 & -1 \\ 0 & 0 & -1 & 1 \end{bmatrix} \left\{ \begin{array}{c} 0 \\ u_2 \\ u_3 \\ u_4 \end{array} \right\} + \left\{ \begin{array}{c} 750 \\ 750 \\ -750 \\ -750 \end{array} \right\} = \left\{ \begin{array}{c} R_{1x} \\ 3\times10^3 \\ 2\times10^3 \\ 0 \end{array} \right\}$$

   By solving the equation, the displacement and reaction forces are computed.
   $u_2 = 0.46\times10^{-3}$m, $u_3 = 0.6\times10^{-3}$m, $u_4 = 0.66\times10^{-3}$m, $R_{1x} = -5$ kN

6. Example 4R.6:

For the axially loaded structure shown in Figure 4.6.6, take $E = 100\text{GPa}$ $(100 \times 10^9$ $\text{N/m}^2)$. A_1 to A_3 are areas of cross section and L_1 to L_3 are the lengths of the elements respectively.

a) Obtain the deflections at all the nodes, 1 to 4, wall reactions, stresses and nodal forces for each element (1 to 3) using FEM.

b) Compare the values from the conventional method.

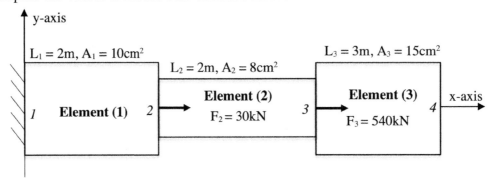

Figure 4.6.6

Solution:

1. The stiffness of each element:

a) Element 1= $k_1 = \dfrac{E_1 A_1}{L_1} = \dfrac{(100 \times 10^9)(10 \times 10^{-4})}{2} = 50 \times 10^6 \, N/m$

b) Element 2= $k_2 = \dfrac{E_2 A_2}{L_2} = \dfrac{(100 \times 10^9)(8 \times 10^{-4})}{2} = 40 \times 10^6 \, N/m$

c) Element 3= $k_3 = \dfrac{E_3 A_3}{L_3} = \dfrac{(100 \times 10^9)(15 \times 10^{-4})}{3} = 50 \times 10^6 \, N/m$

2. The element stiffness matrix:

a) Element 1= $[k_1] = 10^7 \begin{bmatrix} u_1 & u_2 & \\ 5 & -5 & u_1 \\ -5 & 5 & u_2 \end{bmatrix}$

b) Element 2= $[k_2] = 10^7 \begin{bmatrix} u_2 & u_3 & \\ 4 & -4 & u_2 \\ -4 & 4 & u_3 \end{bmatrix}$

c) Element 3= $[k_3] = 10^7 \begin{bmatrix} u_3 & u_4 & \\ 5 & -5 & u_3 \\ -5 & 5 & u_4 \end{bmatrix}$

3. The system stiffness matrix:

$$[k]_{system} = 10^7 \begin{bmatrix} u_1 & u_2 & u_3 & u_4 & \\ 5 & -5 & 0 & 0 & u_1 \\ -5 & 5+4 & -4 & 0 & u_2 \\ 0 & -4 & 4+5 & -5 & u_3 \\ 0 & 0 & -5 & 5 & u_4 \end{bmatrix} = 10^7 \begin{bmatrix} 5 & -5 & 0 & 0 \\ -5 & 9 & -4 & 0 \\ 0 & -4 & 9 & -5 \\ 0 & 0 & -5 & 5 \end{bmatrix}$$

4. Based on the static equilibrium state, the system nodal force vector, stiffness matrix and displacement vector can be expressed as following:

$$\{F\}=[k]\{\delta\}=\{R\}$$

$$\{F\}=10^7\begin{bmatrix} 5 & -5 & 0 & 0 \\ -5 & 9 & -4 & 0 \\ 0 & -4 & 9 & -5 \\ 0 & 0 & -5 & 5 \end{bmatrix}\begin{Bmatrix} u_1=0 \\ u_2 \\ u_3 \\ u_4 \end{Bmatrix}=\begin{Bmatrix} R_{1x} \\ 30\times10^3 \\ 540\times10^3 \\ 0 \end{Bmatrix}$$

By solving the equation, the displacement and reaction forces are computed.
$u_2 = 0.0114\,m$, $u_3 = 0.0249\,m$, $u_4 = 0.0249\,m$, $R_{1x} = -570\,kN$

5. The element stresses:

a) Element 1= $\sigma_1 = \dfrac{E_1}{L_1}\begin{pmatrix} -1 & 1 \end{pmatrix}\begin{Bmatrix} u_1 \\ u_2 \end{Bmatrix} = \dfrac{100\times10^9}{2}\begin{pmatrix} -1 & 1 \end{pmatrix}\begin{Bmatrix} 0 \\ 0.0114 \end{Bmatrix} = 570\times10^6\,N/m^2$

b) Element 2= $\sigma_2 = \dfrac{E_2}{L_2}\begin{pmatrix} -1 & 1 \end{pmatrix}\begin{Bmatrix} u_2 \\ u_3 \end{Bmatrix} = \dfrac{100\times10^9}{2}\begin{pmatrix} -1 & 1 \end{pmatrix}\begin{Bmatrix} 0.0114 \\ 0.0249 \end{Bmatrix} = 675\times10^6\,N/m^2$

c) Element 3= $\sigma_3 = \dfrac{E_3}{L_3}\begin{pmatrix} -1 & 1 \end{pmatrix}\begin{Bmatrix} u_3 \\ u_4 \end{Bmatrix} = \dfrac{100\times10^9}{2}\begin{pmatrix} -1 & 1 \end{pmatrix}\begin{Bmatrix} 0.0249 \\ 0.0249 \end{Bmatrix} = 0$

6. The nodal forces for each element:

a) Element 1= $\{F\}_1 = k_1\begin{pmatrix} 1 & -1 \\ -1 & 1 \end{pmatrix}\begin{Bmatrix} u_1 \\ u_2 \end{Bmatrix} = 5\times10^7\begin{pmatrix} 1 & -1 \\ -1 & 1 \end{pmatrix}\begin{Bmatrix} 0 \\ 0.0114 \end{Bmatrix} = \begin{Bmatrix} -570 \\ 570 \end{Bmatrix}kN$

b) Element 2= $\{F\}_2 = k_2\begin{pmatrix} 1 & -1 \\ -1 & 1 \end{pmatrix}\begin{Bmatrix} u_2 \\ u_3 \end{Bmatrix} = 4\times10^7\begin{pmatrix} 1 & -1 \\ -1 & 1 \end{pmatrix}\begin{Bmatrix} 0.0114 \\ 0.0249 \end{Bmatrix} = \begin{Bmatrix} -540 \\ 540 \end{Bmatrix}kN$

c) Element 3= $\{F\}_3 = k_3\begin{pmatrix} 1 & -1 \\ -1 & 1 \end{pmatrix}\begin{Bmatrix} u_3 \\ u_4 \end{Bmatrix} = 5\times10^7\begin{pmatrix} 1 & -1 \\ -1 & 1 \end{pmatrix}\begin{Bmatrix} 0.0249 \\ 0.0249 \end{Bmatrix} = \begin{Bmatrix} 0 \\ 0 \end{Bmatrix}kN$

7. The conventional method:
a) For Element 1:

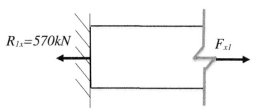

i) $\sum F_x = 0 \rightarrow -570 + F_{x1} = 0 \rightarrow \therefore F_{x1} = 570kN$

ii) $\sigma_{x1} = \dfrac{F_{x1}}{A_1} = \dfrac{570kN}{10\times10^{-4}m^2} = 570\times10^6\,N/m^2$

iii) $\dfrac{u_2 - u_1}{L_1} = \dfrac{\sigma_{x1}}{E} = \dfrac{u_2 - 0}{2} = \dfrac{570 \times 10^6}{100 \times 10^9} N/m^2 \rightarrow \therefore u_2 = 0.0114m$

b) Element 2:

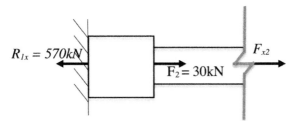

i) $\sum F_x = 0 \rightarrow -570 + 30 + F_{x2} = 0 \rightarrow \therefore F_{x2} = 540kN$

ii) $\sigma_{x2} = \dfrac{F_{x2}}{A_2} = \dfrac{540kN}{8 \times 10^{-4} m^2} = 675 \times 10^6 N/m^2$

iii) $\dfrac{u_3 - u_2}{L_2} = \dfrac{\sigma_{x2}}{E} = \dfrac{u_3 - 0.0114}{2} = \dfrac{675 \times 10^6}{100 \times 10^9} N/m^2 \rightarrow \therefore u_3 = 0.0249m$

c) Element 3:

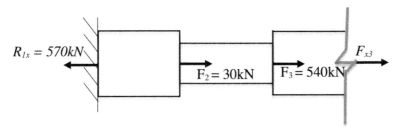

i) $\sum F_x = 0 \rightarrow -570 + 30 + 540 + F_{x3} = 0 \rightarrow \therefore F_{x3} = 0kN$

ii) $\sigma_{x3} = \dfrac{F_{x3}}{A_3} = \dfrac{0kN}{15 \times 10^{-4} m^2} = 0 N/m^2$

iii) $\dfrac{u_4 - u_3}{L_3} = \dfrac{\sigma_{x3}}{E} = \dfrac{u_4 - 0.0249}{3} = \dfrac{0 \times 10^6}{50 \times 10^9} N/m^2 \rightarrow \therefore u_4 = 0.0249m$

7. Example 4R.7:

For the axially loaded structure shown in Figure 4.6.7, obtain the deflections at all the nodes, (1 to 4), wall reactions, stresses and nodal forces for each element (1 to 3) using FEM. Compare the values from the conventional method. Take $E = 100$GPa (100×10^9 N/m²). A_1 to A_3 are areas of cross section and L_1 to L_3 are the lengths of the elements respectively. Calculate the temperature up to which the elements have to be heated such that node 4 touches the boundary CD shown in the Figure *vii*. Coefficient of thermal expansion, $\alpha = 10^{-6}/^{0}$C.

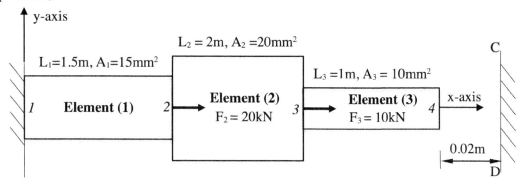

Figure 4.6.7

Solution:

1. The stiffness of each element:

a) Element 1= $k_1 = \dfrac{E_1 A_1}{L_1} = \dfrac{(100 \times 10^9)(15 \times 10^{-6})}{1.5} = 1 \times 10^6 \, N/m$

b) Element 2= $k_2 = \dfrac{E_2 A_2}{L_2} = \dfrac{(100 \times 10^9)(20 \times 10^{-6})}{2} = 1 \times 10^6 \, N/m$

c) Element 3= $k_3 = \dfrac{E_3 A_3}{L_3} = \dfrac{(100 \times 10^9)(10 \times 10^{-6})}{1} = 1 \times 10^6 \, N/m$

2. The element stiffness matrix:

a) Element 1=
$$[k_1] = 10^6 \begin{bmatrix} u_1 & u_2 & \\ 1 & -1 & u_1 \\ -1 & 1 & u_2 \end{bmatrix}$$

b) Element 2=
$$[k_2] = 10^6 \begin{bmatrix} u_2 & u_3 & \\ 1 & -1 & u_2 \\ -1 & 1 & u_3 \end{bmatrix}$$

c) Element 3=
$$[k_3] = 10^6 \begin{bmatrix} u_3 & u_4 & \\ 1 & -1 & u_3 \\ -1 & 1 & u_4 \end{bmatrix}$$

3. The system stiffness matrix:

$$[k]_{system} = 10^6 \begin{bmatrix} u_1 & u_2 & u_3 & u_4 & \\ 1 & -1 & 0 & 0 & u_1 \\ -1 & 1+1 & -1 & 0 & u_2 \\ 0 & -1 & 1+1 & -1 & u_3 \\ 0 & 0 & -1 & 1 & u_4 \end{bmatrix} = 10^6 \begin{bmatrix} 1 & -1 & 0 & 0 \\ -1 & 2 & -1 & 0 \\ 0 & -1 & 2 & -1 \\ 0 & 0 & -1 & 1 \end{bmatrix}$$

4. Based on the static equilibrium state, the system nodal force vector, stiffness matrix and displacement vector can be expressed as following:

$$\{F\}=[k]\{\delta\}=\{R\}$$

$$\{F\}=10^6\begin{bmatrix}1 & -1 & 0 & 0\\ -1 & 2 & -1 & 0\\ 0 & -1 & 2 & -1\\ 0 & 0 & -1 & 1\end{bmatrix}\begin{Bmatrix}u_1=0\\ u_2\\ u_3\\ u_4\end{Bmatrix}=\begin{Bmatrix}R_{1x}\\ 20\times10^3\\ 10\times10^3\\ 0\end{Bmatrix}$$

By solving the equation, the displacement and reaction forces are computed.
$u_2=0.03$ m, $u_3=0.04$ m, $u_4=0.04$ m, $R_{1x}=-30$ kN

5. The element stresses:

a) Element 1=$\sigma_1=\dfrac{E_1}{L_1}\begin{pmatrix}-1 & 1\end{pmatrix}\begin{Bmatrix}u_1\\ u_2\end{Bmatrix}=\dfrac{100\times10^9}{1.5}\begin{pmatrix}-1 & 1\end{pmatrix}\begin{Bmatrix}0\\ 0.03\end{Bmatrix}=2000\times10^6\,N/m^2$

b) Element 2=$\sigma_2=\dfrac{E_2}{L_2}\begin{pmatrix}-1 & 1\end{pmatrix}\begin{Bmatrix}u_2\\ u_3\end{Bmatrix}=\dfrac{100\times10^9}{2}\begin{pmatrix}-1 & 1\end{pmatrix}\begin{Bmatrix}0.03\\ 0.04\end{Bmatrix}=500\times10^6\,N/m^2$

c) Element 3=$\sigma_3=\dfrac{E_3}{L_3}\begin{pmatrix}-1 & 1\end{pmatrix}\begin{Bmatrix}u_3\\ u_4\end{Bmatrix}=\dfrac{100\times10^9}{1}\begin{pmatrix}-1 & 1\end{pmatrix}\begin{Bmatrix}0.04\\ 0.04\end{Bmatrix}=0$

6. The nodal forces for each element:

a) Element 1=$\{F\}_1=k_1\begin{pmatrix}1 & -1\\ -1 & 1\end{pmatrix}\begin{Bmatrix}u_1\\ u_2\end{Bmatrix}=1\times10^6\begin{pmatrix}1 & -1\\ -1 & 1\end{pmatrix}\begin{Bmatrix}0\\ 0.03\end{Bmatrix}=\begin{Bmatrix}-30\\ 30\end{Bmatrix}kN$

b) Element 2=$\{F\}_2=k_2\begin{pmatrix}1 & -1\\ -1 & 1\end{pmatrix}\begin{Bmatrix}u_2\\ u_3\end{Bmatrix}=1\times10^6\begin{pmatrix}1 & -1\\ -1 & 1\end{pmatrix}\begin{Bmatrix}0.03\\ 0.04\end{Bmatrix}=\begin{Bmatrix}-10\\ 10\end{Bmatrix}kN$

c) Element 3= $\{F\}_3=k_3\begin{pmatrix}1 & -1\\ -1 & 1\end{pmatrix}\begin{Bmatrix}u_3\\ u_4\end{Bmatrix}=1\times10^6\begin{pmatrix}1 & -1\\ -1 & 1\end{pmatrix}\begin{Bmatrix}0.04\\ 0.04\end{Bmatrix}=\begin{Bmatrix}0\\ 0\end{Bmatrix}kN$

7. The conventional method:
   a) For Element 1:

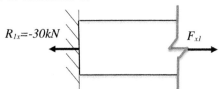

$R_{1x}=-30kN$ F_{x1}

i) $\sum F_x=0\rightarrow-30+F_{x1}=0\rightarrow\therefore F_{x1}=30kN$

ii) $\sigma_{x1}=\dfrac{F_{x1}}{A_1}=\dfrac{30kN}{15\times10^{-6}m^2}=2000\times10^6\,N/m^2$

iii) $\dfrac{u_2-u_1}{L_1}=\dfrac{\sigma_{x1}}{E}=\dfrac{u_2-0}{1.5}=\dfrac{2000\times10^6}{100\times10^9}\,N/m^2\rightarrow\therefore u_2=0.3m$

b) Element 2:

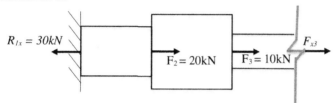

i) $\sum F_x = 0 \rightarrow -30 + 20 + F_{x2} = 0 \rightarrow \therefore F_{x2} = 10 kN$

ii) $\sigma_{x2} = \dfrac{F_{x2}}{A_2} = \dfrac{10kN}{20 \times 10^{-6} m^2} = 500 \times 10^6 \, N/m^2$

iii) $\dfrac{u_3 - u_2}{L_2} = \dfrac{\sigma_{x2}}{E} = \dfrac{u_3 - 0.03}{2} = \dfrac{500 \times 10^6}{100 \times 10^9} \, N/m^2 \rightarrow \therefore u_3 = 0.04m$

c) Element 3:

i) $\sum F_x = 0 \rightarrow -30 + 20 + 10 + F_{x3} = 0 \rightarrow \therefore F_{x3} = 0 kN$

ii) $\sigma_{x3} = \dfrac{F_{x3}}{A_3} = \dfrac{0kN}{10 \times 10^{-6} m^2} = 0 \, N/m^2$

iii) $\dfrac{u_4 - u_3}{L_3} = \dfrac{\sigma_{x3}}{E} = \dfrac{u_4 - 0.04}{1} = \dfrac{0 \times 10^6}{100 \times 10^9} \, N/m^2 \rightarrow \therefore u_4 = 0.04m$

8. The temperature for heating the elements such that node 4 touches the boundary CD is:

$$\{F\} = [k]\{\delta\} + \{F\}_\varepsilon = \{R\} \quad \rightarrow \text{where} \quad \{F_{\varepsilon o}\}^e = \left\{ \begin{array}{c} EA\alpha\Delta T \\ -EA\alpha\Delta T \end{array} \right\}$$

$$10^6 \begin{bmatrix} 1 & -1 & 0 & 0 \\ -1 & 2 & -1 & 0 \\ 0 & -1 & 2 & -1 \\ 0 & 0 & -1 & 1 \end{bmatrix} \left\{ \begin{array}{c} 0 \\ u_2 \\ u_3 \\ 0.02 \end{array} \right\} + 100 \times 10^9 \times 10^{-6} \times \Delta T \left\{ \begin{array}{c} 15 \times 10^{-6} \\ -15 \times 10^{-6} + 20 \times 10^{-6} \\ -20 \times 10^{-4} + 10 \times 10^{-6} \\ -10 \times 10^{-6} \end{array} \right\} = \left\{ \begin{array}{c} R_{1x} \\ 20 \times 10^3 \\ 10 \times 10^3 \\ 0 \end{array} \right\}$$

$$10^6 \left\{ \begin{array}{c} -u_2 \\ 2u_2 - u_3 \\ -u_2 + 2u_3 - 0.02 \\ -u_3 + 0.02 \end{array} \right\} + 0.1 \times \Delta T \left\{ \begin{array}{c} 15 \\ 5 \\ -10 \\ -10 \end{array} \right\} = \left\{ \begin{array}{c} R_{1x} \\ 20 \times 10^3 \\ 10 \times 10^3 \\ 0 \end{array} \right\}$$

By solving the equation, the displacement and reaction forces are computed.
$u_2 = 0.0233$ m, $u_3 = 0.0244$ m, $T = -4444.44^0C$, $R_{1x} = -30$ kN.

4.6.2 Examples of Trusses

1. Example 4T.1

Compute member forces, displacements of nodes and stresses in the two-bar truss shown in Figure 4.6.8 using FEM. For all the truss elements, $A = 10$ cm^2, $E = 100$ GPa, where A = area of cross section, E = modulus of elasticity, and L = length of members. All the joints are hinged. Verify the results using method of joints.

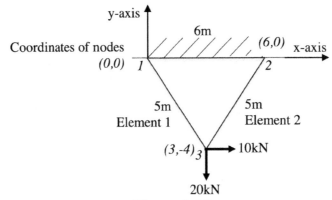

Figure 4.6.8

Solution:

1. The stiffness of each element:

   a) Element 1:

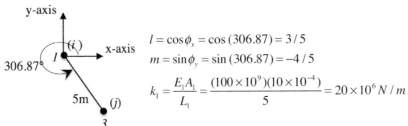

$$l = \cos\phi_x = \cos(306.87) = 3/5$$
$$m = \sin\phi_y = \sin(306.87) = -4/5$$
$$k_1 = \frac{E_1 A_1}{L_1} = \frac{(100\times10^9)(10\times10^{-4})}{5} = 20\times10^6\, N/m$$

Also they can be calculated using the coordinates of the nodes as:
(Taking $i = 1, j = 3$)

$$l = \frac{x_j - x_i}{L} = \frac{x_3 - x_1}{L} = \frac{3}{5} \qquad m = \frac{y_j - y_i}{L} = \frac{y_3 - y_1}{L} = -\frac{4}{5}$$

$$[k_1]_{global} = \frac{E_1 A_1}{L_1}\begin{bmatrix} l^2 & lm & -l^2 & -lm \\ lm & m^2 & -lm & -m^2 \\ -l^2 & -lm & l^2 & lm \\ -lm & -m^2 & lm & m^2 \end{bmatrix} = \frac{20\times10^6}{25}\begin{bmatrix} 9 & -12 & -9 & 12 \\ -12 & 16 & 12 & -16 \\ -9 & 12 & 9 & -12 \\ 12 & -16 & -12 & 16 \end{bmatrix} = 8\times10^5\begin{bmatrix} \overset{u_1}{9} & \overset{v_1}{-12} & \overset{u_3}{-9} & \overset{v_3}{12} \\ -12 & 16 & 12 & -16 \\ -9 & 12 & 9 & -12 \\ 12 & -16 & -12 & 16 \end{bmatrix}\begin{matrix} u_1 \\ v_1 \\ u_3 \\ v_3 \end{matrix}$$

   b) Element 2:

$$l = \cos\phi_x = \cos(233.13) = -3/5$$
$$m = \sin\phi_y = \sin(233.13) = -4/5$$
$$k_2 = \frac{E_2 A_2}{L_2} = \frac{(100\times10^9)(10\times10^{-4})}{5} = 20\times10^6\, N/m$$

(Also, taking $i = 2, j = 3$),

$$l = \frac{x_j - x_i}{L} = \frac{x_3 - x_2}{L} = \frac{3 - 6}{5} = -\frac{3}{5} \qquad m = \frac{y_j - y_i}{L} = \frac{y_3 - y_2}{L} = \frac{-4 - 0}{5} = -\frac{4}{5}$$

$$[k_2]_{global} = \frac{E_2 A_2}{L_2}\begin{bmatrix} l^2 & lm & -l^2 & -lm \\ lm & m^2 & -lm & -m^2 \\ -l^2 & -lm & l^2 & lm \\ -lm & -m^2 & lm & m^2 \end{bmatrix} = \frac{20 \times 10^6}{25}\begin{bmatrix} 9 & 12 & -9 & -12 \\ 12 & 16 & -12 & -16 \\ -9 & -12 & 9 & 12 \\ -12 & -16 & 12 & 16 \end{bmatrix} = 8 \times 10^5 \begin{array}{cccc} u_2 & v_2 & u_3 & v_3 \\ \begin{bmatrix} 9 & 12 & -9 & -12 \\ 12 & 16 & -12 & -16 \\ -9 & -12 & 9 & 12 \\ -12 & -16 & 12 & 16 \end{bmatrix} & & & \begin{array}{c} u_2 \\ v_2 \\ u_3 \\ v_3 \end{array} \end{array}$$

2. The system stiffness matrix:

$$[k]_{system} = 8 \times 10^5 \begin{array}{cccccc} u_1 & v_1 & u_2 & v_2 & u_3 & v_3 \\ \begin{bmatrix} 9 & -12 & 0 & 0 & -9 & 12 \\ -12 & 16 & 0 & 0 & 12 & -16 \\ 0 & 0 & 9 & 12 & -9 & -12 \\ 0 & 0 & 12 & 16 & -12 & -16 \\ -9 & 12 & -9 & -12 & 9+9 & -12+12 \\ 12 & -16 & -12 & -16 & -12+12 & 16+16 \end{bmatrix} & & & & & \begin{array}{c} u_1 \\ v_1 \\ u_2 \\ v_2 \\ u_3 \\ v_3 \end{array} \end{array} = 8 \times 10^5 \begin{bmatrix} 9 & -12 & 0 & 0 & -9 & 12 \\ -12 & 16 & 0 & 0 & 12 & -16 \\ 0 & 0 & 9 & 12 & -9 & -12 \\ 0 & 0 & 12 & 16 & -12 & -16 \\ -9 & 12 & -9 & -12 & 18 & 0 \\ 12 & -16 & -12 & -16 & 0 & 32 \end{bmatrix}$$

3. Based on the static equilibrium state, the system nodal force vector, stiffness matrix and displacement vector can be expressed as following:

$$\{F\} = [k]\{\delta\} = \{R\}$$

$$8 \times 10^5 \begin{bmatrix} 9 & -12 & 0 & 0 & -9 & 12 \\ -12 & 16 & 0 & 0 & 12 & -16 \\ 0 & 0 & 9 & 12 & -9 & -12 \\ 0 & 0 & 12 & 16 & -12 & -16 \\ -9 & 12 & -9 & -12 & 18 & 0 \\ 12 & -16 & -12 & -16 & 0 & 32 \end{bmatrix} \begin{Bmatrix} u_1 = 0 \\ v_1 = 0 \\ u_2 = 0 \\ v_2 = 0 \\ u_3 \\ v_3 \end{Bmatrix} = \begin{Bmatrix} R_{1x} \\ R_{1y} \\ R_{2x} \\ R_{2y} \\ 10000 \\ -20000 \end{Bmatrix}$$

By solving the equation, the displacement and reaction forces are computed.
$u_3 = 0.694 \times 10^{-3}$m, $v_3 = -0.781 \times 10^{-3}$m, $R_{1x} = -12.5$kN, $R_{1y} = 16.66$ kN, $R_{2x} = 2.5$kN, $R_{2y} = 3.33$ kN

4. The nodal forces for each element:
   a) Element 1:

$$\begin{Bmatrix} f_{x1} \\ f_{x3} \end{Bmatrix}_1 = 20 \times 10^6 \begin{pmatrix} 1 & -1 \\ -1 & 1 \end{pmatrix} \begin{bmatrix} 3/5 & -4/5 & 0 & 0 \\ 0 & 0 & 3/5 & -4/5 \end{bmatrix} \begin{Bmatrix} 0 \\ 0 \\ 0.694 \times 10^{-3} \\ -0.781 \times 10^{-3} \end{Bmatrix} = \begin{Bmatrix} -20.824 \\ 20.824 \end{Bmatrix} kN$$

   b) Element 2 =

$$\begin{Bmatrix} f_{x2} \\ f_{x3} \end{Bmatrix}_2 = 20 \times 10^6 \begin{pmatrix} 1 & -1 \\ -1 & 1 \end{pmatrix} \begin{bmatrix} -3/5 & -4/5 & 0 & 0 \\ 0 & 0 & -3/5 & -4/5 \end{bmatrix} \begin{Bmatrix} 0 \\ 0 \\ 0.694 \times 10^{-3} \\ -0.781 \times 10^{-3} \end{Bmatrix} = \begin{Bmatrix} -4.168 \\ 4.168 \end{Bmatrix} kN$$

5. The stresses in each element:
   a) Element 1 =

$$\sigma_1 = \left(\frac{E}{L}\right)_1 \{ -1 \quad 1 \} \begin{bmatrix} l_1 & m_1 & 0 & 0 \\ 0 & 0 & l_1 & m_1 \end{bmatrix} \begin{Bmatrix} u_1 \\ v_1 \\ u_3 \\ v_3 \end{Bmatrix}$$

$$\sigma_1 = \frac{100 \times 10^9}{5} (-1 \quad 1) \begin{Bmatrix} 3/5 & -4/5 & 0 & 0 \\ 0 & 0 & 3/5 & -4/5 \end{Bmatrix} \begin{Bmatrix} 0 \\ 0 \\ 0.694 \times 10^{-3} \\ -0.781 \times 10^{-3} \end{Bmatrix} = \{20.824 \times 10^6\} N/m^2$$
(Tension)

b) Element 2=

$$\sigma_2 = \left(\frac{E}{L}\right)_2 \{ -1 \quad 1 \} \begin{bmatrix} l_2 & m_2 & 0 & 0 \\ 0 & 0 & l_2 & m_2 \end{bmatrix} \begin{Bmatrix} u_2 \\ v_2 \\ u_3 \\ v_3 \end{Bmatrix}$$

$$\sigma_2 = \frac{100 \times 10^9}{5} (-1 \quad 1) \begin{Bmatrix} -3/5 & -4/5 & 0 & 0 \\ 0 & 0 & -3/5 & -4/5 \end{Bmatrix} \begin{Bmatrix} 0 \\ 0 \\ 0.694 \times 10^{-3} \\ -0.781 \times 10^{-3} \end{Bmatrix} = \{4.168 \times 10^6\} N/m^2$$
(Tension)

6. The conventional method:

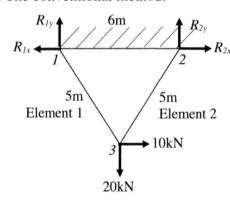

Support Reaction:

$$\sum M_1 = 0 \rightarrow R_{2y}(6) + 10(4) - 20(3) = 0$$
$$\therefore R_{2y} = 3.33 kN$$
$$+\uparrow \sum F_y = 0 \rightarrow R_{1y} + R_{2y} - 20 = 0$$
$$\therefore R_{1y} = 16.67 kN$$

a) At Joint 1:

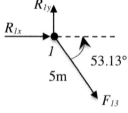

$$+\uparrow \sum F_y = 0 \rightarrow -F_{13}\sin(53.13°) + R_{1y} = 0$$
$$\therefore F_{13} = \frac{16.67 kN}{\sin(53.13°)} = 20.83 kN$$

b) At Joint 2:

$$+\uparrow \sum F_y = 0 \rightarrow -F_{23}\sin(53.13°) + R_{2y} = 0$$
$$\therefore F_{13} = \frac{3.33 kN}{\sin(53.13°)} = 4.16 kN$$

2. Example 4T.2

Determine the reactions, member forces and deflections of the truss members 1-2 and 2-3 with hinged connections at nodes 1, 2 and 3 as in Figure 4.6.9, using FEM. Use one truss element for each element. The coordinates (in meters) are given in the brackets for each node. Compare the member forces from the conventional method (method of joints). Assume area of cross section, A (100 cm²) and modulus of elasticity E (100 GPa) to be the same for all the members.

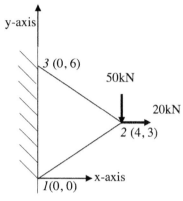

Figure 4.6.9

Solution:

1. The stiffness of each element:

   a) Element 1:

$$l = \cos\phi_x = \cos(36.87) = 4/5 \qquad (= \frac{x_j - x_i}{L} = \frac{4-0}{5} = \frac{4}{5})$$

$$m = \sin\phi_y = \sin(36.87) = 3/5 \qquad (= \frac{y_j - y_i}{L} = \frac{3-0}{5} = \frac{3}{5})$$

$$k_1 = \frac{E_1 A_1}{L_1} = \frac{(100\times10^9)(100\times10^{-4})}{5} = 200\times10^6\,N/m$$

$$[k_1]_{global} = \frac{E_1 A_1}{L_1} \begin{bmatrix} l^2 & lm & -l^2 & -lm \\ lm & m^2 & -lm & -m^2 \\ -l^2 & -lm & l^2 & lm \\ -lm & -m^2 & lm & m^2 \end{bmatrix} = \frac{200\times10^6}{25} \begin{bmatrix} 16 & 12 & -16 & -12 \\ 12 & 9 & -12 & -9 \\ -16 & -12 & 16 & 12 \\ -12 & -9 & 12 & 9 \end{bmatrix} = 8\times10^6 \begin{matrix} u_1 & v_1 & u_2 & v_2 \\ \begin{bmatrix} 16 & 12 & -16 & -12 \\ 12 & 9 & -12 & -9 \\ -16 & -12 & 16 & 12 \\ -12 & -9 & 12 & 9 \end{bmatrix} & \begin{matrix} u_1 \\ v_1 \\ u_2 \\ v_2 \end{matrix} \end{matrix}$$

   b) Element 2:

$$l = \cos\phi_x = \cos(143.13) = -4/5 \qquad (= \frac{x_j - x_i}{L} = \frac{0-4}{5} = -\frac{4}{5})$$

$$m = \sin\phi_y = \sin(143.13) = 3/5 \qquad (= \frac{y_j - y_i}{L} = \frac{6-3}{5} = \frac{3}{5})$$

$$k_2 = \frac{E_2 A_2}{L_2} = \frac{(100\times10^9)(100\times10^{-4})}{5} = 200\times10^6\,N/m$$

$$[k_2]_{global} = \frac{E_2 A_2}{L_2} \begin{bmatrix} l^2 & lm & -l^2 & -lm \\ lm & m^2 & -lm & -m^2 \\ -l^2 & -lm & l^2 & lm \\ -lm & -m^2 & lm & m^2 \end{bmatrix} = \frac{200\times10^6}{25} \begin{bmatrix} 16 & -12 & -16 & 12 \\ -12 & 9 & 12 & -9 \\ -16 & 12 & 16 & -12 \\ 12 & -9 & -12 & 9 \end{bmatrix} = 8\times10^5 \begin{matrix} u_2 & v_2 & u_3 & v_3 \\ \begin{bmatrix} 16 & -12 & -16 & 12 \\ -12 & 9 & 12 & -9 \\ -16 & 12 & 16 & -12 \\ 12 & -9 & -12 & 9 \end{bmatrix} & \begin{matrix} u_2 \\ v_2 \\ u_3 \\ v_3 \end{matrix} \end{matrix}$$

94

2. The system stiffness matrix:

$$[k]_{system} = 8 \times 10^6 \begin{array}{cccccc} & u_1 & v_1 & u_2 & v_2 & u_3 & v_3 \\ \begin{bmatrix} 16 & 12 & -16 & -12 & 0 & 0 \\ 12 & 9 & -12 & -9 & 0 & 0 \\ -16 & -12 & 16+16 & 12-12 & -16 & 12 \\ -12 & -9 & 12-12 & 9+9 & 12 & -9 \\ 0 & 0 & -16 & 12 & 16 & -12 \\ 0 & 0 & 12 & -9 & -12 & 9 \end{bmatrix} & \begin{matrix} u_1 \\ v_1 \\ u_2 \\ v_2 \\ u_3 \\ v_3 \end{matrix} \end{array} = 8 \times 10^6 \begin{bmatrix} 16 & 12 & -16 & -12 & 0 & 0 \\ 12 & 9 & -12 & -9 & 0 & 0 \\ -16 & -12 & 32 & 0 & -16 & 12 \\ -12 & -9 & 0 & 18 & 12 & -9 \\ 0 & 0 & -16 & 12 & 16 & -12 \\ 0 & 0 & 12 & -9 & -12 & 9 \end{bmatrix}$$

3. Based on the static equilibrium state, the system nodal force vector, stiffness matrix and displacement vector can be expressed as following:

$$\{F\} = [k]\{\delta\} = \{R\}$$

$$8 \times 10^6 \begin{bmatrix} 16 & 12 & -16 & -12 & 0 & 0 \\ 12 & 9 & -12 & -9 & 0 & 0 \\ -16 & -12 & 32 & 0 & -16 & 12 \\ -12 & -9 & 0 & 18 & 12 & -9 \\ 0 & 0 & -16 & 12 & 16 & -12 \\ 0 & 0 & 12 & -9 & -12 & 9 \end{bmatrix} \begin{Bmatrix} u_1 = 0 \\ v_1 = 0 \\ u_2 \\ v_2 \\ u_3 = 0 \\ v_3 = 0 \end{Bmatrix} = \begin{Bmatrix} R_{1x} \\ R_{1y} \\ 20000 \\ -50000 \\ R_{3x} \\ R_{3y} \end{Bmatrix}$$

By solving the equation, the displacement and reaction forces are computed.
$u_2 = 7.8125 \times 10^{-5}$ m, $v_2 = -3.4722 \times 10^{-4}$ m, $R_{1x} = 23.13$ kN, $R_{1y} = 17.5$ kN,
$R_{3x} = -43.33$ kN, $R_{3y} = 32.5$ kN

4. The Nodal forces for each element:
   a) Element 1:

$$\begin{Bmatrix} f_{x1} \\ f_{x2} \end{Bmatrix}_1 = k_1 \begin{pmatrix} 1 & -1 \\ -1 & 1 \end{pmatrix} \begin{bmatrix} l_1 & m_1 & 0 & 0 \\ 0 & 0 & l_1 & m_1 \end{bmatrix} \begin{Bmatrix} u_1 \\ v_1 \\ u_2 \\ v_2 \end{Bmatrix}$$

$$\begin{Bmatrix} f_{x1} \\ f_{x2} \end{Bmatrix}_1 = 200 \times 10^6 \begin{pmatrix} 1 & -1 \\ -1 & 1 \end{pmatrix} \begin{bmatrix} 4/5 & 3/5 & 0 & 0 \\ 0 & 0 & 4/5 & 3/5 \end{bmatrix} \begin{Bmatrix} 0 \\ 0 \\ 7.8125 \times 10^{-5} \\ -3.4722 \times 10^{-4} \end{Bmatrix} = \begin{Bmatrix} 29.166 \\ -29.166 \end{Bmatrix} kN$$

   b) Element 2=

$$\begin{Bmatrix} f_{x2} \\ f_{x3} \end{Bmatrix}_2 = k_2 \begin{pmatrix} 1 & -1 \\ -1 & 1 \end{pmatrix} \begin{bmatrix} l_2 & m_2 & 0 & 0 \\ 0 & 0 & l_2 & m_2 \end{bmatrix} \begin{Bmatrix} u_2 \\ v_2 \\ u_3 \\ v_3 \end{Bmatrix}$$

$$\begin{Bmatrix} f_{x2} \\ f_{x3} \end{Bmatrix}_2 = 200 \times 10^6 \begin{pmatrix} 1 & -1 \\ -1 & 1 \end{pmatrix} \begin{bmatrix} -4/5 & 3/5 & 0 & 0 \\ 0 & 0 & -4/5 & 3/5 \end{bmatrix} \begin{Bmatrix} 0 \\ 0 \\ 7.8125 \times 10^{-5} \\ -3.4722 \times 10^{-4} \end{Bmatrix} = \begin{Bmatrix} -54.166 \\ 54.166 \end{Bmatrix} kN$$

The stresses in the members can be obtained either by the expression given in Section 4.2, i.e:

$$\sigma_1 = \left(\frac{E}{L}\right)_1 \left\{\begin{array}{cc} -1 & 1 \end{array}\right\} \left[\begin{array}{cccc} l_1 & m_1 & 0 & 0 \\ 0 & 0 & l_1 & m_1 \end{array}\right] \left\{\begin{array}{c} u_1 \\ v_1 \\ u_3 \\ v_3 \end{array}\right\} = -\frac{-29.166kN}{A_1} = \frac{-29.166 \times 10^3}{100 \times 10^{-4}} = -29.166 \times 10^5 \, N/m^2$$
$$(Compression)$$

$$\sigma_2 = \left(\frac{E}{L}\right)_2 \left\{\begin{array}{cc} -1 & 1 \end{array}\right\} \left[\begin{array}{cccc} l_2 & m_2 & 0 & 0 \\ 0 & 0 & l_2 & m_2 \end{array}\right] \left\{\begin{array}{c} u_2 \\ v_2 \\ u_3 \\ v_3 \end{array}\right\} = \frac{54.66 \times 10^3}{100 \times 10^{-4}} = 54.66 \times 10^5 \, N/m^2$$
$$(Tension)$$

5. The conventional method:

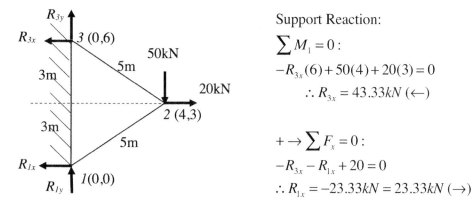

Support Reaction:

$$\sum M_1 = 0:$$

$$-R_{3x}(6) + 50(4) + 20(3) = 0$$

$$\therefore R_{3x} = 43.33kN \; (\leftarrow)$$

$$+ \rightarrow \sum F_x = 0:$$

$$-R_{3x} - R_{1x} + 20 = 0$$

$$\therefore R_{1x} = -23.33kN = 23.33kN \; (\rightarrow)$$

a) At Joint 1:

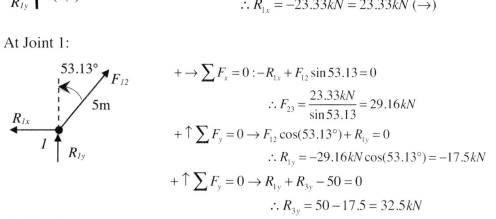

$$+ \rightarrow \sum F_x = 0 : -R_{1x} + F_{12} \sin 53.13 = 0$$

$$\therefore F_{23} = \frac{23.33kN}{\sin 53.13} = 29.16kN$$

$$+ \uparrow \sum F_y = 0 \rightarrow F_{12} \cos(53.13°) + R_{1y} = 0$$

$$\therefore R_{1y} = -29.16kN \cos(53.13°) = -17.5kN$$

$$+ \uparrow \sum F_y = 0 \rightarrow R_{1y} + R_{3y} - 50 = 0$$

$$\therefore R_{3y} = 50 - 17.5 = 32.5kN$$

b) At Joint 3:

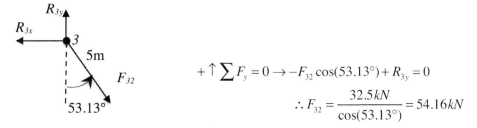

$$+ \uparrow \sum F_y = 0 \rightarrow -F_{32} \cos(53.13°) + R_{3y} = 0$$

$$\therefore F_{32} = \frac{32.5kN}{\cos(53.13°)} = 54.16kN$$

The stresses can be calculated by dividing the axial force in the members by the area of cross-section of the member (+ indicates tension and – indicates compression). These are the same as obtained by FEM above.

96

3. Example 4T.3

Compute member forces, displacements of nodes and stresses in the truss shown in Figure 4.6.10 using FEM. For all the truss elements, $AE/L = 10^6$ kN/m where A = area of cross section, E = modulus of elasticity, and L = length of members: $L_1 = L_2 = 1$m. All the joints are hinged. Verify the results using method of joints.

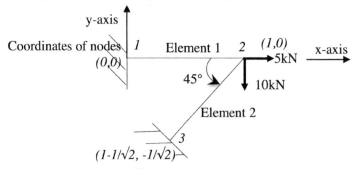

Figure 4.6.10

Solution:

1. The stiffness of each element:

   a) Element 1:

$$l = \cos\phi_x = \cos(0) = 1$$

$$m = \sin\phi_y = \sin(0) = 0$$

$$k_1 = \frac{E_1 A_1}{L_1} = 10^9 \, N/m$$

$$[k_1]_{global} = \frac{E_1 A_1}{L_1} \begin{bmatrix} l^2 & lm & -l^2 & -lm \\ lm & m^2 & -lm & -m^2 \\ -l^2 & -lm & l^2 & lm \\ -lm & -m^2 & lm & m^2 \end{bmatrix} = 10^9 \begin{bmatrix} 1 & 0 & -1 & 0 \\ 0 & 0 & 0 & 0 \\ -1 & 0 & 1 & 0 \\ 0 & 0 & 0 & 0 \end{bmatrix} = 10^9 \begin{matrix} \\ \\ \end{matrix} \begin{array}{cccc} u_1 & v_1 & u_2 & v_2 \\ \end{array} \begin{bmatrix} 1 & 0 & -1 & 0 \\ 0 & 0 & 0 & 0 \\ -1 & 0 & 1 & 0 \\ 0 & 0 & 0 & 0 \end{bmatrix} \begin{matrix} u_1 \\ v_1 \\ u_2 \\ v_2 \end{matrix}$$

   b) Element 2:

$$l = \cos\phi_x = \cos(225) = -\sqrt{2}/2 \qquad (= \frac{x_j - x_i}{L} = -\frac{1}{\sqrt{2}})$$

$$m = \sin\phi_y = \sin(225) = -\sqrt{2}/2 \qquad (= \frac{y_j - y_i}{L} = -\frac{1}{\sqrt{2}})$$

$$k_2 = \frac{E_2 A_2}{L_2} = 10^9 \, N/m$$

$$[k_2]_{global} = \frac{E_2 A_2}{L_2} \begin{bmatrix} l^2 & lm & -l^2 & -lm \\ lm & m^2 & -lm & -m^2 \\ -l^2 & -lm & l^2 & lm \\ -lm & -m^2 & lm & m^2 \end{bmatrix} = \frac{10^9}{4} \begin{bmatrix} 2 & 2 & -2 & -2 \\ 2 & 2 & -2 & -2 \\ -2 & -2 & 2 & 2 \\ -2 & -2 & 2 & 2 \end{bmatrix} = 10^9 \begin{array}{cccc} u_2 & v_2 & u_3 & v_3 \\ \end{array} \begin{bmatrix} 1/2 & 1/2 & -1/2 & -1/2 \\ 1/2 & 1/2 & -1/2 & -1/2 \\ -1/2 & -1/2 & 1/2 & 1/2 \\ -1/2 & -1/2 & 1/2 & 1/2 \end{bmatrix} \begin{matrix} u_2 \\ v_2 \\ u_3 \\ v_3 \end{matrix}$$

2. The system stiffness matrix:

$$[k]_{system} = 10^9 \begin{array}{cccccc} u_1 & v_1 & u_2 & v_2 & u_3 & v_3 \\ \end{array} \begin{bmatrix} 1 & 0 & -1 & 0 & 0 & 0 \\ 0 & 0 & 0 & 0 & 0 & 0 \\ -1 & 0 & 1+1/2 & 0+1/2 & -1/2 & -1/2 \\ 0 & 0 & 0+1/2 & 0+1/2 & -1/2 & -1/2 \\ 0 & 0 & -1/2 & -1/2 & 1/2 & 1/2 \\ 0 & 0 & -1/2 & -1/2 & 1/2 & 1/2 \end{bmatrix} \begin{matrix} u_1 \\ v_1 \\ u_2 \\ v_2 \\ u_3 \\ v_3 \end{matrix} = 10^9 \begin{bmatrix} 1 & 0 & -1 & 0 & 0 & 0 \\ 0 & 0 & 0 & 0 & 0 & 0 \\ -1 & 0 & 3/2 & 1/2 & -1/2 & -1/2 \\ 0 & 0 & 1/2 & 1/2 & -1/2 & -1/2 \\ 0 & 0 & -1/2 & -1/2 & 1/2 & 1/2 \\ 0 & 0 & -1/2 & -1/2 & 1/2 & 1/2 \end{bmatrix}$$

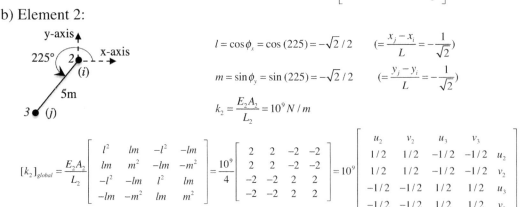

3. Based on the static equilibrium state, the system nodal force vector, stiffness matrix and displacement vector can be expressed as following:

$$\{F\} = [k]\{\delta\} = \{R\}$$

$$10^9 \begin{bmatrix} 1 & 0 & -1 & 0 & 0 & 0 \\ 0 & 0 & 0 & 0 & 0 & 0 \\ -1 & 0 & 3/2 & 1/2 & -1/2 & -1/2 \\ 0 & 0 & 1/2 & 1/2 & -1/2 & -1/2 \\ 0 & 0 & -1/2 & -1/2 & 1/2 & 1/2 \\ 0 & 0 & -1/2 & -1/2 & 1/2 & 1/2 \end{bmatrix} \begin{Bmatrix} u_1 = 0 \\ v_1 = 0 \\ u_2 \\ v_2 \\ u_3 = 0 \\ v_3 = 0 \end{Bmatrix} = \begin{Bmatrix} R_{1x} \\ R_{1y} \\ 5000 \\ -10000 \\ R_{3x} \\ R_{3y} \end{Bmatrix}$$

By solving the equation, the displacement and reaction forces are computed.

$u_2 = 1.5 \times 10^{-5} m$, $v_2 = -3.5 \times 10^{-5} m$, $R_{1x} = -15 kN$, $R_{1y} = 0$ kN, $R_{2x} = 10 kN$, $R_{2y} = 10$ kN.

4. The Nodal forces for each element:
   a) Element 1:

$$\begin{Bmatrix} f_{x1} \\ f_{x2} \end{Bmatrix}_1 = 10^9 \begin{pmatrix} 1 & -1 \\ -1 & 1 \end{pmatrix} \begin{bmatrix} 1 & 0 & 0 & 0 \\ 0 & 0 & 1 & 0 \end{bmatrix} \begin{Bmatrix} 0 \\ 0 \\ 1.5 \times 10^{-5} \\ -3.5 \times 10^{-5} \end{Bmatrix} = \begin{Bmatrix} -15 \\ 15 \end{Bmatrix} kN$$

   b) Element 2=

$$\begin{Bmatrix} f_{x2} \\ f_{x3} \end{Bmatrix}_2 = 10^9 \begin{pmatrix} 1 & -1 \\ -1 & 1 \end{pmatrix} \begin{bmatrix} -\sqrt{2}/2 & -\sqrt{2}/2 & 0 & 0 \\ 0 & 0 & -\sqrt{2}/2 & -\sqrt{2}/2 \end{bmatrix} \begin{Bmatrix} 0 \\ 0 \\ 1.5 \times 10^{-5} \\ -3.5 \times 10^{-5} \end{Bmatrix}$$

$$\begin{Bmatrix} f_{x2} \\ f_{x3} \end{Bmatrix}_2 = \begin{Bmatrix} -10\sqrt{2} \\ 10\sqrt{2} \end{Bmatrix} kN$$

5. The stresses in each element:
   a) Element 1=

$$\sigma_1 = \frac{E}{L}(-1 \quad 1)\begin{Bmatrix} 1 & 0 & 0 & 0 \\ 0 & 0 & 1 & 0 \end{Bmatrix} \begin{Bmatrix} 0 \\ 0 \\ 1.5 \times 10^{-5} \\ -3.5 \times 10^{-5} \end{Bmatrix}$$

$$\sigma_1 = \frac{E}{L}\{1.5 \times 10^{-5}\} N/m^2 \ (Tension)$$

   b) Element 2=

$$\sigma_2 = \frac{E}{L}(-1 \quad 1)\begin{Bmatrix} -\sqrt{2}/2 & -\sqrt{2}/2 & 0 & 0 \\ 0 & 0 & -\sqrt{2}/2 & -\sqrt{2}/2 \end{Bmatrix} \begin{Bmatrix} 0 \\ 0 \\ 1.5 \times 10^{-5} \\ -3.5 \times 10^{-5} \end{Bmatrix}$$

$$\sigma_2 = -\frac{E}{L}\{1.4142 \times 10^{-5}\} N/m^2 \quad (Compression)$$

6. The conventional method:

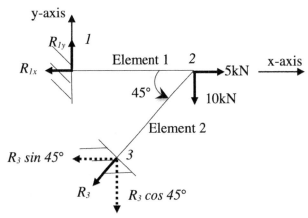

Support Reaction:

$$\sum M_1 = 0:$$

$$-(R_3 \cos 45)(1 - \cos 45) - (R_3 \sin 45)(\sin 45) - 10(1) = 0$$

$$\therefore R_3 = -10\sqrt{2}kN = 10\sqrt{2}kN \ (\text{↗})$$

$+ \rightarrow \sum F_x = 0:$

$-R_{1x} - R_3 \sin 45 + 5 = 0$

$R_{1x} = -(-10\sqrt{2}\sin 45) + 5kN$

$\therefore R_{1x} = 15kN \ (\leftarrow)$

$+ \uparrow \sum F_y = 0:$

$R_{1y} - 10 - R_3 \cos 45 = 0$

$R_{1y} = 10 + (-10\sqrt{2})\cos 45 = 0$

$\therefore R_{1y} = 0kN$

a) At Joint 1:

$+ \rightarrow \sum F_x = 0: \quad -R_{1x} + F_{12} = 0$

$\therefore F_{12} = 15kN \quad \text{(Tension)}$

b) At Joint 2:

$+ \uparrow \sum F_y = 0:$

$F_{23}\sin(45°) - 10 = 0$

$\therefore F_{23} = \dfrac{10kN}{\sin(45°)} = 10\sqrt{2}kN \quad \text{(Compression)}$

4. Example 4T.4

Determine the reactions, member forces and deflections of the truss members BD and CD with hinged connections shown in Figure 4.6.11, using Finite Element Method. Use one truss element for each member. Compare the member forces from the conventional method (method of joints). Assume area of cross section, A (50 cm²) and modulus of elasticity E (200GPa) to be the same for all the members.

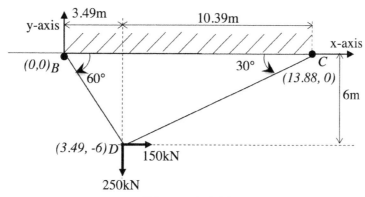

Figure 4.6.11

Solution:

1. The stiffness of each element:

   a) Element BD:

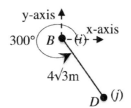

$$l = \cos\phi_x = \cos(300) = 1/2$$

$$m = \sin\phi_y = \sin(300) = -\sqrt{3}/2$$

$$k_1 = \frac{E_1 A_1}{L_1} = \frac{(200\times10^9)(50\times10^{-4})}{4\sqrt{3}} = \frac{25}{\sqrt{3}}\times10^7 N/m$$

$\left(\begin{array}{l}\text{These can also be calculated from the coordinates of the nodes as}\\ \text{illustrated in earlier examples.}\end{array}\right)$

$$[k_1]_{global} = \frac{25}{4\sqrt{3}}\times10^7 \begin{bmatrix} 1 & -\sqrt{3} & -1 & \sqrt{3} \\ -\sqrt{3} & 3 & \sqrt{3} & -3 \\ -1 & \sqrt{3} & 1 & -\sqrt{3} \\ \sqrt{3} & -3 & -\sqrt{3} & 3 \end{bmatrix} = 10^7 \begin{bmatrix} 25/4\sqrt{3} & -25/4 & -25/4\sqrt{3} & 25/4 \\ -25/4 & 25\sqrt{3}/4 & 25/4 & -25\sqrt{3}/4 \\ -25/4\sqrt{3} & 25/4 & 25/4\sqrt{3} & -25/4 \\ 25/4 & -25\sqrt{3}/4 & -25/4 & 25\sqrt{3}/4 \end{bmatrix} \begin{matrix} u_B \\ v_B \\ u_D \\ v_D \end{matrix}$$

(column headers: $u_B \quad v_B \quad u_D \quad v_D$)

   b) Element CD:

$$l = \cos\phi_x = \cos(210) = -\sqrt{3}/2$$

$$m = \sin\phi_y = \sin(210) = -1/2$$

$$k_2 = \frac{E_2 A_2}{L_2} = \frac{(200\times10^9)(50\times10^{-4})}{12} = \frac{25}{3}\times10^7 N/m$$

$$[k_2]_{global} = \frac{25}{3[4]}\times10^7 \begin{bmatrix} 3 & \sqrt{3} & -3 & -\sqrt{3} \\ \sqrt{3} & 1 & -\sqrt{3} & -1 \\ -3 & -\sqrt{3} & 3 & \sqrt{3} \\ -\sqrt{3} & -1 & \sqrt{3} & 1 \end{bmatrix} = 10^7 \begin{bmatrix} 25/4 & 25/4\sqrt{3} & -25/4 & -25/4\sqrt{3} \\ 25/4\sqrt{3} & 25/12 & -25/4\sqrt{3} & -25/12 \\ -25/4 & -25/4\sqrt{3} & 25/4 & 25/4\sqrt{3} \\ -25/4\sqrt{3} & -25/12 & 25/4\sqrt{3} & 25/12 \end{bmatrix} \begin{matrix} u_D \\ v_D \\ u_C \\ v_C \end{matrix}$$

(column headers: $u_D \quad v_D \quad u_C \quad v_C$)

100

2. The system stiffness matrix:

$$[k]_{system} = 10^7 \begin{bmatrix} u_B & v_B & u_D & v_D & u_C & v_C \\ 25/4\sqrt{3} & -25/4 & -25/4\sqrt{3} & 25/4 & 0 & 0 \\ -25/4 & 25\sqrt{3}/4 & 25/4 & -25\sqrt{3}/4 & 0 & 0 \\ -25/4\sqrt{3} & 25/4 & 25/4\sqrt{3}+25/4 & -25/4+25/4\sqrt{3} & -25/4 & -25/4\sqrt{3} \\ 25/4 & -25\sqrt{3}/4 & -25/4+25/4\sqrt{3} & 25\sqrt{3}/4+25/12 & -25/4\sqrt{3} & -25/12 \\ 0 & 0 & -25/4 & -25/4\sqrt{3} & 25/4 & 25/4\sqrt{3} \\ 0 & 0 & -25/4\sqrt{3} & -25/12 & 25/4\sqrt{3} & 25/12 \end{bmatrix} \begin{matrix} u_B \\ v_B \\ u_D \\ v_D \\ u_C \\ v_C \end{matrix}$$

$$[k]_{system} = 10^7 \begin{bmatrix} 3.608 & -6.25 & -3.608 & 6.25 & 0 & 0 \\ -6.25 & 10.825 & 6.25 & -10.825 & 0 & 0 \\ -3.608 & 6.25 & 9.858 & -2.642 & -6.25 & -3.608 \\ 6.25 & -10.825 & -2.642 & 12.909 & -3.608 & -2.083 \\ 0 & 0 & -6.25 & -3.608 & 6.25 & 3.608 \\ 0 & 0 & -3.608 & -2.083 & 3.608 & 2.083 \end{bmatrix}$$

3. Based on the static equilibrium state, the system nodal force vector, stiffness matrix and displacement vector can be expressed as following:

$$\{F\} = [k]\{\delta\} = \{R\}$$

$$10^7 \begin{bmatrix} 3.608 & -6.25 & -3.608 & 6.25 & 0 & 0 \\ -6.25 & 10.825 & 6.25 & -10.825 & 0 & 0 \\ -3.608 & 6.25 & 9.858 & -2.642 & -6.25 & -3.608 \\ 6.25 & -10.825 & -2.642 & 12.909 & -3.608 & -2.083 \\ 0 & 0 & -6.25 & -3.608 & 6.25 & 3.608 \\ 0 & 0 & -3.608 & -2.083 & 3.608 & 2.083 \end{bmatrix} \begin{Bmatrix} u_B = 0 \\ v_B = 0 \\ u_D \\ v_D \\ u_C = 0 \\ v_C = 0 \end{Bmatrix} = \begin{Bmatrix} R_{Bx} \\ R_{By} \\ 150000 \\ -250000 \\ R_{Cx} \\ R_{Cy} \end{Bmatrix}$$

By solving the equation, the displacement and reaction forces are computed.
$u_D = 1.0608 \times 10^{-3}$m, $v_D = -1.720 \times 10^{-3}$m, $R_{Bx} = -145.77$kN, $R_{By} = 252.49$ kN, $R_{Cx} = -4.24$kN, $R_{Cy} = -2.44$ kN

4. The Nodal forces for each element:
   a) Element 1:

$$\begin{Bmatrix} f_{xB} \\ f_{xD} \end{Bmatrix}_1 = \frac{25}{\sqrt{3}} \times 10^7 \begin{pmatrix} 1 & -1 \\ -1 & 1 \end{pmatrix} \begin{bmatrix} 1/2 & -\sqrt{3}/2 & 0 & 0 \\ 0 & 0 & 1/2 & -\sqrt{3}/2 \end{bmatrix} \begin{Bmatrix} 0 \\ 0 \\ 1.0608 \times 10^{-3} \\ -1.720 \times 10^{-3} \end{Bmatrix} = \begin{Bmatrix} -291.56 \\ 291.56 \end{Bmatrix} kN \atop (Tension)$$

   b) Element 2=

$$\begin{Bmatrix} f_{xC} \\ f_{xD} \end{Bmatrix}_2 = \frac{25}{3} \times 10^7 \begin{pmatrix} 1 & -1 \\ -1 & 1 \end{pmatrix} \begin{bmatrix} -\sqrt{3}/2 & -1/2 & 0 & 0 \\ 0 & 0 & -\sqrt{3}/2 & -1/2 \end{bmatrix} \begin{Bmatrix} 0 \\ 0 \\ 1.0608 \times 10^{-3} \\ -1.720 \times 10^{-3} \end{Bmatrix} = \begin{Bmatrix} 4.89 \\ -4.89 \end{Bmatrix} kN \atop (Compression)$$

The stresses in the members can be directly calculated (also by using the relevant expressions) by dividing the member force by the area of cross-section: Accordingly,

$$\sigma_1 = \text{Stress in element 1} = \frac{291.56 \times 10^3}{50 \times 10^{-4}} = \frac{58.3 \times 10^6 \, N/m^2}{(Tension)}$$

$$\sigma_2 = \text{Stress in element 2} = -\frac{4.89 \times 10^3}{50 \times 10^{-4}} = \frac{0.978 \times 10^6 \, N/m^2}{(\textit{Compression})}$$

5. The conventional method:

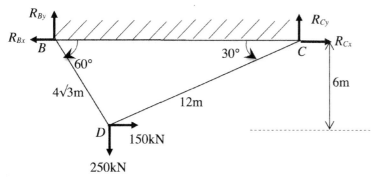

Support Reaction:

$$\sum M_B = 0: \quad -(250)(4\sqrt{3}\cos 60) + (150)(6) + (R_{cy})(4\sqrt{3}\cos 60 + 12\cos 30) = 0$$

$$\therefore R_{cy} = -2.45 \, kN = 2.45 \, kN \, (\downarrow)$$

$$\frac{R_{cy}}{R_{cx}} = \tan 30° \quad \rightarrow \therefore R_{cx} = \frac{-2.45 \, kN}{\tan 30°} = -4.24 \, kN = 4.24 \, kN (\leftarrow)$$

$$+\uparrow \sum F_y = 0: \quad R_{By} + R_{cy} - 250 = 0$$

$$R_{By} - 2.45 - 250 = 0$$

$$\therefore R_{By} = 252.45 \, kN$$

$$+\rightarrow \sum F_x = 0: \quad -R_{Bx} + R_{Cx} + 150 = 0$$

$$R_{Bx} = 150 - 4.24$$

$$\therefore R_{Bx} = 145.76 \, kN$$

a) At Joint B:

$$+\rightarrow \sum F_x = 0: \quad -R_{Bx} + F_{BD}\cos 60 = 0$$

$$\therefore F_{BD} = 291.52 \, kN \quad (\text{Tension})$$

b) At Joint C:

$$+\uparrow \sum F_y = 0: \quad F_{CD}\sin(30°) - R_{Cy} = 0$$

$$\therefore F_{CD} = \frac{2.45 \, kN}{\sin(30°)} = 4.9 \, kN \quad (\text{Compression})$$

5. Example 4T.5

Find the forces in the bars of the hinged right angled truss 123 as shown in Figure 4.6.12 and mark the forces as compression and tension (C and T) and show the directions by means of appropriate arrows. Use FEM. Take 12, 23 and 31 as element 1, 2 and 3. Check the reactions at 1 and 2 from strength of materials approach. E and A are same for all bars.

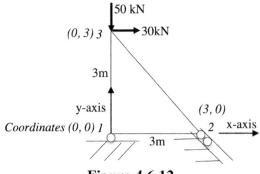

Figure 4.6.12

Solution:

1. The stiffness of each element:

   a) Element 1:

$$l = \cos\phi_x = \cos(0) = 1$$
$$m = \sin\phi_y = \sin(0) = 0$$
$$k_1 = \frac{E_1 A_1}{L_1} = \frac{EA}{3} N/m$$

(The direction cosines can also be calculated from the coordinates of the nodes as illustrated in earlier examples.)

$$[k_1]_{global} = \frac{E_1 A_1}{L_1}\begin{bmatrix} l^2 & lm & -l^2 & -lm \\ lm & m^2 & -lm & -m^2 \\ -l^2 & -lm & l^2 & lm \\ -lm & -m^2 & lm & m^2 \end{bmatrix} = \frac{EA}{3}\begin{bmatrix} 1 & 0 & -1 & 0 \\ 0 & 0 & 0 & 0 \\ -1 & 0 & 1 & 0 \\ 0 & 0 & 0 & 0 \end{bmatrix} = \frac{EA}{3}\begin{bmatrix} \overset{u_1}{1} & \overset{v_1}{0} & \overset{u_2}{-1} & \overset{v_2}{0} \\ 0 & 0 & 0 & 0 \\ -1 & 0 & 1 & 0 \\ 0 & 0 & 0 & 0 \end{bmatrix}\begin{matrix} u_1 \\ v_1 \\ u_2 \\ v_2 \end{matrix}$$

   b) Element 2:

$$l = \cos\psi_x = \cos(135) = -1/\sqrt{2}$$
$$m = \sin\phi_y = \sin(135) = 1/\sqrt{2}$$
$$k_2 = \frac{E_2 A_2}{L_2} = \frac{EA}{3\sqrt{2}} N/m$$

$$[k_2]_{global} = \frac{EA}{(2)3\sqrt{2}}\begin{bmatrix} 1 & -1 & -1 & 1 \\ -1 & 1 & 1 & -1 \\ -1 & 1 & 1 & -1 \\ 1 & -1 & -1 & 1 \end{bmatrix} = \frac{EA}{3}\begin{bmatrix} \overset{u_2}{1/2\sqrt{2}} & \overset{v_2}{-1/2\sqrt{2}} & \overset{u_3}{-1/2\sqrt{2}} & \overset{v_3}{1/2\sqrt{2}} \\ -1/2\sqrt{2} & 1/2\sqrt{2} & 1/2\sqrt{2} & -1/2\sqrt{2} \\ -1/2\sqrt{2} & 1/2\sqrt{2} & 1/2\sqrt{2} & -1/2\sqrt{2} \\ 1/2\sqrt{2} & -1/2\sqrt{2} & -1/2\sqrt{2} & 1/2\sqrt{2} \end{bmatrix}\begin{matrix} u_2 \\ v_2 \\ u_3 \\ v_3 \end{matrix}$$

   c) Element 3:

$$l = \cos\phi_x = \cos(90) = 0$$
$$m = \sin\phi_y = \sin(90) = 1$$
$$k_3 = \frac{E_3 A_3}{L_3} = \frac{EA}{3} N/m$$

$$[k_3]_{global} = \frac{EA}{3}\begin{bmatrix} \overset{u_3}{0} & \overset{v_3}{0} & \overset{u_1}{0} & \overset{v_1}{0} \\ 0 & 1 & 0 & -1 \\ 0 & 0 & 0 & 0 \\ 0 & -1 & 0 & 1 \end{bmatrix}\begin{matrix} u_3 \\ v_3 \\ u_1 \\ v_1 \end{matrix}$$

2. The system stiffness matrix:

$$[k]_{system} = \frac{EA}{3}\begin{bmatrix} u_1 & v_1 & u_2 & v_2 & u_3 & v_3 & \\ 1+0 & 0+0 & -1 & 0 & 0+0 & 0+0 & u_1 \\ 0+0 & 0+1 & 0 & 0 & 0+0 & 0-1 & v_1 \\ -1 & 0 & 1+1/2\sqrt{2} & 0-1/2\sqrt{2} & -1/2\sqrt{2} & 1/2\sqrt{2} & u_2 \\ 0 & 0 & 0-1/2\sqrt{2} & 0+1/2\sqrt{2} & 1/2\sqrt{2} & -1/2\sqrt{2} & v_2 \\ 0+0 & 0+0 & -1/2\sqrt{2} & 1/2\sqrt{2} & 1/2\sqrt{2}+0 & -1/2\sqrt{2}+0 & u_3 \\ 0+0 & 0-1 & 1/2\sqrt{2} & -1/2\sqrt{2} & -1/2\sqrt{2}+0 & 1/2\sqrt{2}+1 & v_3 \end{bmatrix}$$

$$[k]_{system} = \frac{EA}{3}\begin{bmatrix} 1 & 0 & -1 & 0 & 0 & 0 \\ 0 & 1 & 0 & 0 & 0 & -1 \\ -1 & 0 & (4+\sqrt{2})/4 & -1/2\sqrt{2} & -1/2\sqrt{2} & 1/2\sqrt{2} \\ 0 & 0 & -1/2\sqrt{2} & 1/2\sqrt{2} & 1/2\sqrt{2} & -1/2\sqrt{2} \\ 0 & 0 & -1/2\sqrt{2} & 1/2\sqrt{2} & 1/2\sqrt{2} & -1/2\sqrt{2} \\ 0 & -1 & 1/2\sqrt{2} & -1/2\sqrt{2} & -1/2\sqrt{2} & (4+\sqrt{2})/4 \end{bmatrix}$$

3. Based on the static equilibrium state, the system nodal force vector, stiffness matrix and displacement vector can be expressed as following:

$$\{F\} = [k]\{\delta\} = \{R\}$$

$$\frac{EA}{3}\begin{bmatrix} 1 & 0 & -1 & 0 & 0 & 0 \\ 0 & 1 & 0 & 0 & 0 & -1 \\ -1 & 0 & (4+\sqrt{2})/4 & -1/2\sqrt{2} & -1/2\sqrt{2} & 1/2\sqrt{2} \\ 0 & 0 & -1/2\sqrt{2} & 1/2\sqrt{2} & 1/2\sqrt{2} & -1/2\sqrt{2} \\ 0 & 0 & -1/2\sqrt{2} & 1/2\sqrt{2} & 1/2\sqrt{2} & -1/2\sqrt{2} \\ 0 & -1 & 1/2\sqrt{2} & -1/2\sqrt{2} & -1/2\sqrt{2} & (4+\sqrt{2})/4 \end{bmatrix}\begin{Bmatrix} u_1 = 0 \\ v_1 = 0 \\ u_2 = -\delta/\sqrt{2} \\ v_2 = \delta/\sqrt{2} \\ u_3 \\ v_3 \end{Bmatrix} = \begin{Bmatrix} R_{1x} \\ R_{1y} \\ R_2\cos 45 \\ R_2\sin 45 \\ 30000 \\ -50000 \end{Bmatrix}$$

By solving the equation, the displacement and reaction forces are computed.
$\delta = -(180000\sqrt{2}/EA)$ m, $u_3 = (554558.44/EA)$ m, $v_3 = -(60000/EA)$ m,
$R_{1x} = -60$kN, $R_{1y} = 20$ kN, $R_2 = 30\sqrt{2}$kN.

4. The Nodal forces for each element:

a) Element 1:

$$\begin{Bmatrix} f_{x1} \\ f_{x2} \end{Bmatrix}_1 = \frac{EA}{3}\begin{pmatrix} 1 & -1 \\ -1 & 1 \end{pmatrix}\begin{bmatrix} 1 & 0 & 0 & 0 \\ 0 & 0 & 1 & 0 \end{bmatrix}\begin{Bmatrix} 0 \\ 0 \\ 180000/EA \\ -180000/EA \end{Bmatrix} = \begin{Bmatrix} -60 \\ 60 \end{Bmatrix}kN$$

(*Tension*)

b) Element 2=

$$\begin{Bmatrix} f_{x2} \\ f_{x3} \end{Bmatrix}_2 = \frac{EA}{3\sqrt{2}}\begin{pmatrix} 1 & -1 \\ -1 & 1 \end{pmatrix}\begin{bmatrix} -1/\sqrt{2} & 1/\sqrt{2} & 0 & 0 \\ 0 & 0 & -1/\sqrt{2} & 1/\sqrt{2} \end{bmatrix}\begin{Bmatrix} 180000/EA \\ -180000/EA \\ 554558.44/EA \\ -60000/EA \end{Bmatrix} = \begin{Bmatrix} 30\sqrt{2} \\ -30\sqrt{2} \end{Bmatrix}kN$$

(*Compression*)

c) Element 3=

$$\begin{Bmatrix} f_{x3} \\ f_{x1} \end{Bmatrix}_3 = \frac{EA}{3}\begin{pmatrix} 1 & -1 \\ -1 & 1 \end{pmatrix}\begin{bmatrix} 0 & 1 & 0 & 0 \\ 0 & 0 & 0 & 1/ \end{bmatrix}\begin{Bmatrix} 554558.44/EA \\ -60000/EA \\ 0 \\ 0 \end{Bmatrix} = \begin{Bmatrix} -20 \\ 20 \end{Bmatrix}kN$$

(*Tension*)

104

The stresses in each member can be directly calculated (or by relevant expression) by dividing the member force by the area of cross-section. Accordingly,

$$\sigma_1 = \frac{60 \times 10^3}{A} N/m^2 (Tension)$$

$$\sigma_2 = \frac{-30\sqrt{2} \times 10^3}{A} N/m^2 (Compression)$$

$$\sigma_3 = \frac{20 \times 10^3}{A} N/m^2 (Tension)$$

(*A is in m² units.*)

5. The forces and directions of the case:

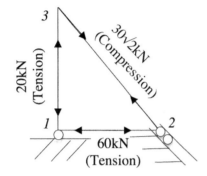

6. The conventional method:

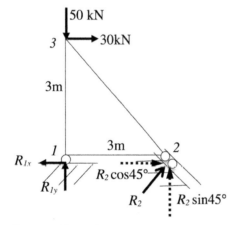

Support Reaction:

$$\sum M_2 = 0: \quad -30(3) + 50(3) - R_{1y}(3) = 0$$

$$\therefore R_{1y} = 20kN$$

$$+\uparrow \sum F_y = 0: \quad -50 + R_{1y} + R_2 \sin 45 = 0$$

$$\therefore R_2 = 30\sqrt{2}kN$$

$$+\rightarrow \sum F_x = 0: \quad -R_{1x} + R_2 \cos 45 + 30 = 0$$

$$\therefore R_{1x} = 60kN$$

4.6.3   Examples of Beams

1.   Example 4B.1:

Compute the deflections at the node 2 and reactions at node 1 of the cantilever beam shown in Figure 4.6.13 using FEM (use one beam element). Draw the BMD and SFD for the beam. Assume $EI = 10^5$ kNm^2, where E = modulus of elasticity, kN/m^2 and I = moment of inertia, m^4. Check the results with conventional method.

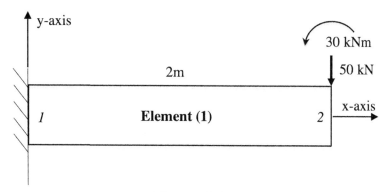

Figure 4.6.13

Solution:

1. The element stiffness matrix:

   a) Element 1=

$$[k]_1 = \frac{E_1 I_1}{L_1^3} \begin{bmatrix} 12 & 6L & -12 & 6L \\ 6L & 4L^2 & -6L & 2L^2 \\ -12 & -6L & 12 & -6L \\ 6L & 2L^2 & -6L & 4L^2 \end{bmatrix} = \frac{10^8}{8} \begin{matrix} & v_1 & \theta_1 & v_2 & \theta_2 & \\ & \begin{bmatrix} 12 & 12 & -12 & 12 \\ 12 & 16 & -12 & 8 \\ -12 & -12 & 12 & -12 \\ 12 & 8 & -12 & 16 \end{bmatrix} & \begin{matrix} v_1 \\ \theta_1 \\ v_2 \\ \theta_2 \end{matrix} \end{matrix}$$

$$[k]_1 = 10^8 \begin{bmatrix} 1.5 & 1.5 & -1.5 & 1.5 \\ 1.5 & 2 & -1.5 & 1 \\ -1.5 & -1.5 & 1.5 & -1.5 \\ 1.5 & 1 & -1.5 & 2 \end{bmatrix}$$

2. Based on the static equilibrium state, the system nodal force vector, stiffness matrix and displacement vector can be expressed as following:

$$\{F\} = [k]\{\delta\} = \{R\}$$

$$\{F\} = 10^8 \begin{bmatrix} 1.5 & 1.5 & -1.5 & 1.5 \\ 1.5 & 2 & -1.5 & 1 \\ -1.5 & -1.5 & 1.5 & -1.5 \\ 1.5 & 1 & -1.5 & 2 \end{bmatrix} \begin{Bmatrix} v_1 = 0 \\ \theta_1 = 0 \\ v_2 \\ \theta_2 \end{Bmatrix} = \begin{Bmatrix} R_{1y} \\ M_{1z} \\ -50000 \\ 30000 \end{Bmatrix}$$

By solving the equation, the displacement and reaction forces are computed.
$v_2 = -7.333 \times 10^{-4}$ m, $\theta_2 = -4 \times 10^{-4}$ rad, $R_{1y} = 50$ kN, $M_{1z} = 70$ kNm

106

3. The conventional method:

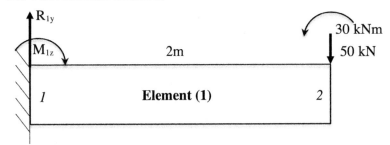

Support Reactions:

$$+\uparrow \sum F_y = 0 : \qquad R_{1y} - 50 = 0$$

$$\therefore R_{1y} = 50 \ kN$$

$$\sum M_1 = 0 : \qquad -50(2) + 30 - M_{1z} = 0$$

$$\therefore M_{1z} = -70kN = 70kN \ (\curvearrowleft)$$

4. The shear force and bending moment diagrams for the beam:

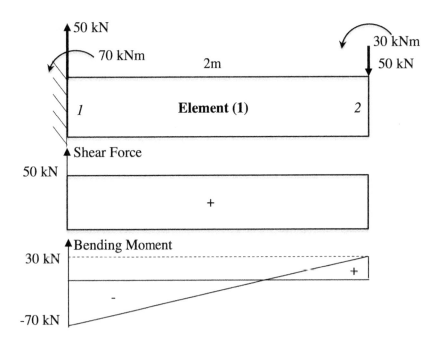

2. Example 4B.2:

Compute the deflections at the nodes (1, 2 and 3) and reactions on the beam shown in Figure 4.6.14 using two beam elements 1 and 2. Draw the BMD and SFD for the beam. Assume EI = constant, units (E = modulus of elasticity, kN/m^2, I = moment of inertia, m^4).

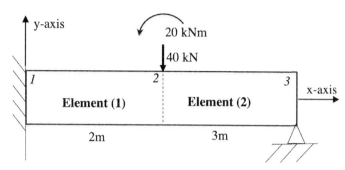

Figure 4.6.14

Solution:

1. The element stiffness matrix:

a) Element 1=

$$[k]_1 = \frac{E_1 I_1}{L_1^3} \begin{bmatrix} 12 & 6L & -12 & 6L \\ 6L & 4L^2 & -6L & 2L^2 \\ -12 & -6L & 12 & -6L \\ 6L & 2L^2 & -6L & 4L^2 \end{bmatrix} = \frac{EI}{8} \begin{matrix} v_1 & \theta_1 & v_2 & \theta_2 \\ \begin{bmatrix} 12 & 12 & -12 & 12 \\ 12 & 16 & -12 & 8 \\ -12 & -12 & 12 & -12 \\ 12 & 8 & -12 & 16 \end{bmatrix} \begin{matrix} v_1 \\ \theta_1 \\ v_2 \\ \theta_2 \end{matrix} \end{matrix} = EI \begin{bmatrix} 1.5 & 1.5 & -1.5 & 1.5 \\ 1.5 & 2 & -1.5 & 1 \\ -1.5 & -1.5 & 1.5 & -1.5 \\ 1.5 & 1 & -1.5 & 2 \end{bmatrix}$$

b) Element 2=

$$[k]_2 = \frac{EI}{27} \begin{matrix} v_2 & \theta_2 & v_3 & \theta_3 \\ \begin{bmatrix} 12 & 18 & -12 & 18 \\ 18 & 36 & -18 & 18 \\ -12 & -18 & 12 & -18 \\ 18 & 18 & -18 & 36 \end{bmatrix} \begin{matrix} v_2 \\ \theta_2 \\ v_3 \\ \theta_3 \end{matrix} \end{matrix} = EI \begin{bmatrix} 4/9 & 2/3 & -4/9 & 2/3 \\ 2/3 & 4/3 & -2/3 & 2/3 \\ -4/9 & -2/3 & 4/9 & -2/3 \\ 2/3 & 2/3 & -2/3 & 4/3 \end{bmatrix}$$

2. The system stiffness matrix:

$$[k]_{system} = EI \begin{matrix} v_1 & \theta_1 & v_2 & \theta_2 & v_3 & \theta_3 \\ \begin{bmatrix} 1.5 & 1.5 & -1.5 & 1.5 & 0 & 0 \\ 1.5 & 2 & -1.5 & 1 & 0 & 0 \\ -1.5 & -1.5 & 1.5+4/9 & -1.5+2/3 & -4/9 & 2/3 \\ 1.5 & 1 & -1.5+2/3 & 2+4/3 & -2/3 & 2/3 \\ 0 & 0 & -4/9 & -2/3 & 4/9 & -2/3 \\ 0 & 0 & 2/3 & 2/3 & -2/3 & 4/3 \end{bmatrix} \begin{matrix} v_1 \\ \theta_1 \\ v_2 \\ \theta_2 \\ v_3 \\ \theta_3 \end{matrix} \end{matrix}$$

$$[k]_{system} = EI \begin{bmatrix} 1.5 & 1.5 & -1.5 & 1.5 & 0 & 0 \\ 1.5 & 2 & -1.5 & 1 & 0 & 0 \\ -1.5 & -1.5 & 35/18 & -5/6 & -4/9 & 2/3 \\ 1.5 & 1 & -5/6 & 10/3 & -2/3 & 2/3 \\ 0 & 0 & -4/9 & -2/3 & 4/9 & -2/3 \\ 0 & 0 & 2/3 & 2/3 & -2/3 & 4/3 \end{bmatrix}$$

108

3. Based on the static equilibrium state, the system nodal force vector, stiffness matrix and displacement vector can be expressed as following:

$$\{F\}=[k]\{\delta\}=\{R\}$$

$$\{F\}= EI\begin{bmatrix} 1.5 & 1.5 & -1.5 & 1.5 & 0 & 0 \\ 1.5 & 2 & -1.5 & 1 & 0 & 0 \\ -1.5 & -1.5 & 35/18 & -5/6 & -4/9 & 2/3 \\ 1.5 & 1 & -5/6 & 10/3 & -2/3 & 2/3 \\ 0 & 0 & -4/9 & -2/3 & 4/9 & -2/3 \\ 0 & 0 & 2/3 & 2/3 & -2/3 & 4/3 \end{bmatrix} \begin{Bmatrix} v_1=0 \\ \theta_1=0 \\ v_2 \\ \theta_2 \\ v_3=0 \\ \theta_3 \end{Bmatrix} = \begin{Bmatrix} R_{1y} \\ M_{1z} \\ -40 \\ 20 \\ R_{3y} \\ 0 \end{Bmatrix}$$

By solving the equation, the displacement and reaction forces are computed.
$v_2 = -27.84/EI$ m, $\theta_2 = -4.16/EI$ rad, $\theta_3 = 16/EI$ rad,
$R_{1y} = 35$ kN, $M_{1z} = 35$ kNm, $R_{3y} = 5$ kN.

4. The shear force and bending moment diagrams for the beam:

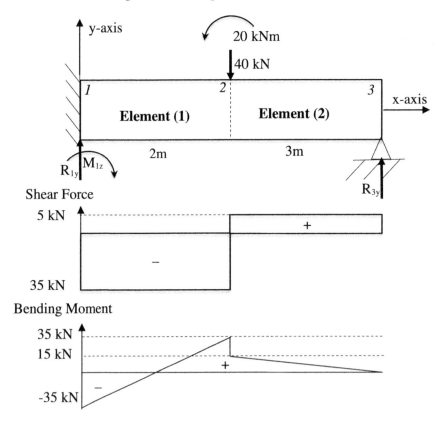

3.   Example 4B.3:

Compute the deflections at the nodes (1, 2 and 3) and reactions on the beam shown in Figure 4.6.15 using two beam elements 1 and 2. Draw the BMD and SFD for the beam. Assume EI = constant, units (E = modulus of elasticity, kN/m², I = moment of inertia, m⁴).

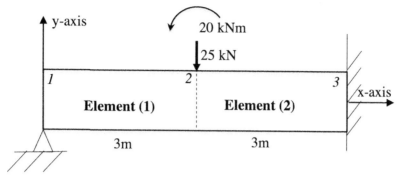

Figure 4.6.15

Solution:

1. The element stiffness matrix:
   a) Element 1=

$$[k]_1 = \frac{E_1 I_1}{L_1^3} \begin{bmatrix} 12 & 6L & -12 & 6L \\ 6L & 4L^2 & -6L & 2L^2 \\ -12 & -6L & 12 & -6L \\ 6L & 2L^2 & -6L & 4L^2 \end{bmatrix} = \frac{EI}{27} \begin{matrix} & v_1 & \theta_1 & v_2 & \theta_2 & \\ & 12 & 18 & -12 & 18 & v_1 \\ & 18 & 36 & -18 & 18 & \theta_1 \\ & -12 & -18 & 12 & -18 & v_2 \\ & 18 & 18 & -18 & 36 & \theta_2 \end{matrix}$$

   b) Element 2=

$$[k]_2 = \frac{E_2 I_2}{L_2^3} \begin{bmatrix} 12 & 6L & -12 & 6L \\ 6L & 4L^2 & -6L & 2L^2 \\ -12 & -6L & 12 & -6L \\ 6L & 2L^2 & -6L & 4L^2 \end{bmatrix} = \frac{EI}{27} \begin{matrix} & v_2 & \theta_2 & v_3 & \theta_3 & \\ & 12 & 18 & -12 & 18 & v_2 \\ & 18 & 36 & -18 & 18 & \theta_2 \\ & -12 & -18 & 12 & -18 & v_3 \\ & 18 & 18 & -18 & 36 & \theta_3 \end{matrix}$$

2. The system stiffness matrix:

$$[k]_{system} = \frac{EI}{27} \begin{matrix} v_1 & \theta_1 & v_2 & \theta_2 & v_3 & \theta_3 & \\ 12 & 18 & -12 & 18 & 0 & 0 & v_1 \\ 18 & 36 & -18 & 18 & 0 & 0 & \theta_1 \\ -12 & -18 & 12+12 & -18+18 & -12 & 18 & v_2 \\ 18 & 18 & -18+18 & 36+36 & -18 & 18 & \theta_2 \\ 0 & 0 & -12 & -18 & 12 & -18 & v_3 \\ 0 & 0 & 18 & 18 & -18 & 36 & \theta_3 \end{matrix}$$

$$[k]_{system} = \frac{EI}{27} \begin{bmatrix} 12 & 18 & -12 & 18 & 0 & 0 \\ 18 & 36 & -18 & 18 & 0 & 0 \\ -12 & -18 & 24 & 0 & -12 & 18 \\ 18 & 18 & 0 & 72 & -18 & 18 \\ 0 & 0 & -12 & -18 & 12 & -18 \\ 0 & 0 & 18 & 18 & -18 & 36 \end{bmatrix}$$

110

3. Based on the static equilibrium state, the system nodal force vector, stiffness matrix and displacement vector can be expressed as following:

$$\{F\}=[k]\{\delta\}=\{R\}$$

$$\{F\}=\frac{EI}{27}\begin{bmatrix} 12 & 18 & -12 & 18 & 0 & 0 \\ 18 & 36 & -18 & 18 & 0 & 0 \\ -12 & -18 & 24 & 0 & -12 & 18 \\ 18 & 18 & 0 & 72 & -18 & 18 \\ 0 & 0 & -12 & -18 & 12 & -18 \\ 0 & 0 & 18 & 18 & -18 & 36 \end{bmatrix}\begin{Bmatrix} v_1=0 \\ \theta_1 \\ v_2 \\ \theta_2 \\ v_3=0 \\ \theta_3=0 \end{Bmatrix}=\begin{Bmatrix} R_{1y} \\ 0 \\ -25 \\ 20 \\ R_{3y} \\ M_{3z} \end{Bmatrix}$$

By solving the equation, the displacement and reaction forces are computed.
$\theta_1 = -35.625/EI$ rad, $v_2 = -54.844/EI$ m, $\theta_2 = 16.406/EI$ rad,
$R_{1y} = 11.562$ kN, $R_{3y} = 13.438$ kN, $M_{3z} = -25.625$ kNm.

4. The shear force and bending moment diagrams for the beam:

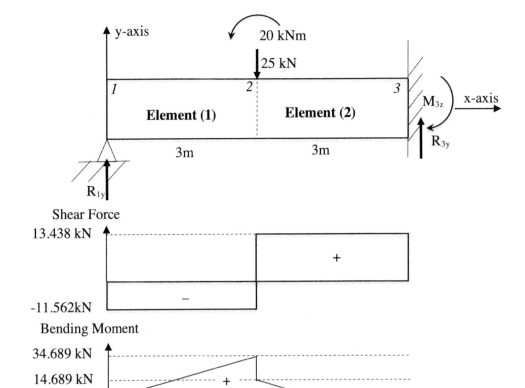

4.6.4 Examples of Elastic Solids

1. Example 4E.1:

Obtain the global stiffness matrix, nodal displacements, support reactions (hinged at A and C) and stresses for the plane stress plate using two triangular elements 1 and 2 as shown in Figure 4.6.16. $E = 200$ MN/m², $v = 0$, Thickness $t = $ 1m. All the nodal coordinates are in meters.

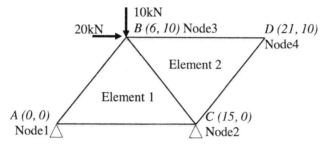

Figure 4.6.16

Solution:

1. The stiffness of each element:

   a) Element 1:

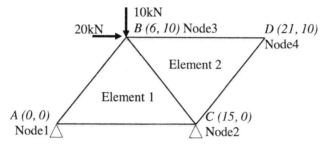

$a_i = x_j y_m - x_m y_j = 15(10) - (6)(0) = 150 m^2$

$a_j = x_m y_i - x_i y_m = 6(0) - (0)(10) = 0$

$a_m = x_i y_j - x_j y_i = 0(15) - (0)(0) = 0$

$2\Delta = a_1 + a_2 + a_3 = 150 m^2$

$b_i = y_j - y_m = 0 - 10 = -10m$ $c_i = x_m - x_j = 6 - 15 = -9m$

$b_j = y_m - y_i = 10 - 0 = 10m$ $c_j = x_i - x_m = 0 - 6 = -6m$

$b_m = y_i - y_j = 0 - 0 = 0m$ $c_m = x_j - x_i = 15 - 0 = 15$

$$[B] = \frac{1}{2\Delta}\begin{bmatrix} b_i & 0 & b_j & 0 & b_m & 0 \\ 0 & c_i & 0 & c_j & 0 & c_m \\ c_i & b_i & c_j & b_j & c_m & b_m \end{bmatrix} = \frac{1}{150}\begin{bmatrix} -10 & 0 & 10 & 0 & 0 & 0 \\ 0 & -9 & 0 & -6 & 0 & 15 \\ -9 & -10 & -6 & 10 & 15 & 0 \end{bmatrix}m^{-1}$$

$$[D] = \frac{E}{(1-v^2)}\begin{bmatrix} 1 & v & 0 \\ v & 1 & 0 \\ 0 & 0 & (1-v)/2 \end{bmatrix} = \frac{200\times10^6}{(1-0^2)}\begin{bmatrix} 1 & 0 & 0 \\ 0 & 1 & 0 \\ 0 & 0 & (1-0)/2 \end{bmatrix} = 2\times10^8\begin{bmatrix} 1 & 0 & 0 \\ 0 & 1 & 0 \\ 0 & 0 & 0.5 \end{bmatrix}N/m^2$$

$$[K] = [B]^T[D][B]t\Delta$$

$$[K]_1 = \frac{2\times10^8}{150}\times\frac{1\times75}{150}\begin{bmatrix} -10 & 0 & -9 \\ 0 & -9 & -10 \\ 10 & 0 & -6 \\ 0 & -6 & 10 \\ 0 & 0 & 15 \\ 0 & 15 & 0 \end{bmatrix}\begin{bmatrix} 1 & 0 & 0 \\ 0 & 1 & 0 \\ 0 & 0 & 0.5 \end{bmatrix}\begin{bmatrix} -10 & 0 & 10 & 0 & 0 & 0 \\ 0 & -9 & 0 & -6 & 0 & 15 \\ -9 & -10 & -6 & 10 & 15 & 0 \end{bmatrix}$$

112

$$\therefore [K]_1 = \frac{2\times 10^6}{3}\begin{bmatrix} u_A & v_A & u_C & v_C & u_B & v_B \\ 140.5 & 45 & -73 & -45 & -67.5 & 0 \\ 45 & 131 & 30 & 4 & -75 & -135 \\ -73 & 30 & 118 & -30 & -45 & 0 \\ -45 & 4 & -30 & 86 & 75 & -90 \\ -67.5 & -75 & -45 & 75 & 112.5 & 0 \\ 0 & -135 & 0 & -90 & 0 & 225 \end{bmatrix}\begin{matrix} u_A \\ v_A \\ u_C \\ v_C \\ u_B \\ v_B \end{matrix}$$

b) Element 2:

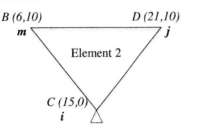

$$a_i = x_j y_m - x_m y_j = 21(10)-(6)(10) = 150m^2$$
$$a_j = x_m y_i - x_i y_m = 6(0)-(15)(10) = -150m^2$$
$$a_m = x_i y_j - x_j y_i = 15(10)-(0)(21) = 150m^2$$
$$2\Delta = a_1 + a_2 + a_3 = 150m^2$$

$$b_i = y_j - y_m = 10-10 = 0m \qquad\qquad c_i = x_m - x_j = 6-21 = -15m$$
$$b_j = y_m - y_i = 10-0 = 10m \qquad\qquad c_j = x_i - x_m = 15-6 = 9m$$
$$b_m = y_i - y_j = 0-10 = -10m \qquad\qquad c_m = x_j - x_i = 21-15 = 6m$$

$$[B] = \frac{1}{2\Delta}\begin{bmatrix} b_i & 0 & b_j & 0 & b_m & 0 \\ 0 & c_i & 0 & c_j & 0 & c_m \\ c_i & b_i & c_j & b_j & c_m & b_m \end{bmatrix} = \frac{1}{150}\begin{bmatrix} 0 & 0 & 10 & 0 & -10 & 0 \\ 0 & -15 & 0 & 9 & 0 & 6 \\ -15 & 0 & 9 & 10 & 6 & -10 \end{bmatrix}m^{-1}$$

$$[K] = [B]^T[D][B]t\Delta$$

$$[K]_2 = \frac{2\times 10^8}{150}\times\frac{1\times 75}{150}\begin{bmatrix} 0 & 0 & -15 \\ 0 & -15 & 0 \\ 10 & 0 & 9 \\ 0 & 9 & 10 \\ -10 & 0 & 6 \\ 0 & 6 & -10 \end{bmatrix}\begin{bmatrix} 1 & 0 & 0 \\ 0 & 1 & 0 \\ 0 & 0 & 0.5 \end{bmatrix}\begin{bmatrix} 0 & 0 & 10 & 0 & -10 & 0 \\ 0 & -15 & 0 & 9 & 0 & 6 \\ -15 & 0 & 9 & 10 & 6 & -10 \end{bmatrix}$$

$$\therefore [K]_2 = \frac{2\times 10^6}{3}\begin{bmatrix} u_C & v_C & u_D & v_D & u_B & v_B \\ 112.5 & 0 & -67.5 & -75 & -45 & 75 \\ 0 & 225 & 0 & -135 & 0 & -90 \\ -67.5 & 0 & 140.5 & 45 & -73 & -45 \\ -75 & -135 & 45 & 131 & 30 & 4 \\ -45 & 0 & -73 & 30 & 118 & -30 \\ 75 & -90 & -45 & 4 & -30 & 86 \end{bmatrix}\begin{matrix} u_C \\ v_C \\ u_D \\ v_D \\ u_B \\ v_B \end{matrix}$$

2. The global stiffness matrix:

$$[k]_{system} = \frac{2\times 10^6}{3}\begin{bmatrix} u_A & v_A & u_C & v_C & u_B & v_B & u_D & v_D \\ 140.5 & 45 & -73 & -45 & -67.5 & 0 & 0 & 0 \\ 45 & 131 & 30 & 4 & -75 & -135 & 0 & 0 \\ -73 & 30 & 118+112.5 & -30+0 & -45-45 & 0+75 & -67.5 & -75 \\ -45 & 4 & -30+0 & 86+225 & 75+0 & -90-90 & 0 & -135 \\ -67.5 & -75 & -45-45 & 75+0 & 112.5+118 & 0-30 & -73 & 30 \\ 0 & -135 & 0+75 & -90-90 & 0-30 & 225+86 & -45 & 4 \\ 0 & 0 & -67.5 & 0 & -73 & -45 & 140.5 & 45 \\ 0 & 0 & -75 & -135 & 30 & 4 & 45 & 131 \end{bmatrix}\begin{matrix} u_A \\ v_A \\ u_C \\ v_C \\ u_B \\ v_B \\ u_D \\ v_D \end{matrix}$$

113

$$[k]_{system} = \frac{2 \times 10^6}{3} \begin{bmatrix} 140.5 & 45 & -73 & -45 & -67.5 & 0 & 0 & 0 \\ 45 & 131 & 30 & 4 & -75 & -135 & 0 & 0 \\ -73 & 30 & 230.5 & -30 & -90 & 75 & -67.5 & -75 \\ -45 & 4 & -30 & 311 & 75 & -180 & 0 & -135 \\ -67.5 & -75 & -90 & 75 & 230.5 & -30 & -73 & 30 \\ 0 & -135 & 75 & -180 & -30 & 311 & -45 & 4 \\ 0 & 0 & -67.5 & 0 & -73 & -45 & 140.5 & 45 \\ 0 & 0 & -75 & -135 & 30 & 4 & 45 & 131 \end{bmatrix}$$

3. Based on the static equilibrium state, the system nodal force vector, stiffness matrix and displacement vector can be expressed as following:

$$\{F\} = [k]\{\delta\} = \{R\}$$

$$\frac{2 \times 10^6}{3} \begin{bmatrix} 140.5 & 45 & -73 & -45 & -67.5 & 0 & 0 & 0 \\ 45 & 131 & 30 & 4 & -75 & -135 & 0 & 0 \\ -73 & 30 & 230.5 & -30 & -90 & 75 & -67.5 & -75 \\ -45 & 4 & -30 & 311 & 75 & -180 & 0 & -135 \\ -67.5 & -75 & -90 & 75 & 230.5 & -30 & -73 & 30 \\ 0 & -135 & 75 & -180 & -30 & 311 & -45 & 4 \\ 0 & 0 & -67.5 & 0 & -73 & -45 & 140.5 & 45 \\ 0 & 0 & -75 & -135 & 30 & 4 & 45 & 131 \end{bmatrix} \begin{Bmatrix} u_A = 0 \\ v_A = 0 \\ u_C = 0 \\ v_C = 0 \\ u_B \\ v_B \\ u_D \\ v_D \end{Bmatrix} = \begin{Bmatrix} R_{Ax} \\ R_{Ay} \\ R_{Cx} \\ R_{Cy} \\ 20000 \\ -10000 \\ 0 \\ 0 \end{Bmatrix}$$

By solving the equation, the displacement and reaction forces are computed.
$u_B = 1.73 \times 10^{-4}$ m, $v_B = -1.465 \times 10^{-5}$ m, $u_D = 1.099 \times 10^{-4}$ m, $v_D = -7.69 \times 10^{-5}$ m,
$R_{Ax} = -7785$ kN, $R_{Ay} = -7331.5$ kN, $R_{3x} = -12213$ kN, $R_{3y} = 17329$ kN

4. The stress in the element:

$$\{\sigma\} = [D][B]\{\delta\}$$

$$\{\sigma\}_1 = \frac{2 \times 10^8}{150} \begin{bmatrix} 1 & 0 & 0 \\ 0 & 1 & 0 \\ 0 & 0 & 0.5 \end{bmatrix} \begin{bmatrix} -10 & 0 & 10 & 0 & 0 & 0 \\ 0 & -9 & 0 & -6 & 0 & 15 \\ -9 & -10 & -6 & 10 & 15 & 0 \end{bmatrix} \begin{Bmatrix} 0 \\ 0 \\ 0 \\ 0 \\ 1.73 \times 10^{-4} \\ -1.465 \times 10^{-5} \end{Bmatrix} = \begin{Bmatrix} 0 \\ -293 \\ 1730 \end{Bmatrix} N / m^2$$

$$\{\sigma\}_2 = \frac{2 \times 10^8}{150} \begin{bmatrix} 1 & 0 & 0 \\ 0 & 1 & 0 \\ 0 & 0 & 0.5 \end{bmatrix} \begin{bmatrix} 0 & 0 & 10 & 0 & -10 & 0 \\ 0 & -15 & 0 & 9 & 0 & 6 \\ -15 & 0 & 9 & 10 & 6 & -10 \end{bmatrix} \begin{Bmatrix} 0 \\ 0 \\ 1.099 \times 10^{-4} \\ -7.69 \times 10^{-5} \\ 1.73 \times 10^{-4} \\ -1.465 \times 10^{-5} \end{Bmatrix} = \begin{Bmatrix} -841.33 \\ -1040 \\ 936.4 \end{Bmatrix} N / m^2$$

2. Example 4E.2:

Calculate the deflections at the nodes (coordinates in meters are shown in brackets), reactions at the supports (hinged at 1 and rollers at 2), and the stresses at point D (5,5) of the elastic plane stress triangular element shown in Figure 4.6.17. Thickness, $t = 1m$, modulus of elasticity $= E$, and Poisson's Ratio, $v = 0$. Check the values of reactions from elementary theory. Assume the loads applied along the boundaries to be shared equally between the two end nodes.

What will be the equivalent forces at the nodes if the displacements at nodes (in $\mu m = 10^{-6}m$) are: $u_1, v_1 = 10, 5$; $u_2, v_2 = 5, 10$; $u_3, v_3 = 5, 5$? Use one element only.

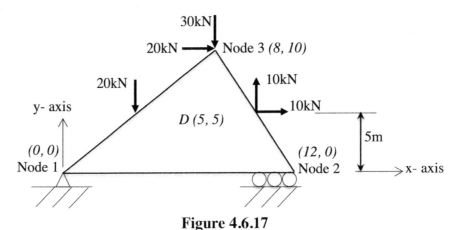

Figure 4.6.17

Solution:

Lumping the loads on the boundaries equally at the end nodes. The equivalent applied nodal loads $\{R\}$ are shown below.

1. The stiffness of the element:

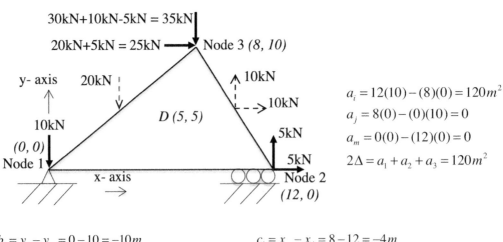

$$a_i = 12(10) - (8)(0) = 120m^2$$
$$a_j = 8(0) - (0)(10) = 0$$
$$a_m = 0(0) - (12)(0) = 0$$
$$2\Delta = a_1 + a_2 + a_3 = 120m^2$$

$$b_i = y_j - y_m = 0 - 10 = -10m \qquad c_i = x_m - x_j = 8 - 12 = -4m$$
$$b_j = y_m - y_i = 10 - 0 = 10m \qquad c_j = x_i - x_m = 0 - 8 = -8m$$
$$b_m = y_i - y_j = 0 - 0 = 0m \qquad c_m = x_j - x_i = 12 - 0 = 12m$$

$$[B] = \frac{1}{2\Delta}\begin{bmatrix} b_i & 0 & b_j & 0 & b_m & 0 \\ 0 & c_i & 0 & c_j & 0 & c_m \\ c_i & b_i & c_j & b_j & c_m & b_m \end{bmatrix} = \frac{1}{120}\begin{bmatrix} -10 & 0 & 10 & 0 & 0 & 0 \\ 0 & -4 & 0 & -8 & 0 & 12 \\ -4 & -10 & -8 & 10 & 12 & 0 \end{bmatrix} m^{-1}$$

$$[D] = \frac{E}{(1-v^2)}\begin{bmatrix} 1 & v & 0 \\ v & 1 & 0 \\ 0 & 0 & (1-v)/2 \end{bmatrix} = \frac{E}{(1-0^2)}\begin{bmatrix} 1 & 0 & 0 \\ 0 & 1 & 0 \\ 0 & 0 & (1-0)/2 \end{bmatrix} = E\begin{bmatrix} 1 & 0 & 0 \\ 0 & 1 & 0 \\ 0 & 0 & 0.5 \end{bmatrix} N/m^2$$

$$[K] = [B]^T[D][B]t\Delta$$

$$[K] = \frac{E}{120}\times\frac{1\times60}{120}\begin{bmatrix} -10 & 0 & -4 \\ 0 & -4 & -10 \\ 10 & 0 & -8 \\ 0 & -8 & 10 \\ 0 & 0 & 12 \\ 0 & 12 & 0 \end{bmatrix}\begin{bmatrix} 1 & 0 & 0 \\ 0 & 1 & 0 \\ 0 & 0 & 0.5 \end{bmatrix}\begin{bmatrix} -10 & 0 & 10 & 0 & 0 & 0 \\ 0 & -4 & 0 & -8 & 0 & 12 \\ -4 & -10 & -8 & 10 & 12 & 0 \end{bmatrix}$$

$$\therefore [K] = \frac{E}{240}\begin{bmatrix} 108 & 20 & -84 & -20 & -24 & 0 \\ 20 & 66 & 40 & -18 & -60 & -48 \\ -84 & 40 & 132 & -40 & -48 & 0 \\ -20 & -18 & -40 & 114 & 60 & -96 \\ -24 & -60 & -48 & 60 & 72 & 0 \\ 0 & -48 & 0 & -96 & 0 & 144 \end{bmatrix}$$

2. Based on the static equilibrium state, the system nodal force vector, stiffness matrix and displacement vector can be expressed as following:

$$\{F\} = [k]\{\delta\} = \{R\}$$

$$\{F\} = \frac{E}{240}\begin{bmatrix} 108 & 20 & -84 & -20 & -24 & 0 \\ 20 & 66 & 40 & -18 & -60 & -48 \\ -84 & 40 & 132 & -40 & -48 & 0 \\ -20 & -18 & -40 & 114 & 60 & -96 \\ -24 & -60 & -48 & 60 & 72 & 0 \\ 0 & -48 & 0 & -96 & 0 & 144 \end{bmatrix}\begin{Bmatrix} u_1 = 0 \\ v_1 = 0 \\ u_2 \\ v_2 = 0 \\ u_3 \\ v_3 \end{Bmatrix} = \begin{Bmatrix} R_{1x} \\ R_{1y} - 10\times10^3 \\ 5\times10^3 \\ R_{2y} + 5\times10^3 \\ 25\times10^3 \\ -35\times10^3 \end{Bmatrix}$$

By solving the equation, the displacement and reaction forces are computed.
$u_2 = 52000/E$ m, $u_3 = 118000/E$ m, $v_3 = -175000/3E$ m,
$R_{1x} = -30000$ N, $R_{1y} = 2500/3$ N, $R_{2y} = 117500/3$ N.

3. The stress in the element:

$$\{\sigma\} = [D][B]\{\delta\}$$

$$\{\sigma\} = \frac{E}{120}\begin{bmatrix} 1 & 0 & 0 \\ 0 & 1 & 0 \\ 0 & 0 & 0.5 \end{bmatrix}\begin{bmatrix} -10 & 0 & 10 & 0 & 0 & 0 \\ 0 & -4 & 0 & -8 & 0 & 12 \\ -4 & -10 & -8 & 10 & 12 & 0 \end{bmatrix}\begin{Bmatrix} 0 \\ 0 \\ 52000/E \\ 0 \\ 118000/E \\ -175000/3E \end{Bmatrix}$$

$$\{\sigma\} = \left\{ \begin{array}{c} 4333.33 \\ -5833.33 \\ 4166.67 \end{array} \right\} N/m^2$$

4. The conventional method:

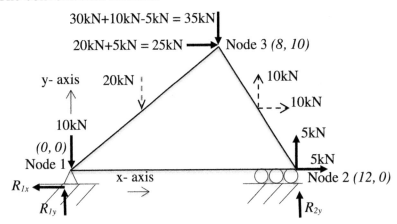

Support Reaction:

$$\sum M_1 = 0:$$

$$R_{2y}(12) + 5(12) - 35(8) - 25(10) = 0$$

$$\therefore R_{2y} = 39.167 kN \ (\uparrow)$$

$$+ \uparrow \sum F_y = 0:$$

$$R_{1y} - 10 - 35 + 5 + R_{2y} = 0$$

$$\therefore R_{1y} = 0.833 kN \ (\uparrow)$$

$$+ \rightarrow \sum F_x = 0:$$

$$-R_{1x} + 25 + 5 = 0$$

$$\therefore R_{1x} = 30 kN \ (\leftarrow)$$

5. The equivalent forces at the nodes for the specified displacements:

$$\{F\} = \frac{E}{240} \begin{bmatrix} 108 & 20 & -84 & -20 & -24 & 0 \\ 20 & 66 & 40 & -18 & -60 & -48 \\ -84 & 40 & 132 & -40 & -48 & 0 \\ -20 & -18 & -40 & 114 & 60 & -96 \\ -24 & -60 & -48 & 60 & 72 & 0 \\ 0 & -48 & 0 & -96 & 0 & 144 \end{bmatrix} \left\{ \begin{array}{c} 10 \times 10^{-6} \\ 5 \times 10^{-6} \\ 5 \times 10^{-6} \\ 10 \times 10^{-6} \\ 5 \times 10^{-6} \\ 5 \times 10^{-6} \end{array} \right\} = \left\{ \begin{array}{c} 1.833E \\ 0.0417E \\ -2.583E \\ 1.958E \\ 0.75E \\ -2E \end{array} \right\} \times 10^{-6} N$$

3. Example 4E.3:

For a Constant Strain Triangle element in a finite element model, the nodal coordinates and external forces condition are indicated in the Figure 4.6.18. The element thickness is 1m and the material properties are $E = 100$ GPa and $v = 0$. Upon the loading, determine:

a) The element stiffness matrix
b) The nodal displacements and support reactions
c) The stress in the element

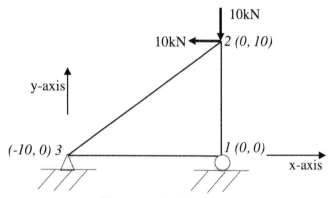

Figure 4.6.18

Solution:

1. The element stiffness matrix:

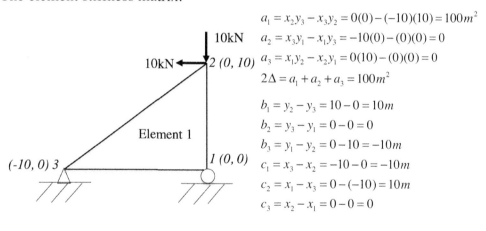

$$a_1 = x_2 y_3 - x_3 y_2 = 0(0) - (-10)(10) = 100 m^2$$
$$a_2 = x_3 y_1 - x_1 y_3 = -10(0) - (0)(0) = 0$$
$$a_3 = x_1 y_2 - x_2 y_1 = 0(10) - (0)(0) = 0$$
$$2\Delta = a_1 + a_2 + a_3 = 100 m^2$$

$$b_1 = y_2 - y_3 = 10 - 0 = 10 m$$
$$b_2 = y_3 - y_1 = 0 - 0 = 0$$
$$b_3 = y_1 - y_2 = 0 - 10 = -10 m$$
$$c_1 = x_3 - x_2 = -10 - 0 = -10 m$$
$$c_2 = x_1 - x_3 = 0 - (-10) = 10 m$$
$$c_3 = x_2 - x_1 = 0 - 0 = 0$$

$$[B] = \frac{1}{2\Delta} \begin{bmatrix} b_1 & 0 & b_2 & 0 & b_3 & 0 \\ 0 & c_1 & 0 & c_2 & 0 & c_3 \\ c_1 & b_1 & c_2 & b_2 & c_3 & b_3 \end{bmatrix} = \frac{1}{100} \begin{bmatrix} 10 & 0 & 0 & 0 & -10 & 0 \\ 0 & -10 & 0 & 10 & 0 & 0 \\ -10 & 10 & 10 & 0 & 0 & -10 \end{bmatrix} m^{-1}$$

$$[B]^T = \frac{1}{100} \begin{bmatrix} 10 & 0 & -10 \\ 0 & -10 & 10 \\ 0 & 0 & 10 \\ 0 & 10 & 0 \\ -10 & 0 & 0 \\ 0 & 0 & -10 \end{bmatrix} m^{-1}$$

118

$$[D] = \frac{E}{2(1-v^2)}\begin{bmatrix} 2 & 2v & 0 \\ 2v & 2 & 0 \\ 0 & 0 & (1-v) \end{bmatrix} = \frac{100 \times 10^9}{2(1-0^2)}\begin{bmatrix} 2 & 2(0) & 0 \\ 2(0) & 2 & 0 \\ 0 & 0 & (1-0) \end{bmatrix} = 50 \times 10^9 \begin{bmatrix} 2 & 0 & 0 \\ 0 & 2 & 0 \\ 0 & 0 & 1 \end{bmatrix} N/m^2$$

$$[K] = [B]^T[D][B]t\Delta$$

$$[K] = \frac{50 \times 10^9}{100} \times \frac{1}{100} \times \frac{100}{2} \begin{bmatrix} 10 & 0 & -10 \\ 0 & -10 & 10 \\ 0 & 0 & 10 \\ 0 & 10 & 0 \\ -10 & 0 & 0 \\ 0 & 0 & -10 \end{bmatrix} \begin{bmatrix} 2 & 0 & 0 \\ 0 & 2 & 0 \\ 0 & 0 & 1 \end{bmatrix} \begin{bmatrix} 10 & 0 & 0 & 0 & -10 & 0 \\ 0 & -10 & 0 & 10 & 0 & 0 \\ -10 & 10 & 10 & 0 & 0 & -10 \end{bmatrix}$$

$$[K] = 10^9 \begin{bmatrix} 75 & -25 & -25 & 0 & -50 & 25 \\ -25 & 75 & 25 & -50 & 0 & -25 \\ -25 & 25 & 25 & 0 & 0 & -25 \\ 0 & -50 & 0 & 50 & 0 & 0 \\ -50 & 0 & 0 & 0 & 50 & 0 \\ 25 & -25 & -25 & 0 & 0 & 25 \end{bmatrix}$$

2. Based on the static equilibrium state, the system nodal force vector, stiffness matrix and displacement vector can be expressed as following:

$$\{F\} = [k]\{\delta\} = \{R\}$$

$$\{F\} = 10^9 \begin{bmatrix} 75 & -25 & -25 & 0 & -50 & 25 \\ -25 & 75 & 25 & -50 & 0 & -25 \\ -25 & 25 & 25 & 0 & 0 & -25 \\ 0 & -50 & 0 & 50 & 0 & 0 \\ -50 & 0 & 0 & 0 & 50 & 0 \\ 25 & -25 & -25 & 0 & 0 & 25 \end{bmatrix} \begin{Bmatrix} u_1 \\ v_1 = 0 \\ u_2 \\ v_2 \\ v_3 = 0 \\ \theta_3 = 0 \end{Bmatrix} = \begin{Bmatrix} 0 \\ R_1 \\ -10 \times 10^3 \\ -10 \times 10^3 \\ R_{3x} \\ R_{3y} \end{Bmatrix}$$

By solving the equation, the displacement and reaction forces are computed.
$u_1 = -2 \times 10^{-7}$m, $u_2 = -6 \times 10^{-7}$m, $v_2 = -2 \times 10^{-7}$m,
$R_1 = 0$ kN, $R_{3x} = 10$ kN, $R_{3y} = 10$ kN

3. The stress in the element:

$$\{\sigma\} = [D][B]\{\delta\}$$

$$\{\sigma\} = \frac{50 \times 10^9}{100} \begin{bmatrix} 2 & 0 & 0 \\ 0 & 2 & 0 \\ 0 & 0 & 1 \end{bmatrix} \begin{bmatrix} 10 & 0 & 0 & 0 & -10 & 0 \\ 0 & -10 & 0 & 10 & 0 & 0 \\ -10 & 10 & 10 & 0 & 0 & -10 \end{bmatrix} \begin{Bmatrix} -2 \times 10^{-7} \\ 0 \\ -6 \times 10^{-7} \\ -2 \times 10^{-7} \\ 0 \\ 0 \end{Bmatrix}$$

$$\{\sigma\} = \begin{Bmatrix} -2000 \\ -2000 \\ -2000 \end{Bmatrix} N/m^2$$

4. Example 4E.4:

a. Calculate the deflections at the nodes (coordinates in meters are shown in brackets), reactions at the supports (hinged at 1 and rollers at 2), and the stresses at point D (5,5) of the elastic plane stress triangular element shown in Figure 4.6.19. Thickness, $t = 1m$, modulus of elasticity $= E$, and Poisson's Ratio, $v = 0$. Check the values of reactions from elementary theory. Assume the loads applied along the boundaries to be shared equally between the two end nodes.

b. What will be the equivalent forces at the nodes if the displacements at nodes (in $\mu m = 10^{-6} m$) are: $u_1, v_1 = 10, 5$; $u_2, v_2 = 5, 10$; $u_3, v_3 = 5, 5$? Use one element only.

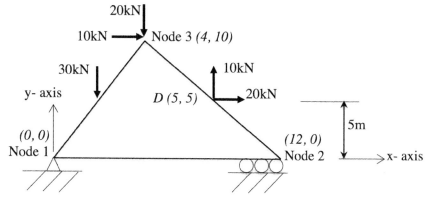

Figure 4.6.19

Solution:

After lumping the loads applied on the boundaries, equally at the end nodes, the equivalent applied nodal loads $\{R\}$ are shown below.

1. The stiffness of the element:

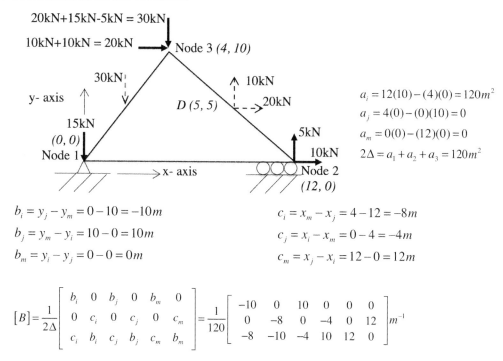

$$a_i = 12(10) - (4)(0) = 120m^2$$
$$a_j = 4(0) - (0)(10) = 0$$
$$a_m = 0(0) - (12)(0) = 0$$
$$2\Delta = a_1 + a_2 + a_3 = 120m^2$$

$b_i = y_j - y_m = 0 - 10 = -10m$ $c_i = x_m - x_j = 4 - 12 = -8m$

$b_j = y_m - y_i = 10 - 0 = 10m$ $c_j = x_i - x_m = 0 - 4 = -4m$

$b_m = y_i - y_j = 0 - 0 = 0m$ $c_m = x_j - x_i = 12 - 0 = 12m$

$$[B] = \frac{1}{2\Delta}\begin{bmatrix} b_i & 0 & b_j & 0 & b_m & 0 \\ 0 & c_i & 0 & c_j & 0 & c_m \\ c_i & b_i & c_j & b_j & c_m & b_m \end{bmatrix} = \frac{1}{120}\begin{bmatrix} -10 & 0 & 10 & 0 & 0 & 0 \\ 0 & -8 & 0 & -4 & 0 & 12 \\ -8 & -10 & -4 & 10 & 12 & 0 \end{bmatrix} m^{-1}$$

$$[D] = \frac{E}{(1-v^2)} \begin{bmatrix} 1 & v & 0 \\ v & 1 & 0 \\ 0 & 0 & (1-v)/2 \end{bmatrix} = \frac{E}{(1-0^2)} \begin{bmatrix} 1 & 0 & 0 \\ 0 & 1 & 0 \\ 0 & 0 & (1-0)/2 \end{bmatrix} = E \begin{bmatrix} 1 & 0 & 0 \\ 0 & 1 & 0 \\ 0 & 0 & 0.5 \end{bmatrix} N/m^2$$

$$[K] = [B]^T [D][B] t \Delta$$

$$[K] = \frac{E}{120} \times \frac{1 \times 60}{120} \begin{bmatrix} -10 & 0 & -8 \\ 0 & -8 & -10 \\ 10 & 0 & -4 \\ 0 & -4 & 10 \\ 0 & 0 & 12 \\ 0 & 12 & 0 \end{bmatrix} \begin{bmatrix} 1 & 0 & 0 \\ 0 & 1 & 0 \\ 0 & 0 & 0.5 \end{bmatrix} \begin{bmatrix} -10 & 0 & 10 & 0 & 0 & 0 \\ 0 & -8 & 0 & -4 & 0 & 12 \\ -8 & -10 & -4 & 10 & 12 & 0 \end{bmatrix}$$

$$\therefore [K] = \frac{E}{240} \begin{bmatrix} 132 & 40 & -84 & -40 & -48 & 0 \\ 40 & 114 & 20 & -18 & -60 & -96 \\ -84 & 20 & 108 & -20 & -24 & 0 \\ -40 & -18 & -20 & 66 & 60 & -48 \\ -48 & -60 & -24 & 60 & 72 & 0 \\ 0 & -96 & 0 & -48 & 0 & 144 \end{bmatrix}$$

2. Based on the static equilibrium state, the system nodal force vector, stiffness matrix and displacement vector can be expressed as following:

$$\{F\} = [k]\{\delta\} = \{R\}$$

$$\{F\} = \frac{E}{240} \begin{bmatrix} 132 & 40 & -84 & -40 & -48 & 0 \\ 40 & 114 & 20 & -18 & -60 & -96 \\ -84 & 20 & 108 & -20 & -24 & 0 \\ -40 & -18 & -20 & 66 & 60 & -48 \\ -48 & -60 & -24 & 60 & 72 & 0 \\ 0 & -96 & 0 & -48 & 0 & 144 \end{bmatrix} \begin{Bmatrix} u_1 = 0 \\ v_1 = 0 \\ u_2 \\ v_2 = 0 \\ u_3 \\ v_3 \end{Bmatrix} = \begin{Bmatrix} R_{1x} \\ R_{1y} - 15 \times 10^3 \\ 10 \times 10^3 \\ R_{2y} + 5 \times 10^3 \\ 20 \times 10^3 \\ -30 \times 10^3 \end{Bmatrix}$$

By solving the equation, the displacement and reaction forces are computed.
$u_2 = 40000/E$ m, $u_3 = 80000/E$ m, $v_3 = -50000/E$ m,
$R_{1x} = -30000$ N, $R_{1y} = 18333.33$ N, $R_{2y} = 21666.67$ N.

3. The stress in the element:

$$\{\sigma\} = [D][B]\{\delta\}$$

$$\{\sigma\} = \frac{E}{120} \begin{bmatrix} 1 & 0 & 0 \\ 0 & 1 & 0 \\ 0 & 0 & 0.5 \end{bmatrix} \begin{bmatrix} -10 & 0 & 10 & 0 & 0 & 0 \\ 0 & -8 & 0 & -4 & 0 & 12 \\ -8 & -10 & -4 & 10 & 12 & 0 \end{bmatrix} \begin{Bmatrix} 0 \\ 0 \\ 40000/E \\ 0 \\ 80000/E \\ -50000/E \end{Bmatrix}$$

$$\{\sigma\} = \begin{Bmatrix} 3333.33 \\ -5000 \\ 3333.33 \end{Bmatrix} N/m^2$$

4. The conventional method:

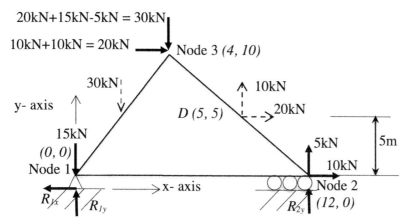

20kN+15kN-5kN = 30kN

10kN+10kN = 20kN → Node 3 *(4, 10)*

30kN

10kN

y- axis

D (5, 5) 20kN

15kN

5kN 5m

(0, 0)

Node 1 10kN

R_{1x} R_{1y} Node 2 *(12, 0)*

x- axis R_{2y}

Support Reaction:

$\sum M_1 = 0$:

$R_{2y}(12) + 5(12) - 35(8) - 25(10) = 0$

$\therefore R_{2y} = 39.167 kN \ (\uparrow)$

$+\uparrow \sum F_y = 0$:

$R_{1y} - 10 - 35 + 5 + R_{2y} = 0$

$\therefore R_{1x} = 0.833 kN \ (\leftarrow)$

$+\rightarrow \sum F_x = 0$:

$-R_{1x} + 20 + 10 = 0$

$\therefore R_{1x} = 30 kN \ (\leftarrow)$

5. The equivalent forces at the nodes for the specified displacements:

$$\{F\} = \frac{E}{240} \begin{bmatrix} 132 & 40 & -84 & -40 & -48 & 0 \\ 40 & 114 & 20 & -18 & -60 & -96 \\ -84 & 20 & 108 & -20 & -24 & 0 \\ -40 & -18 & -20 & 66 & 60 & -48 \\ -48 & -60 & -24 & 60 & 72 & 0 \\ 0 & -96 & 0 & -48 & 0 & 144 \end{bmatrix} \begin{Bmatrix} 10 \times 10^{-6} \\ 5 \times 10^{-6} \\ 5 \times 10^{-6} \\ 10 \times 10^{-6} \\ 5 \times 10^{-6} \\ 5 \times 10^{-6} \end{Bmatrix} = \begin{Bmatrix} 1.9167E \\ 0.4583E \\ -2.1667E \\ 0.5417E \\ 0.25E \\ -1E \end{Bmatrix} \times 10^{-6} N$$

122

5. Example 4E.5:

a. Calculate the deflections at the nodes (coordinates in meters are shown in brackets), reactions at the supports (hinged at 1 and rollers at 2), and the stresses at point D, the center of gravity of the elastic plane stress triangular element shown in Figure 4.6.20. Thickness, t, Modulus of elasticity = E, and Poisson's Ratio, $v = 0$.

b. Check the values of reactions from elementary theory. Assume the loads applied along the boundaries to be shared equally between the two end nodes.

c. What will be the equivalent forces at the nodes if the displacements at nodes (in μm = 10^{-6} m) are: $u_1, v_1 = 12, 15$; $u_2, v_2 = 8, 6$; $u_3, v_3 = 4, 8$?
Use one element only.

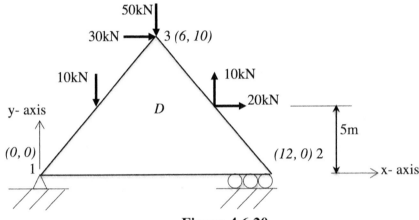

Figure 4.6.20

Solution:

After lumping the loads applied on the boundaries equally at the end nodes, the equivalent applied nodal loads $\{R\}$ are shown below.

1. The stiffness of the element:

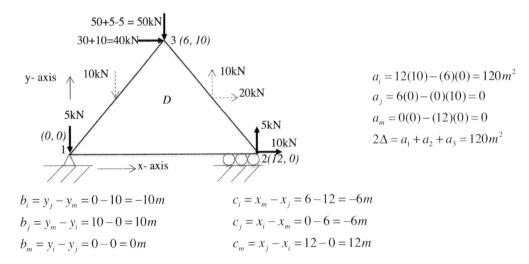

$$a_i = 12(10) - (6)(0) = 120\,m^2$$
$$a_j = 6(0) - (0)(10) = 0$$
$$a_m = 0(0) - (12)(0) = 0$$
$$2\Delta = a_1 + a_2 + a_3 = 120\,m^2$$

$$b_i = y_j - y_m = 0 - 10 = -10m \qquad c_i = x_m - x_j = 6 - 12 = -6m$$
$$b_j = y_m - y_i = 10 - 0 = 10m \qquad c_j = x_i - x_m = 0 - 6 = -6m$$
$$b_m = y_i - y_j = 0 - 0 = 0m \qquad c_m = x_j - x_i = 12 - 0 = 12m$$

123

$$[B] = \frac{1}{2\Delta} \begin{bmatrix} b_i & 0 & b_j & 0 & b_m & 0 \\ 0 & c_i & 0 & c_j & 0 & c_m \\ c_i & b_i & c_j & b_j & c_m & b_m \end{bmatrix} = \frac{1}{120} \begin{bmatrix} -10 & 0 & 10 & 0 & 0 & 0 \\ 0 & -6 & 0 & -6 & 0 & 12 \\ -6 & -10 & -6 & 10 & 12 & 0 \end{bmatrix} m^{-1}$$

$$[D] = \frac{E}{(1-v^2)} \begin{bmatrix} 1 & v & 0 \\ v & 1 & 0 \\ 0 & 0 & (1-v)/2 \end{bmatrix} = \frac{E}{(1-0^2)} \begin{bmatrix} 1 & 0 & 0 \\ 0 & 1 & 0 \\ 0 & 0 & (1-0)/2 \end{bmatrix} = E \begin{bmatrix} 1 & 0 & 0 \\ 0 & 1 & 0 \\ 0 & 0 & 0.5 \end{bmatrix} N/m^2$$

$$[K] = [B]^T [D][B] t \Delta$$

$$[K] = \frac{E}{120} \times \frac{1 \times 60}{120} \begin{bmatrix} -10 & 0 & -6 \\ 0 & -6 & -10 \\ 10 & 0 & -6 \\ 0 & -6 & 10 \\ 0 & 0 & 12 \\ 0 & 12 & 0 \end{bmatrix} \begin{bmatrix} 1 & 0 & 0 \\ 0 & 1 & 0 \\ 0 & 0 & 0.5 \end{bmatrix} \begin{bmatrix} -10 & 0 & 10 & 0 & 0 & 0 \\ 0 & -6 & 0 & -6 & 0 & 12 \\ -6 & -10 & -6 & 10 & 12 & 0 \end{bmatrix}$$

$$\therefore [K] = \frac{E}{240} \begin{bmatrix} 118 & 30 & -82 & -30 & -36 & 0 \\ 30 & 86 & 30 & -14 & -60 & -72 \\ -82 & 30 & 118 & -30 & -36 & 0 \\ -30 & -14 & -30 & 86 & 60 & -72 \\ -36 & -60 & -36 & 60 & 72 & 0 \\ 0 & -72 & 0 & -72 & 0 & 144 \end{bmatrix}$$

2. Based on the static equilibrium state, the system nodal force vector, stiffness matrix and displacement vector can be expressed as following:

$$\{F\} = [k]\{\delta\} = \{R\}$$

$$\{F\} = \frac{E}{240} \begin{bmatrix} 118 & 30 & -82 & -30 & -36 & 0 \\ 30 & 86 & 30 & -14 & -60 & -72 \\ -82 & 30 & 118 & -30 & -36 & 0 \\ -30 & -14 & -30 & 86 & 60 & -72 \\ -36 & -60 & -36 & 60 & 72 & 0 \\ 0 & -72 & 0 & -72 & 0 & 144 \end{bmatrix} \begin{Bmatrix} u_1 = 0 \\ v_1 = 0 \\ u_2 \\ v_2 = 0 \\ u_3 \\ v_3 \end{Bmatrix} = \begin{Bmatrix} R_{1x} \\ R_{1y} - 5 \times 10^3 \\ 10 \times 10^3 \\ R_{2y} + 5 \times 10^3 \\ 40 \times 10^3 \\ -50 \times 10^3 \end{Bmatrix}$$

By solving the equation, the displacement and reaction forces are computed.
$u_2 = 72000/E$ m, $u_3 = 508000/3E$ m, $v_3 = -250000/3E$ m,
$R_{1x} = -50000$ N, $R_{1y} = -3333.33$ N, $R_{2y} = 53333.33$ N.

3. The stress in the element:

$$\{\sigma\} = [D][B]\{\delta\}$$

$$\{\sigma\} = \frac{E}{120} \begin{bmatrix} 1 & 0 & 0 \\ 0 & 1 & 0 \\ 0 & 0 & 0.5 \end{bmatrix} \begin{bmatrix} -10 & 0 & 10 & 0 & 0 & 0 \\ 0 & -6 & 0 & -6 & 0 & 12 \\ -6 & -10 & -6 & 10 & 12 & 0 \end{bmatrix} \begin{Bmatrix} 0 \\ 0 \\ -50000/E \\ 0 \\ -3333.33/E \\ 53333.33/E \end{Bmatrix}$$

$$\{\sigma\} = \left\{ \begin{array}{c} -4166.67 \\ 5333.33 \\ 1083.33 \end{array} \right\} N/m^2$$

4. The conventional method:

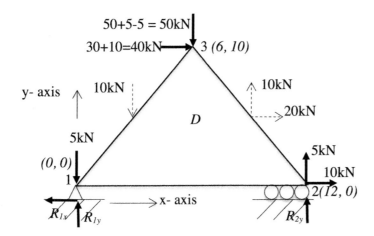

Support Reaction:

$\sum M_1 = 0:$

$R_{2y}(12) + 5(12) - 50(6) - 40(10) = 0$

$\quad \therefore R_{2y} = 53.33kN \ (\uparrow)$

$+\uparrow \sum F_y = 0:$

$R_{1y} - 5 - 50 + 5 + R_{2y} = 0$

$\therefore R_{1y} = -3.33kN = 3.33kN \ (\downarrow)$

$+\rightarrow \sum F_x = 0:$

$-R_{1x} + 40 + 10 = 0$

$\therefore R_{1x} = 50kN \ (\leftarrow)$

5. The equivalent forces at the nodes for the specified displacements:

$$\{F\} = \frac{E}{240} \begin{bmatrix} 118 & 30 & -82 & -30 & -36 & 0 \\ 30 & 86 & 30 & -14 & -60 & -72 \\ -82 & 30 & 118 & -30 & -36 & 0 \\ -30 & -14 & -30 & 86 & 60 & -72 \\ -36 & -60 & -36 & 60 & 72 & 0 \\ 0 & -72 & 0 & -72 & 0 & 144 \end{bmatrix} \left\{ \begin{array}{c} 12\times10^{-6} \\ 15\times10^{-6} \\ 8\times10^{-6} \\ 6\times10^{-6} \\ 4\times10^{-6} \\ 8\times10^{-6} \end{array} \right\} = \left\{ \begin{array}{c} 3.692E \\ 4.125E \\ 0.358E \\ -2.625E \\ -4.05E \\ -1.5E \end{array} \right\} \times10^{-6} N$$

4.7 Assignment Problems

The assignment problems for Chapter 4 are given topic wise in Section 4.7 i.e. Rods (Section 4.7.1), Trusses (Section 4.7.2), Beams (Section 4.7.3) and Elastic Solids (Section 4.7.4) respectively.

4.7.1 Rods

1a) For the axially loaded structure shown in Figure 4.7.1, obtain the deflections at all the nodes, 1 to 4, wall reactions, stresses and nodal forces for each element (1 to 3) using FEM. Compare the values from the conventional method. Take $E = 150$ GPa (150×10^9 N/m²). A_1 to A_3 are areas of cross section, and L_1 to L_3 are the lengths of the elements respectively.

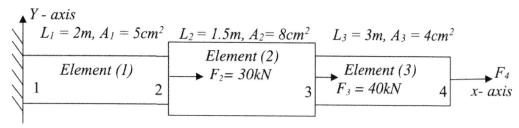

Figure 4.7.1

b) Redo the problem 1a) if the entire rod is heated up by 200^0C taking the coefficient of thermal expansion, $\alpha = 35\mu /^0$ Celcius ($\mu = 10^{-6}$).

c) What should be the force F_4 to be applied at node 4 to restrict the displacement as 0.5m in problems 1a) and b)?

2a) For the axially loaded structure shown in Figure 4.7.2, obtain the deflections at all the nodes, 1 to 4, wall reactions, stresses and nodal forces for each element (1 to 3) using FEM. Compare the values from the conventional method. Take $E = 250$ GPa (150×10^9 N/m²). A_1 to A_3 are areas of cross section, and L_1 to L_3 are the lengths of the elements respectively.

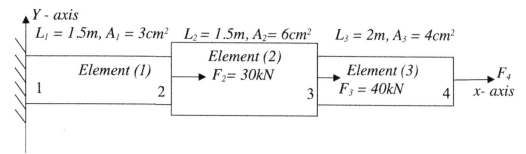

Figure 4.7.2

b) Redo the problem 2a) if the entire rod is heated up by 100°C taking the coefficient of thermal expansion, $\alpha = 30\mu /^\circ$ Celcius ($\mu = 10^{-6}$).

c) What should be the force F_4 to be applied at node 4 to restrict the displacement as 0.3m in problems 2a) and 2b)?

3a) For the axially loaded structure shown in Figure 4.7.3, obtain the deflections at all the nodes, 1 to 4, wall reactions, stresses and nodal forces for each element (1 to 3) using FEM. Compare the values from the conventional method. Take $E = 150$ GPa (150×10^9 N/m²). A_1 to A_3 are areas of cross section, and L_1 to L_3 are the lengths of the elements respectively.

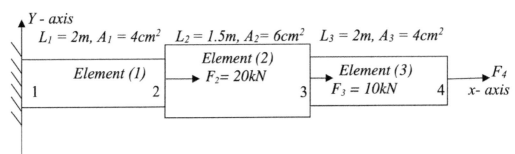

Figure 4.7.3

b) Redo the problem 3a) if the entire rod is heated up by 100°C taking the coefficient of thermal expansion, $\alpha = 20\mu /^\circ$ Celcius ($\mu = 10^{-6}$).

c) What should be the force F_4 to be applied at node 4 to restrict the displacement as 0.3m in problems 3a) and 3b)?

4a) For the axially loaded structure shown in Figure 4.7.4, obtain the deflections at all the nodes, 1 to 4, wall reactions, stresses and nodal forces for each element (1 to 3) using FEM. Compare the values from the conventional method. Take $E = 100$ GPa (100×10^9 N/m²). A_1 to A_3 are areas of cross section, and L_1 to L_3 are the lengths of the elements respectively.

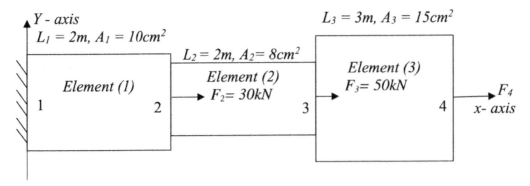

Figure 4.7.4

127

b) Redo the problem 4a) if the entire rod is heated up by 100°C taking the coefficient of thermal expansion, $\alpha = 15\mu / ^\circ$ Celcius ($\mu = 10^{-6}$).

c) What should be the force F_4 to be applied at node 4 to restrict the displacement as 0.05m in problems 4a) and b)?

5a) For the axially loaded structure shown in Figure 4.7.5, obtain the deflections at all the nodes, 1 to 4, wall reactions, stresses and nodal forces for each element (1 to 3) using FEM. Compare the values from the conventional method. Take E = 200 GPa (200 x 10^9 N/m²). A_1 to A_3 are areas of cross section, and L_1 to L_3 are the lengths of the elements respectively.

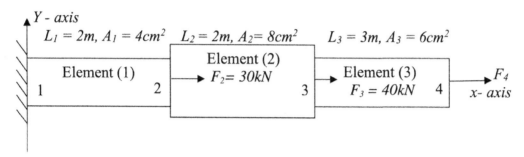

Figure 4.7.5

b) Redo the problem 5a) if the entire rod is heated up by 200°C taking the coefficient of thermal expansion, $\alpha = 25\mu / ^\circ$ Celcius ($\mu = 10^{-6}$).

c) What should be the force F_4 to be applied at node 4 to restrict the displacement as 0.1m in problems 5a) and b)?

4.7.2 Trusses

1) For the truss in Figure 4.7.6, let $L = 500$ mm, $P = 2$ kN, and for each member $E = 70$ GPa, and the cross-section area $A = 950$ mm^2.
   a) Using the finite element method, determine the deflections of the pins, the reaction forces, the nodal forces and the element stresses.
   b) Verify the results of part 1a) using
      i) method of joints
      ii) method of sections

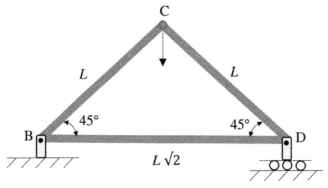

Figure 4.7.6

2) For the truss in Figure 4.7.7, all members have a cross-sectional area of 500 mm^2 and a modulus of elasticity of 210 GPa. Other than member (5) all members have a length of 400 mm.
   a) Using the finite element method determine the global displacements of each node, all support force reactions, and the element nodal forces and stress in element (5).
   b) Verify the results of part 2a) using
      i) method of joints
      ii) method of sections

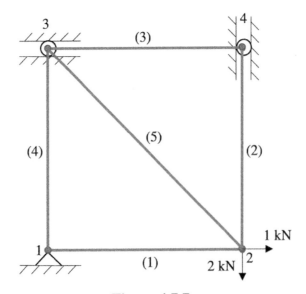

Figure 4.7.7

3) The i and j nodes of an element in a truss structure have the coordinates $(2.5, -2.5, 0)$ and $(5, 2.5, -5)$ cm. The element has an area of 1.3 cm^2 and the modulus of elasticity is 70 GPa. The global displacements are found to be:

$$\{u_i \quad v_i \quad w_i \quad u_j \quad v_j \quad w_j\}^T = \{-6 \quad 5 \quad 5 \quad 4 \quad 3 \quad 7\}^T \; (10^{-2}) \text{ in}$$

4) For the two-dimensional truss structure shown in Figure 4.7.8, determine the nodal deflections and the stress and nodal forces of each element. The support reaction forces R_{1x}, R_{1y} and R_{3x} are also indicated in the figure next to the supports. Members (1) and (2) each have a length of 2m. Each member has a cross-sectional area of 80 mm^2 and a modulus of elasticity of 200 GPa. Ignore the possibility of buckling in compression members.

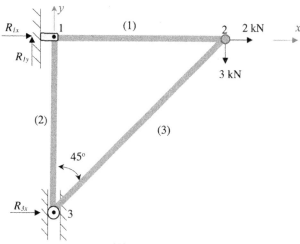

Figure 4.7.8

5a) Calculate the reactions and forces in all the members and the displacements at the joints of the trusses shown in Figure 4.7.9 using FEM.

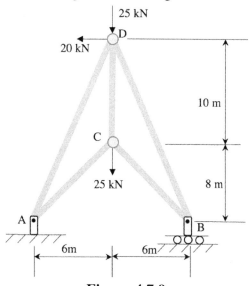

Figure 4.7.9

b) Redo the 5a) if the joint A settles vertically by 0.3m and horizontally by 0.4 m.

130

6a) Calculate the reactions and forces in all the members and the displacements at the joints of the trusses shown in Figure 4.7.10 using FEM.

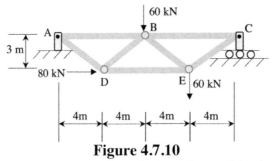

Figure 4.7.10

b) Redo the 6a) if the joint A settles vertically by 0.3m and horizontally by 0.4 m.

7a) Calculate the reactions and forces in all the members and the displacements at the joints of the trusses shown in Figure 4.7.11 using FEM.

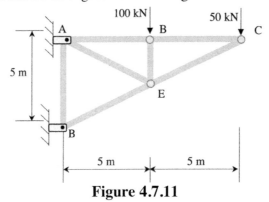

Figure 4.7.11

b) Redo the 7a) if the joint A moves vertically by 0.2m and horizontally by 0.5 m.

8a) Calculate the reactions and forces in all the members and the displacements at the joints of the trusses shown in Figure 4.7.12 using FEM.

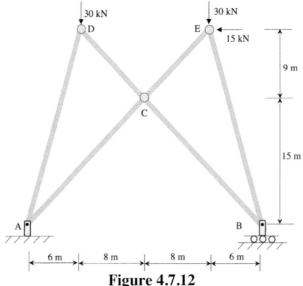

Figure 4.7.12

b) Redo the 8a) if the joint A moves vertically by 0.2m and horizontally by 0.5 m.

131

4.7.3 Beams

1) Determine the slope and deflections at nodes 2 and 3 for the beam shown in Figure 4.7.13. The modulus of elasticity of the beam is E and $I_1 = 2\,I_2$.

   HINT:

   When dealing with beam systems without numerical parameters the element stiffness equation can be put into the form:

$$\begin{Bmatrix} V_i \\ M_i \\ V_j \\ M_j \end{Bmatrix} = \frac{EI}{L^3} \begin{bmatrix} 12 & 6L & -12 & 6L \\ 6L & 4L^2 & -6L & 2L^2 \\ -12 & -6L & 12 & -6L \\ 6L & 2L^2 & -6L & 4L^2 \end{bmatrix} \begin{Bmatrix} v_i \\ \theta_i \\ v_j \\ \theta_j \end{Bmatrix}^e + \begin{Bmatrix} F_i \\ M_i \\ F_j \\ M_j \end{Bmatrix}^e_p + \{F\}^e_{\varepsilon_0}$$

   or as:

$$\begin{Bmatrix} V_i \\ M_i/L \\ V_j \\ M_j/L \end{Bmatrix}_e = \left(\frac{EI}{L^3}\right)_e \begin{bmatrix} 12 & 6 & -12 & 6 \\ 6 & 4 & -6 & 2 \\ -12 & -6 & 12 & -6 \\ 6 & 2 & -6 & 4 \end{bmatrix}_e \begin{Bmatrix} v_i \\ \theta_i L \\ v_j \\ \theta_j L \end{Bmatrix}_e + \begin{Bmatrix} F_i \\ M_i \\ F_j \\ M_j \end{Bmatrix}^e_p + \{F\}^e_{\varepsilon_0}$$

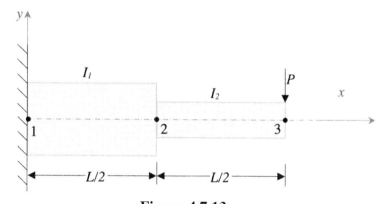

Figure 4.7.13

Care must be exercised, however, when applying a moment to a node (see Problem 2).

2) Repeat Problem 1) by replacing the force P with a clockwise moment M on node 3.

3) Repeat Problem 1) using non-uniform cross section with I_1 for element 1 and $I_2 = 2\,I_1$.

4) Repeat Problem 1) with both ends simply supported at nodes 1 and 3. Compare the values of deflections with standard values available in books on Strength of Materials.

5) Repeat Problem 1) with both ends fixed at nodes 1 and 3. Compare the values of deflections with standard values available in books on Strength of Materials.

6) Repeat Problem 1) with simply supported end at node 1 and fixed end at node 3. Compare the values of deflections with standard values available in books on Strength of Materials.

7) Repeat Problem 1) using uniform cross section as element 1 i.e. $I_1 = I_2 = I$.

4.7.4 Elastic Solids

1a) Calculate the deflections at the nodes, reactions at the supports, and the stresses at the Point E (8,6) of the triangular elastic element shown in Figure 4.7.14. Thickness, $t = 1$ m, Modulus of elasticity, $E = 350$ GN/m², and Poisson's ratio, $v = 0.2$. The supports at A and C are hinged.

Calculate the additional forces and stresses at the nodes if the element is heated to 40^0 C above room temperature (Coefficient of thermal expansion, $\alpha = 3 \times 10^{-6}$ per degree).

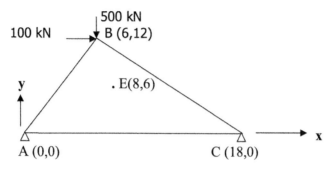

Figure 4.7.14

b) Redo the problem when support C sinks vertically down by .05m and horizontally by 0.03m.

2) The thin plate structure shown in Figure 4.7.15 is to be modelled with the two elements i.e. Triangular Element 1 with nodes 1-4-5 and Triangular element 2 with 1-2-5. Compute the stresses and displacements at point 3 and at the Centres of gravity of the elements 1-4-5 and 1-2-5. Assume $E = 150$ GPa, $v = 0.1$, $w = 8$m, $h = 5$m, t (thickness) = 12mm, $P_x = 60$ kN and $P_x = 120$ kN.

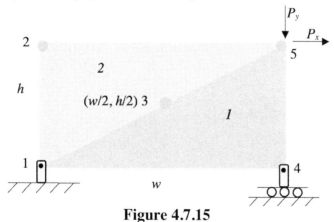

Figure 4.7.15

3) The thin plate structure shown in Figure 4.7.16 is to be modelled with only Element 1 with nodes 1-2-3 with hinged supports at 1 and 2. Compute the reactions at supports, stresses and displacements at point 3 and 4 shown in figure with the force $P_x = 104$ kN and $P_y = 52$ kN. Compare the results using conventional analysis. Assume $E = 150$ GPa, $v = 0.1$, $w = 8$m, $h = 5$m, t (thickness) = 12mm.

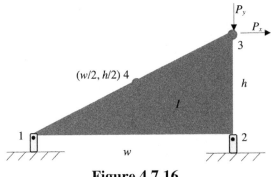

Figure 4.7.16

4) The thin plate structure shown in Figure 4.7.17 is to be modelled with the Triangular Element 1 and Triangular element 2. Compute the stresses and displacements at point 3 and at the Centres of gravity of the Triangular Elements 1 and 2. Assume $E = 250$ GPa, $v = 0.2$, $w = 10$m, $h = 4$m, t (thickness) = 20mm $P_x = 60$ kN and $P_x = 120$ kN.

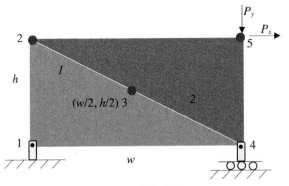

Figure 4.7.17

5) Compute the reactions at supports, stresses and displacements at points 2 and 3 shown in Figure 4.7.18 with the force $P_x = 52$kN and $P_y = 104$ kN applied at Node 2. Compare the results using conventional analysis. Assume $E = 150$ GPa, $v = 0.1$, $w = 8$m, $h = 5$m, t (thickness) = 12mm.

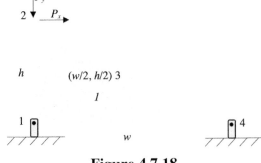

Figure 4.7.18

6) Consider only Element 1 with nodes 1-4-2 with hinged supports at 1 and 4. Compute the reactions at supports, stresses and displacements at point 2 and 3 shown in Figure 4.7.18 with the force $P_x = 80$ kN and $P_y = 150$ kN applied at Node 2. Compare the results using conventional analysis.

Chapter 5
FEM for Foundation Analysis and Design

5.1 Soils, Foundations, and Superstructures

Foundations transfer the loads coming from the superstructures such as buildings, bridges, dams, etc. and for that matter every engineering structure. Generally, that part of the structure above the foundation and extending above the ground level is referred to as superstructure. The foundations in turn are supported by soil medium below. Thus, soil is also the foundation for the structure and bears the entire load coming from above. Hence, the structural foundation and the soil together are also referred to as "substructure". The substructure is generally below the superstructure and refers to that part of the system that is below ground level. Thus, the structural foundation interfaces the superstructure and the soil below as shown in Figures 5.1 and 5.2. The soil supporting the entire structure above is also referred to as subsoil and (or) subgrade. For a satisfactory performance of the superstructure, a proper foundation is essential.

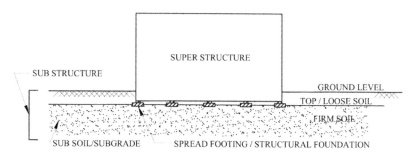

Figure 5.1 Building with spread foundations

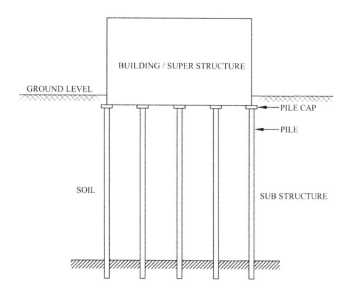

Figure 5.2 Superstructure with pile foundations

The manmade superstructures or facilities / utilities are expected to become very intricate and complex depending on creativity, architecture and infinite scope in modern times. However, the soil medium is mother earth which is a natural element and very little can be manipulated to achieve the desirable engineering properties to carry the large loads transmitted by the superstructure through the interfacing structural foundation (which is usually referred to as foundation). Further, almost all problems involving soils need detailed analysis of all components (Kameswara Rao, 2011) since soils have very complex behaviour as follows:

1. Natural soil media are usually not linear and do not have a unique constitutive (stress-strain) relationship.
2. Soil is generally non homogeneous, anisotropic and location dependent.
3. Soil behaviour is influenced by environment, pressure, time and several other parameters.
4. The soil being below ground, its prototype behaviour can not be seen in its entirety and has to be estimated on the basic of small samples taken from random locations (as per provisions and guidelines).
5. Most soils are very sensitive to disturbances due to sampling. Accordingly, their predicted behaviour as per laboratory samples could be very much different from the in-situ soil.

Thus, foundation design becomes a challenging task to provide a safe interface between the manmade superstructure and the natural soil media whose characteristics have limited scope for manipulation. Hence, the above factors make every foundation or soil problem very unique which may have several solutions.

The generally insufficient and conflicting soil data, selection of proper design parameters for design, the anticipated mode for design, the perception of a proper solution etc require a high degree of intuition- i.e., engineering judgement. Thus, foundation engineering is a complex blend of soil mechanics as a science and its practice through foundation engineering as an Art. This may be also referred to as "Geotechnique" or "Geotechnical Engineering".

5.2   Classification of Foundations

Foundations are classified as shallow and deep foundations based on the depth at which the load is transmitted to the underlying and (or) surrounding soil by the foundation as follows.

Shallow foundation:
A typical shallow foundation is shown Figure 5.3 (a). If $D_f/B \leq 1$, the foundations are called shallow foundations, where D_f = depth of foundation below ground level (GL) and B = width of the foundation (least dimension). Common types of shallow foundations are continuous wall footing, spread footing, combined footing, strap footing, grillage foundation, raft or mat foundation etc. These are shown in Figure 5.4.

Design and analysis considerations of shallow foundations are discussed extensively in several books, codes, etc. (Kameswara Rao, 2011) The shallow foundations are thus used to "spread" the load / pressure coming from the column or superstructure (which is several times the safe bearing pressure of supporting soil), so that, it is transmitted at a level that the soil can safely support. These are used when the natural soil at the site has reasonable safe bearing capacity, acceptable compressibility and the column loads are not very high.

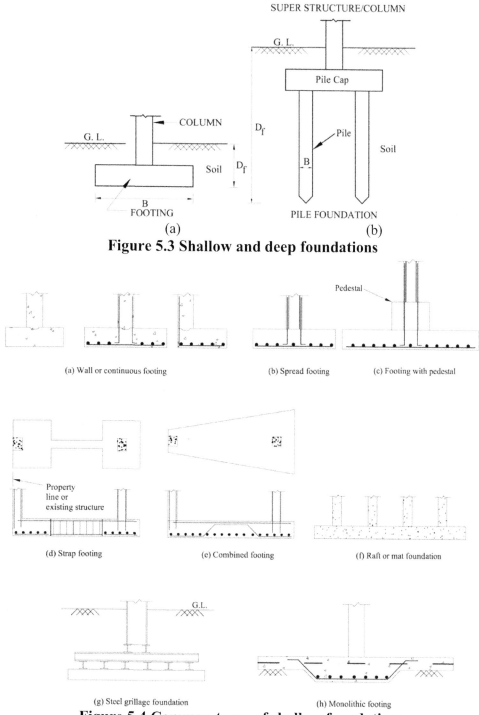

(a) (b)

Figure 5.3 Shallow and deep foundations

(a) Wall or continuous footing   (b) Spread footing   (c) Footing with pedestal

(d) Strap footing   (e) Combined footing   (f) Raft or mat foundation

(g) Steel grillage foundation   (h) Monolithic footing

Figure 5.4 Common types of shallow foundations

Deep foundations:

A typical deep foundation is shown in Figure 5.3 (b). If $D_f/B \geq 1$, the foundations are called deep foundations such as in piles, drilled piers/ caissons, well foundations, large diameter piers, pile raft systems. The details of analysis and design of such foundations are very elaborate and more practice oriented and are available in several books and reports (Kameswara Rao, 2011).

Deep foundations are similar to shallow foundations except that the load coming from columns or superstructure is transferred deeper into the soil vertically. These are used when column loads are very large, the top soils are weak, and soil with good strength and compressibility characteristics are at a reasonable depth below ground level. Further, earth retaining structures are also classified under deep foundations.

Foundations can be classified in terms of the materials used for their construction and (or) fabrication. Usually reinforced concrete (RCC) is used for the construction of foundations. Plain concrete, stone and brick pieces are also used for wall footings if the loads transmitted to the soil are relatively small. Engineers also use other materials such as steel beams and sections (such as in grillage foundations and pile foundations), wood as piles (for temporary structures), steel sheets (for temporary retaining structures and cofferdams) and other composite materials.

Sometimes, these are also encased in concrete depending on the load and strength requirements (Tomlinson, 2001 and Bowles, 1996).

Selection of type of foundation:

While engineering judgement and cost play a very important role in selecting a proper foundation for design, there are general guidelines available in literature including codes etc. (Kameswara Rao, 2011).

5.3 Modelling, Parameters, and Analysis

All practical problems need to be reduced to physical models representing the behaviour by corresponding analytical equations. The physical parameters of the system form the inputs in the mathematical equations for computing the responses. The models used should be simple enough so that the physical parameters needed for computations are accurately and reliably determined using inexpensive test procedures. For example, in a foundation-soil system, the foundation can be modelled as rigid, while soil may be assumed to be elastic. The physical parameters needed in such a model are the elasticity parameters of the soil, i.e. Young's modulus of elasticity, E and Poisson's ratio, v of soil. Obviously E and v have to be accurately determined for the soil under consideration as they will be needed for the computation of the responses of the system. Thus modelling, evaluation of parameters and analysis are closely linked and the accuracy of the solutions obtained are highly dependent on all these aspects.

The responses thus obtained have to be judged using appropriate design criteria specified either by codes or evolved out of practice and / or experience. The design process necessarily has two vital components namely, the methods of analysis and experimental data which have to be integrated to yield accurate results. However, both the methods and data depend entirely on the mechanism chosen for mathematical idealization of the system components. It is at this juncture, engineering judgment and experience will be very useful. It may be noted that optimum accuracy in analysis and design can be achieved only by proper match of data and analytical methods used. It is also obvious that any improvement either in the data alone or sophistication in analytical methods alone can even reduce the accuracy of the results/predictions (Lambe, 1973).

5.4 Analysis of Shallow Foundations and Deep Foundations

Foundation structures are customarily divided as 'shallow' or 'deep' on the basis of their depth in relation to their width, i.e. $D_f B \leq 1$ for shallow foundations and $D_f B > 1$ for deep foundations (Figures 1.3 a, b).

The difference between 'shallow' and 'deep' foundation is based on the structural response as well as the depth up to which the foundation is taken. Thus bending (flexure) is the predominant structural action in the case of shallow foundations. The behavior of deep foundation could result in axial and lateral loads besides bending moments and torsional moments. The deep foundation soil interaction needs a detailed analysis, practical and empirical considerations, economical and constructional expertise, and large computational resources for carrying out FEM analysis if necessary. Hence, they are not included in the present book. Shallow foundations can be of the following types (Figure 5.4).

Footings can be classified further as:
1. Continuous (or Strip Footings):
   These types of footings are primarily used for load bearing walls and generally of rectangular cross sections.
2. Independent (Isolated or Spread) Footings:
   These types of footings are generally used for individual columns and can be rectangular or trapezoidal, square or circular in shape.
3. Strip Footings:
   These footings support more than one column or wall.
4. Combined Footings:
   These types of footings are used for two or more columns in one row. These are generally rectangular, trapezoidal, cantilever (two interconnected footings) in shape.
5. Mat Foundations:
   These foundations support two dimensional arrays (regular or irregular) of columns. Rafts are generally used for two or more columns in several rows. These can be rectangular, square, circular, annular or octagonal in shape. Rafts also may have to be used if the allowable design soil (contact) pressure of soil is very low.
6. Others:
   Grillage, grids etc.

5.5   Conventional Design and Rational Design

In the conventional design of footings, the soil pressure is assumed to be uniform or linearly varying depending upon whether the foundation supports symmetric or eccentric loading (Figure 5.5). However, the actual contact pressure distribution, which is the result of the soil foundation interaction, can be far from the assumed uniform or linear distribution. The contact pressure distribution for flexible footings could be uniform for both clay and sand. The contact pressure is maximum at the edges for rigid footings on clay while it is minimum at the edges for rigid footings on sand. The typical distributions of immediate settlement and contact pressure for flexible and rigid footings are shown in Figure 5.6.

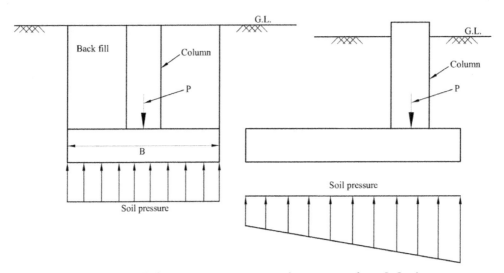

Figure 5.5 Soil contact pressures in conventional design

Hence the assumption of uniform contact pressure distribution will result in a slightly unsafe design for rigid footings on clays as the maximum bending moment at the center is underestimated. It will give a conservative design for rigid footings on sandy soils, as the maximum bending moment is overestimated. Similarly, the actual bending moments and shear forces in flexible footings could be at considerable variance with the design values obtained with the assumption of uniform contact pressure distribution.

Hence the necessity for developing effective and safe design for foundations procedure based on realistic distribution of soil pressure, obtained by a rational interaction analysis, referred to as "flexible" or "elastic" design, arises from the above drawbacks in conventional design (Kameswara Rao, 2011). However, most design codes are focused on conventional design.

In ether of the above analysis and design methods, the soil reaction is incorporated as an external load or as an internal load on the structural foundation and rest of the procedure follows standard structural analysis using any of the methods (Kameswara Rao, 2011) i.e.   i. Conventional methods ii. Analytical methods iii. Numerical methods iv. Finite element method (FEM) v. Other methods as may be required.

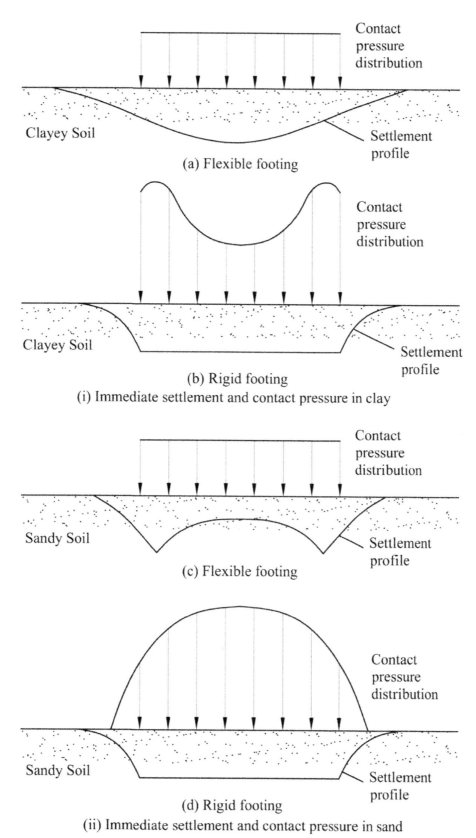

(a) Flexible footing

(b) Rigid footing

(i) Immediate settlement and contact pressure in clay

(c) Flexible footing

(d) Rigid footing

(ii) Immediate settlement and contact pressure in sand

Figure 5.6 Typical distribution of immediate settlement and contact pressure in soils

5.6   Procedures for the Analysis and Design of Footings

Footings may be designed as outlined below (Kameswara Rao,2011):

1. Calculate the loads applied at top of footings. Two types of loads are necessary, one for bearing capacity determination and the other for settlement analysis.
2. Sketch a soil profile or soil profiles showing the soil stratification at the site. Draw an outline of the proposed foundation scheme on the soil profile of the site.
3. Mark the maximum water level from the borehole data.
4. Determine minimum depth of footings.
5. Determine the bearing capacity of supporting stratum.
6. Proportion the footing sizes.
7. Check for danger of overstressing the soil strata at greater depths.
8. Predict the total and differential settlements.
9. Check stability due to eccentric loading.
10. Check uplift on individual footings and basement slabs, footings on slopes.
11. Analyze and Design the footings to comply with design criteria. This includes the structural design.
12. Check for foundation drains, waterproofing or damp proofing and other building code requirements.

5.7   Conventional Design of Footings

In practice all individual and wall footings and rafts are designed on the assumption that the distribution of the soil pressure against the bottom of the footing is linear or planar. Thus, when the load is applied at the centroid of the footing area, the unit pressure is equal to the total load divided by the footing area. In case of eccentric load, the pressure may be calculated by an appropriate procedure assuming linearly varying planar distribution of contact pressure (Kameswara Rao, 2011).

By far the majority of footings are constructed of concrete and the design of such footings should follow the concrete codes and design criteria prescribed. The structural design is a specialized subject and has to be pursued after the analysis is carried out. Several books and codes are available in literature on structural design of foundations (Kameswara Rao, 2011).

5.8   Modelling Soil Structure Interaction for Rational Design of Foundations

As summarized in Section 5.5, the contact pressure is taken as a uniform / linear / planar pressure for the conventional design of foundation. While all other requirements and precautions outlined in Section 5.6 are essentially the same for both conventional and elastic/flexible/rational design of foundations, the use of a realistic soil - structure interaction model can make the design more rational. The foundation and soil can either be modelled as solids (as continua), or the footing can be modeled as a beam (one dimensional) or a plate or a shell (two dimensional) and classical bending theories can be used for representing their response, and, the soil reaction has to be incorporated in the integrated analysis of soil-structure interaction equation by modeling the soil appropriately using different discrete models. (Vlasov and Leontev, 1966, Kameswara Rao, 2011).

The soil structure interaction involves three components to be modelled for analysis. The modelling options generally used for analysis are summarized below.
1. Soil medium:
   a. It can be modelled as a continuum / solid with appropriate constitutive relations such as elastic, plastic, linear, nonlinear, homogeneous /non- homogenous, isotropic / anisotropic, with or without tension, and other models and combinations of the above. However, the material properties needed for the analysis of the chosen model need to be evaluated using proper experimentation.
   b. It can be modelled using discrete representation of the solid soil medium using discrete models such as closely spaced loose springs referred to as Winkler's model, and other improved models as discussed in the following sections.
   c. Others such as multi-phase materials etc.

2. Footing:
   a. It can be modelled as a solid with all the variations as mentioned above for soil medium.
   b. It can be modelled as a beam, plate, shell or a combination of these thus representing as a solid with approximate theories such as bending theory to facilitate easy analysis while focusing on the design parameters needed for structural design.
   c. It can be modelled using other practical models as may be required.

3. Soil – structure interface / contact surface:
   a. It can be modelled as smooth, frictional (partial or full), bonded, slippery etc.
   b. It can be modelled using provisions as may be required.

However, the simple models of the above are discussed below with focus on FEM analysis. For the analysis for general models (mentioned above), users may refer to Zienkiewicz and Taylor, (1989), Rao, S.S, (1982), Desai and Abel, (1972), Rajasekaran (1993) and other books available in literature.

The modelling of soil structure interaction analysis using soil as a discrete elastic medium represented by discrete models, and the structure as a beam or plate in bending is discussed below. The interface / contact surface between the soil and structure is assumed to be fully bonded for the following discussion which is the one most commonly adopted in practice.

5.8.1 Soil – Structure Interaction Equations

The foundation – soil system subjected to external loads is shown in Figures 5.7 and 5.8 depending on the geometry of the foundation i.e., beam or a plate. Most of the footings can be considered either as beams (one dimensional) or plates (two dimensional– rectangular, squares, circular, annular or other shapes).

With the above soil structure interaction model, the differential equation of bending of the beam or plate on an elastic foundation can be written as follows.

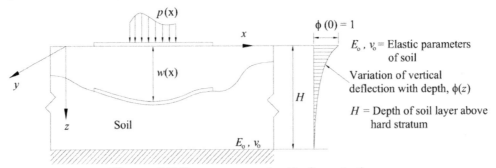

Figure 5.7 Beam on an elastic foundation

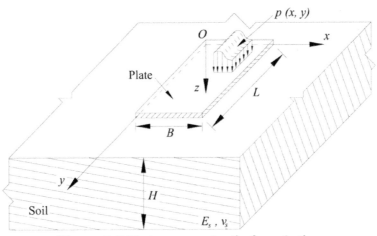

Figure 5.8 Plate on an elastic foundation

5.8.2　Beams on Elastic Foundations (BEF):

Neglecting friction between beam and the soil medium since the interface is assumed to be fully bonded, the governing equation (referring to Figure 5.7), can be written from bending theory as (Kameswara Rao, 2011):

$$EI\frac{d^4w}{dx^4} = p(x) - q(x) \tag{5.1}$$

EI = flexural rigidity of the beam

E　= modulus of elasticity of beam material

I　= moment of Inertia of the beam cross section

$$= \frac{h^3}{12(1-v_p^2)}\text{ for unit strip in case of plane strain structure (strip footing)} \tag{5.2}$$

v_p = Poisson's ratio of the strip / plane strain structure

$p(x)$ = external load applied on the footing (in units $-FL^{-1}$)

$q(x)$ = reaction from the supporting soil (same units as $p(x)$ i.e. FL^{-1})

w　= vertical deflection along z axis

x, y, z = right-handed coordinate system

E_s = modulus of elasticity of soil

v_s = Poisson's ratio of soil

E_0, v_0 = modified elastic parameters of soil to be used for PEF and strips (plane strain case) on elastic foundations and all three dimensional problems (shown in Figure 8) and are defined as (Vlasov and Leontev, 1966, Kameswara Rao,2011),

$$E_0 = \frac{E_s}{1-v_s^2} \text{ and } v_0 = \frac{v_s}{1-v_s}$$

The other parameters that can be defined for beams using classical bending theory are as follows:

$$w' = \theta = \text{slope} = \frac{dw}{dx}$$

$$M = \text{bending moment} = -EI\frac{d^2w}{dx^2} \qquad (5.3)$$

$$Q = \text{shear force} = -EI\frac{d^3w}{dx^3}$$

$q(x)$ = soil reaction or contact pressure

$$= EI\frac{d^4w}{dx^4} - p(x)$$

The convention from bending theory for bending moment (BM) and shear forces (SF) are shown below (Figure 5.9).

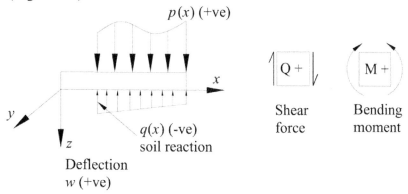

Figure 5.9 Convention sketch for bending theory of beams

5.8.3 Plates on Elastic Foundations (PEF):

Consider a rectangular plate on elastic foundation as shown in Figure 5.8. The assumptions usually made in the theory of bending of thin plates will be deemed to apply to this case. Friction and adhesion between the plate and the surface of the elastic foundation is neglected since complete bond between soil and foundation is assumed.

The differential equation of bending of the plate, in Cartesian coordinates, is

$$D\nabla^2\nabla^2 w(x,y) = p(x,y) - q(x,y) \qquad (5.4)$$

(∇^2 denotes the Laplace operator) or, in expanded form:

146

$$DV^4w = D\left(\frac{\partial^4 w}{\partial x^4} + \frac{\partial^4 w}{\partial x^2 \partial y^2} + \frac{\partial^4 w}{\partial y^4}\right) = p(x,y) - q(x,y) \quad\quad (5.5)$$

where $w = w(x, y)$ = vertical displacements of the plate surface,

$p = p(x, y)$ = distributed load applied on the plate (units , FL^{-2})

$q(x,y)$ = soil reaction (units , FL^{-2})

$$D = \frac{E_p h^3}{12(1-v_p^2)} = \text{flexural rigidity of the plate} \quad\quad (5.6)$$

in which E_p = modulus of elasticity of plate material

v_p = Poisson's ratio of plate material

h = thickness of the plate

Although Eq. (5.4) is known as the equation of bending of thin plates, it can be applied to the analysis of most rectangular plates. Further, soil properties of elastic foundation shown in Figure 5.8 are:

$$E_0 = \frac{E_s}{1-v_s^2}, \quad v_0 = \frac{v_s}{1-v_s} \quad\quad (5.7)$$

where E_s and v_s are respectively the modulus of elasticity and Poisson's ratio for the material of the soil.

After $w(x,y)$ has been determined from Eq. (5.4) and the boundary conditions, the reactions $q(x,y)$ can be evaluated as discussed below. The moments and shearing forces in the plate (Figure 5.10) can be computed using formulae of the theory of bending of plates as follows (Timoshenko and Kreiger, 1959).

$$M_x = -D\left(\frac{\partial^2 w}{\partial x^2} + v_p \frac{\partial^2 w}{\partial y^2}\right)$$

$$M_y = -D\left(\frac{\partial^2 w}{\partial y^2} + v_p \frac{\partial^2 w}{\partial x^2}\right)$$

$$M_{xy} = H = H_x = -H_y = -D(1-v_p)\frac{\partial^2 w}{\partial x \partial y} \quad\quad (5.8)$$

$$N_x = -D\frac{\partial}{\partial x}\left(\frac{\partial^2 w}{\partial x^2} + \frac{\partial^2 w}{\partial y^2}\right)$$

$$N_y = -D\frac{\partial}{\partial y}\left(\frac{\partial^2 w}{\partial x^2} + \frac{\partial^2 w}{\partial y^2}\right)$$

Following Kirchhoff (Timoshenko and Kreiger, 1959 and Vlasov and Leontev, 1966), the shearing forces N_x, N_y, and the torque H at the plate edges are usually replaced by the reduced shearing forces Q_x and Q_y which for a rectangular plate, are:

$$Q_x = -D\left(\frac{\partial^3 w}{\partial x^3} + (2-v_p)\frac{\partial^3 w}{\partial x \partial y^2}\right)$$

$$Q_y = -D\left(\frac{\partial^3 w}{\partial y^3} + (2 - \nu_p)\frac{\partial^3 w}{\partial x^2 \partial y}\right)$$ (5.9)

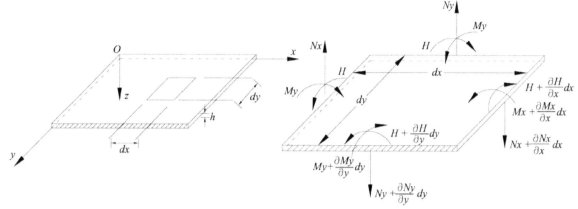

Figure 5.10 Convention sketch for plate bending

Equations (5.4) and (5.5) are also valid for plates with other geometries such as circular, annular, etc., since Laplacean operator is invariant except that its expansion in other coordinate systems has to be taken for solving the equation. For example, for circular plates or annular plates, the Laplacean operator ∇^2 has the expanded form in r, θ coordinate system as (Figure 5.11)

$$\nabla^2 = \frac{\partial^2}{\partial r^2} + \frac{1}{r}\frac{\partial}{\partial r} + \frac{1}{r^2}\frac{\partial^2}{\partial \theta^2}$$ (5.10a)

If the load is axisymmetric, then θ coordinate can be omitted in Eq. (4.23a) resulting in:

$$\nabla^2 = \frac{\partial^2}{\partial r^2} + \frac{1}{r}\frac{\partial}{\partial r}$$ (5.10b)

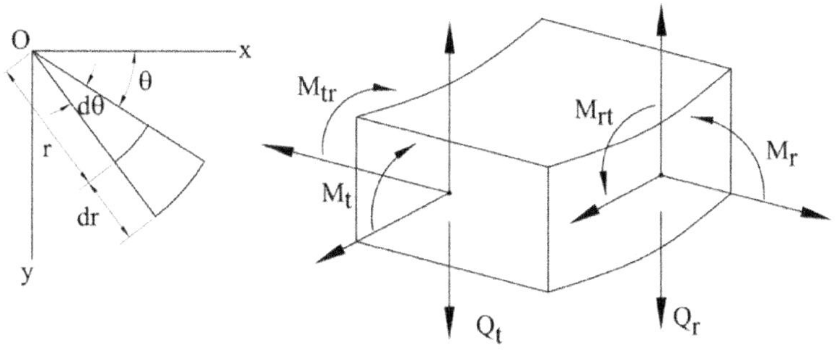

(a)   (b)

Figure 5.11 Convention sketch for circular and annular plates in polar coordinates

However, the bending moments and shear forces take the following forms in such a situation (Timoshenko and Kreiger, 1959). The convention sketch is shown in Figure 5.11.

148

Bending Moments:

$$M_r = -D\left(\frac{\partial^2 w}{\partial x^2} + v_p \frac{\partial^2 w}{\partial y^2}\right)_{\theta=0} = -D\left[\frac{\partial^2 w}{\partial r^2} + v_p\left(\frac{1}{r}\frac{\partial w}{\partial r} + \frac{1}{r^2}\frac{\partial^2 w}{\partial \theta^2}\right)\right] \quad (5.10c)$$

$$M_t = -D\left(\frac{\partial^2 w}{\partial y^2} + v_p \frac{\partial^2 w}{\partial x^2}\right)_{\theta=0} = -D\left[\frac{1}{r}\frac{\partial w}{\partial r} + \frac{1}{r^2}\frac{\partial^2 w}{\partial \theta^2} + v_p \frac{\partial^2 w}{\partial r^2}\right]$$

$$M_{rt} = (1 - v_p)D\left(\frac{\partial^2 w}{\partial x \partial y}\right)_{\theta=0} = (1 - v_p)D\left(\frac{1}{r}\frac{\partial^2 w}{\partial r \partial \theta} - \frac{1}{r^2}\frac{\partial w}{\partial \theta}\right)$$

The expressions for shear forces are as follows.

$$Q_r = -D\frac{\partial}{\partial r}\left(\nabla^2 w\right)$$

$$Q_t = -\frac{D}{r}\frac{\partial}{\partial \theta}\left(\nabla^2 w\right) \quad (5.10d)$$

It can be noted that solution of soil structure interaction Equations (5.1) and (5.5) is essentially the same as the corresponding structural analysis problem with the inclusion of contact pressure, q as a load applied either externally or internally on the structure.

5.8.4 Soil Reaction $q(x) / q(x,y)$ - Contact pressure

Hence, to solve the final form of soil structure interaction Equations (5.1) and (5.5), the soil reaction $q(x) / q(x,y)$ has to be incorporated in those equations which are dependent on the beam / plate and soil characteristics and the bond at the interface. The rest of the solution follows the standard structural analysis problem as can be noted from the above equations.

The various options for modelling soil medium for analyzing soil structure interaction problems are summarized above in Section 5.8. In case the soil is modelled as a 3-D or 2D solid medium, the soil reaction gets incorporated automatically in the governing equations which are very difficult to solve either analytically or numerically (Kameswara Rao, 2011). However, for such 2-D or 3-D models, FEM can be used for analysis using computer programs or packages (Zienkiwicz and Taylor, 1989 etc.). Due to complexity in such elaborate models, alternative approaches using discrete models for soil medium and simple solid models for the foundation/ structure such as beams and plates in bending, are preferred for analysis and design and the soil reaction can be incorporatd easily in the above Eqns. (5.1) and (5.5). The rest of the solution procedure is the same as that of the structural analysis of the structural component adopted for the footing/foundation.

Assuming frictionless contact, and complete bond at the interface between the beam / plate and the soil, the soil reaction can be expressed in terms of soil displacements (mainly vertical displacement for vertical loads) using different discrete foundation models. A review of these models is given by Reissner, (1937), Kameswara Rao, (2011) and others. The important features of these models are summarized below.

5.9 Elastic Foundations

The theory of elastic foundations has attracted considerable attention due to its useful application in various technical disciplines besides foundation engineering. The problems of elastically supported structures are of interest in solid propellant rocket motors, aerospace structures, construction projects in cold regions, and several other fields. While in some problems, the structure and the elastic support, generally referred to as "foundation", can be physically identified, in many others the concept of structure and foundation may be of an abstract nature.

The problem of foundation-structure interaction is generally solved by incorporating the reaction from the foundation, into the response mechanism of the structure, by idealizing the foundation by a suitable mathematical model. Even if the foundation medium happens to be complex in some problems as mentioned in Section 5.8, in a majority of cases, the response of the structure at the contact surface is of prime interest and hence, it would be of immense help in the analysis, if the foundation can be represented by a simple mathematical model, without foregoing the desired accuracy. To accomplish this objective, many foundation models have been proposed and a comprehensive review pertaining to these has been given by Reissner, (1937), Kameswara Rao, (2011). These are briefly described below.

5.9.1 Brief Review of the Foundation Models

The earliest formulation of the foundation model was due to Winkler, (1988), who assumed the foundation model to consist of closely spaced independent linear springs, as shown in Figure 5.12. If such a foundation is subjected to a partially distributed surface loading, q, the springs will not be affected beyond the loaded region. For such a situation, an actual foundation is observed to have the surface deformation as shown in Figure 5.13. Hence by comparing the behavior of theoretical model and actual foundation, it can be seen that this model essentially suffers from a complete "lack of continuity" in the supporting medium. The load-deflection equation for this case can be written as

$$q = kw \tag{5.11}$$

where q is the contact pressure at the interface between soil and foundation, k is the spring constant of the soil and is often referred to as "foundation modulus", and w is the vertical deflection of the contact surface / foundation as both are assumed to be completely bonded. It can be observed that Eq. (5.11) is exactly satisfied by an elastic plate floating on the surface of a liquid and carrying some load which causes it to deflect. The pressure distribution under such a plate will be equivalent to the force of buoyancy, k being the specific weight of the liquid. With this analogy in view, the first solution for the bending of plates on Winkler-type foundation was presented by Hertz, (1884).

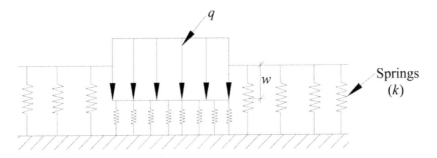

Figure 5.12 Load on Winkler's foundation

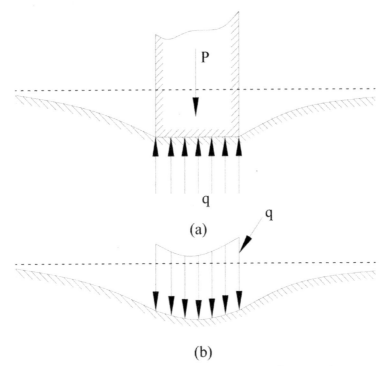

(a)

(b)

Figure 5.13 Deformation of actual foundation

Also, in such a foundation model the displacements of the loaded region will be constant whether the foundation is subjected to a rigid stamp or a uniform load as can be seen from Figure 5.12. However, the displacements for these cases are quite different in actual foundations as can be noted from Figure 5.13 (a) and (b). Though this model leads to some inconsistencies, being the simplest, it is amenable to an easy analysis. Through the years a large variety of solutions have been presented on this basis (Winkler, 1888, Reissner, 1937, Timoshenko and Kreiger, 1959, Hetenyi, 1946, 1950, Kameswara Rao, 2011 etc.).

Another approach is to assume the foundation medium to be a continuous elastic solid. Though this hypothesis closely simulates the physical behavior of an actual foundation, it makes the analysis unduly complex. Despite several mathematical complexities, solutions were presented on these bases (Kameswara Rao,2011), which, however, were limited to relatively simple cases. Also, it was observed that the foundation performance as predicted by this theory differed from the actual behavior,

151

probably due to the questionable assumptions of elasticity, homogeneity and isotropy of the materials, inherent in this hypothesis. As an example, in soils it has been observed that the surface displacements away from the loaded region decreased more rapidly than predicted by this theory (Kameswara Rao, 2011), and, materials like soils and foam rubber hardly satisfy the basic assumptions stipulated above.

The need for bridging the gap between these two extremes and limiting cases and to arrive at a physically close and mathematically simple foundation model has been felt for some time. Several authors have proposed foundation models which involve more than one parameter for the characterization for the supporting medium.

One such attempt has been presented by Filonenko-Borodich (Reissner,1937), who modified the Winkler foundation by providing some continuity by connection top ends of the springs by a stretched elastic membrane subjected to a constant tension field 'T', as shown in Figure 5.14. The equilibrium in the vertical direction yields the equation

$$q = kw - T\nabla^2 w \tag{5.12}$$

where q is the distributed vertical load applied on the surface of the soil, w is the vertical deflection of the surface, ∇^2 is the Laplace operator, k and T are the two parameters characterizing the foundation.

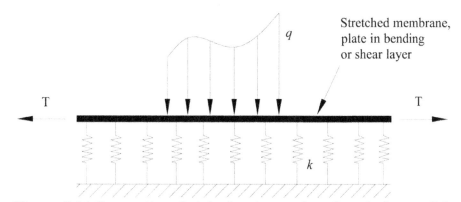

Figure 5.14 Convention sketch showing various foundation models

Hetenyi (1946, 1950) achieved the continuity in the Winklers's foundation model by embedding, an elastic beam in the two-dimensional case, and an elastic plate in three dimensional case, (Figure 5.14), with the stipulation that the hypothetical beam or plate deforms in bending only. In this case relation between the load q, and the deflection of the surface w, can be expressed as

$$q = kw + D\nabla^4 w \tag{5.13}$$

where D is the flexural rigidity of the embedded beam or plate and ∇^4 is the bi-harmonic operator.

By providing for shear interaction between the Winkler's spring elements, Pasternak (Kameswara Rao, 2011) presented a foundation model as shown in Figure 5.14. The shear interaction between the springs has been achieved by connecting the ends of the springs to a beam or a plate (as the case may be), consisting of incompressible vertical elements, which hence deform in transverse shear only. The corresponding equation relating the load q, and deflection w, can be derived as

$$q = kw - \mu \nabla^2 w \qquad (5.14)$$

where μ is the shear modulus of the foundation material and ∇^2 is the Laplace operator. It can be seen that this foundation model also consists of two parameters k and μ, and is equivalent to the models proposed by Filonenko-Borodich (Eq. 5.12) and Wieghardt (Kameswara Rao, 2011).

Pasternak proposed another foundation model (Kameswara Rao, 2011) consisting of two layers of springs connected by shear layer in between as shown in Figure 5.15. The relation between the load q, and the deflection w of the surface of the foundation can be expressed as

$$\left(1 + \frac{k}{c}\right) q - \frac{\mu}{c} \nabla^2 q = kw - \mu \nabla^2 w \qquad (5.15)$$

where c and k are the spring constants of the upper and lower layers of springs and μ is the shear modulus of the shear layer. Several contributions based on Pasternak-type foundations are available in literature.

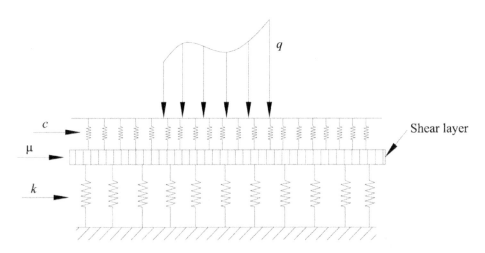

Figure 5.15 Pasternak's modified foundation model

In all the above models, the Winkler's model has been modified by providing for some interaction between the spring elements and hence assuring the continuity of the foundation to some degree. In contrast to them, starting from the elastic-continuum theory, and introducing simplifying assumptions with respect to the expected stresses and (or) displacements, some models were proposed. One such contribution was from Reissner (Kameswara Rao, 2011), who assumed that the in-plane stresses σ_{xx}, σ_{yy} and τ_{xy} are negligible throughout the foundation layer. Also, the horizontal displacements at the upper and lower surfaces of the foundation layer were assumed to be zero. Proceeding with these assumptions and solving the elastic-continuum equations, the equation relating the applied distributed surface load q, and the resulting surface displacement w, has been derived as

$$c_1 w - c_2 \nabla^2 w = q - \frac{c_2}{4c_1} \nabla^2 q \qquad (5.16)$$

where $c_1 = \dfrac{E}{H}$ and $c_2 = \dfrac{Hv}{3}$

E and v are the elastic constants of the foundation material and H is the thickness of the foundation layer. It can be seen that Eqns. (5.15) and (5.16) are similar. Also, for constant and linearly varying loads, this equation can be seen to be mathematically equivalent to Eqns. (5.12) and (5.14), thus establishing the similarity of the models. In this case, neglecting the in-plane stresses, it can be shown that shear stresses τ_{zx} and τ_{zy} are constant throughout the depth of the foundation for a given surface point, which is inconsistent with the actual foundation performance, especially for thick foundation layers.

Vlasov and Leontev (1966) have developed a foundation model starting from elastic-continuum theory and neglecting the horizontal displacements of the supporting medium. Using Vlasov's "general variational method" the load-displacement relation can be derived as

$$q = kw - 2t_1 \nabla^2 w \qquad (5.17a)$$

where q is the distributed surface load and w is the vertical deflection. k and t_1 are the two parameters characterizing the foundation and can be expressed in terms of the elastic constants of the material and geometric properties of the foundation layer (Vlasov and Leontev,1966). In Eq. (5.17a), these parameters can be obtained as,

$$k = \frac{E_0 B}{1 - v_0^2} \int_0^H \psi^2(z)\, dz \qquad (5.17b)$$

$$t_1 = \frac{E_0 B}{1 - v_0^2} \int_0^H \psi^2(z)\, dz \qquad (5.17c)$$

where $E_0, v_0 = E_s, v_s$ of the soil respectively to be used for beams on elastic foundation problems (plane stress problems)

$$E_0 = \frac{E_s}{1 - v_s^2}, v_0 = \frac{v_s}{1 - v_s} \qquad (5.17d)$$

for plates and plane strain case (strips and three dimensional problems)

B = width of the foundation

$\psi(z)$ is the assumed distribution function of vertical displacement with depth (preferably a function with $\psi(0) = 1$ and easy to integrate (Figure 5.7). $\psi(z)$ can be chosen as $e^{-\beta z}$ for infinite soil layers (or layers with large depth), where β is the soil parameter. β can be chosen between 0.5 and 2.5 (0.5 for clayey soils and 1 to 2.5 for sandy soils). $\psi(z)$ can be chosen as $\dfrac{H - z}{H}$ i.e. linearly decreasing with depth for soil layers of finite depth.

Comparing Eq. (5.17a) and Eqns. (5.12) and (5.14), it can be observed that, this model is equivalent to the models proposed by Filonenko-Borodich, Pasternak and Wieghardt.

A close examination of the various models reviewed above, reveals the fact that these methods fall short for direct application to practical problems, either because the analysis is cumbersome, or, because the assumptions made for natural foundation media cannot be fully justified and often lead to some inconsistencies.

To overcome the above inconsistencies of the soil behaviour, Kameswara Rao (2011) modified Vlasov and Leontev's model to account for the horizontal displacements in the elastic foundation and basic equations have been presented using Vlasov's general variational method. The resulting model is very close to elastic-continuum hypothesis and is easy for mathematical analysis.

5.9.2 Winkler's Model

After reviewing the various foundation models as outlined above (Section 5.9.1), it can be seen that Winkler's model is the simplest both in terms of representation of the soil reaction at the footing soil interface as well as analysis of the resulting soil structure interaction Equations (5.1) and (5.5), though it has inherent deficiencies as outlined therein. It has an added advantage that the soil parameter used for expressing the soil reaction i.e. k (Eq. 5.11), referred to as spring constant (of the idealized springs of the Winkler's model as shown in Figure 5.12 is relatively easy to evaluate from laboratory and field experiments as explained in the following Section 5.9.3. Hence Winkler's model is preferred for rational analysis and design by most of the researchers and designers and is presented in this book. Thus, using Winkler's model for representing the soil and using Eq. (5.11) for soil reaction q, the soil-structure interaction equation can be expressed as follows:

Beams on elastic foundations: (Eq. 5.1)

$$EI\frac{d^4w(x)}{dx^4} + kw(x) = p(x) \tag{5.18}$$

For plates on elastic foundation: (Eq. 5.5)

$$D\nabla^4 w(x, y) + kw(x, y) = p(x, y) \tag{5.19}$$

in which k = spring constant of the soil medium, to be evaluated from suitable laboratory and field tests as explained below (Section 5.9.3).

5.9.3 Evaluation of Spring Constant k, in Winkler's Soil Model and Elastic Properties E and v of Soil

The spring constant of the soil k is related to modulus of subgrade reaction, k_s, (also referred to as coefficient of elastic uniform compression, C_u) as

$$k = k_s A = C_u A \tag{5.20}$$

where A is the area of contact between soil and footing. Laboratory tests such as CBR test, consolidation test, triaxial test etc. and (or) field tests (Kameswara Rao, 2011 etc.)

are most commonly used to evaluate the soil parameters needed for the soil structure interaction analysis. However, field plate load test is widely used for the evaluation of the above soil parameters and is described below.

5.9.4 Evaluation of k_s / C_u using Plate Load Test In-Situ

The idea of modeling soil as an elastic medium was first introduced by Winkler and this principle is now referred to as the Winkler soil model. The modulus of subgrade reaction at any point along the beam is assumed to be directly proportional to the vertical displacement of the beam at that point. In other words, the soil is assumed to be elastic and obeys Hooke's Law. Hence, the modulus of subgrade reaction (k_s) for the soil is given by

$$k_s = \frac{q}{w} = C_u \tag{5.21}$$

where q is the bearing pressure at a point along the beam, and w is the vertical displacement of the beam at that point. k_s is also referred to as coefficient of elastic uniform compression C_u.

The main difficulty in applying the Winkler soil model is that of quantifying the modulus of subgrade reaction (k_s) to be used in the analysis, as soil is a very variable material. In practical terms, k_s can be found only by carrying out in-situ plate load tests or relating it in some way to elastic characteristics of the soil. The plate load test is widely used and is described in BS 1377: Part 9: 1990, IS: 1888, 1982 (Kameswara Rao, 2011), etc. Plate load test is described in detail in the Appendix 5A at the end of the chapter. Evaluation of modulus of subgrade reaction, k_s and elastic constants, modulus of elasticity, E and Poisson's ratio, v of the soil are explained in Appendix 5A.

5.10 Foundation Analysis using FEM

As mentioned in Sections. 5.8 and 5.9, the two approaches for the analysis and design of foundations are:
1. Conventional approach which assumes the foundation to be rigid and the contact pressure at the interface to be planar.
2. Rational approach which incorporates the flexibility of the footing as well as the soil contact pressure based on elastic theories using modulus of subgrade reaction.

The footing and soil can be idealized either as:
i. solids (elastic or other feasible solid models) along with required interface making the analysis very difficult and complex, or,
ii. the footing as a beam (spread footings, combined footings, strap footings, wall footings, etc.) or as a plate (mat or raft foundations, circular footings, annular or ring footings and footings of general shape which are two dimensional in plan supporting several loads from columns, walls, etc.), and soil as a discrete model with fully bonded interface which is relatively easy to analyze and
iii. any combination of models mentioned in i and ii stated above which also may need complex analysis.

Option ii is most commonly used in view of its feasibility for relatively easy analysis resulting in design parameters compatible with design procedures prescribed in codes.

Analysis using option ii above is essentially a structural analysis problem with the soil reaction as an additional force applied on the structure. While several models for incorporating soil reaction in the soil–structure interaction equations are described in Sections. 5.8 and 5.9, the Winkler's model is extensively used for modelling the soil in practice. As mentioned in Section 5.8, the methods of solutions for rational analysis of soil structure systems are: a. Analytical methods b. Numerical methods c. FEM d. other methods. Extensive literature is available for reference to these methods (Kameswara Rao, 2011) and only FEM is discussed in the following sections in view of its versatility, realistic modelling, flexibility, modularity, computational efficiency and accuracy as may be required. While noting that most of the codes prescribe conventional analysis to be used in practice, a rational analysis becomes necessary for the following reasons.

a. Verification of the results from conventional analysis.
b. Diagnosis of the behavior of the system for safety, quality control, rehabilitation in case of structural failure etc.
c. Accurate assessment of the response at critical zones and locations of the system.
d. Validation of the input parameters obtained from expensive and difficult laboratory and field tests in general and in cases of questionable reliability of the experimental results.
e. Simulation of the behavior and sensitivity studies.
f. Very important systems needing total safety.
g. Research, design and development, and practice.
h. Other reasons, such as statutory requirements etc.

Accordingly, FEM analysis of the soil-structure systems using beams and plates as structural components, and Winkler's model representation for the soil with total bond at the interface is discussed below.

5.10.1 Review of FEM Procedure

The details of the Finite Element Method are presented in Chapters 2 and 3. Some relevant details are reviewed below for continuity and easy understanding. FEM can be divided into three phases:

(a) Structural idealization
(b) Evaluation of the element characteristics
(c) Structural analysis of element assemblage

The structural idealization is the process of subdivision of the original system into an assemblage of discrete segments of proper sizes and shapes. Judgement is required as results can be valid only to the extent that the behavior of the substitute structure simulates the actual structure. In general, better results can be achieved by finer subdivision.

The objective of the second phase is to find the stiffness or the flexibility of the element, which is an important step in the analysis and explained in Chapter3 and discussed further in the following sections.

The third phase is a standard structural problem as has been briefly outlined earlier in Chapters 2 and 3 in general and Chapter 4 in particular for structural analysis of different types of structures including beams. The individual element configurations are of no concern. The same techniques apply to systems of one-, two- or three-dimensional elements or any combinations of these. The essential problem is to satisfy the three conditions of equilibrium, compatibility and force-deflections relationships. Any of the basic approaches of structural analysis known as "force method" and "displacement method" or "hybrid methods" can be used. However, the displacement method is preferred as it is simpler for formulation and computer programming. The fundamental steps involved are summarized as follows.

1. Evaluation of the stiffness properties of the individual structural elements, expressed in any convenient local (element) coordinate system. The procedure includes the evaluation of the element stiffness matrix, $[k]^e$ of the structural element and relating it to the applied external and internal forces as described in Chapters 2 and 3.

   Assuming elastic behavior, the characteristic relationship between the forces and the displacements of the element e will be of the form

   $$\{R\}^e = [k]^e\{\delta\}^e + \{F\}^e_p + \{F\}^e_{\varepsilon_0} \tag{5.22}$$

   where $\{R\}^e$ = externally applied forces at the nodes of the element

   $[k]^e$ = stiffness matrix of the element

   $\{\delta\}^e$ = nodal displacement matrix of the element

   $\{F\}^e_p$ = equivalent nodal forces due internal distributed forces

   $\{F\}^e_{\varepsilon_0}$ = equivalent nodal forces due to any other internal causes such as
   initial strain, lack of fitness etc.

   Similarly stresses, $\{\sigma\}^e$ at any specified point or points of the element can be expressed in terms of nodal displacements as

   $$\{\sigma\}^e = [S]^e\{\delta\}^e + \{\sigma\}^e_p + \{\sigma\}^e_{\varepsilon_0} \tag{5.23}$$

   where $\{\sigma\}^e$ = stress matrix at any point of the element

$[S]^e$ = stress matrix of the element

$\{\delta\}^e$ = nodal displacement matrix of the element

$\{\sigma\}^e_p$ = stress due to internal distributed forces

$\{\sigma\}^e_{\varepsilon_0}$ = stress due to other internal causes such as initial strain, lack of fitness etc.

All the above quantities can be evaluated using expressions derived from FEM stiffness analysis as presented in Chapter 3. Alternatively, equivalent nodal forces acting on the element (external or internal) can be calculated using lumped method for simplicity which is a common practice in analysis and design (explained in Chapter 4)

2. Transformation of the element stiffness matrices from the local coordinates (as may be necessary) to global coordinates system of the complete structural assemblage. This can be done as explained in Chapters 2 and 3.

3. To get the complete solution of the structural assembly, (Chapter 2 and 3), the two conditions to be satisfied are

    (a) Displacement compatibility
    (b) Equilibrium

    By listing the nodal displacements for all the elements of the structural assembly, the first condition of compatibility of nodal displacements is taken care of as only unique solutions of displacements are being solved for. The second condition is to satisfy the equilibrium of each element as well as the total system / assemblage of all elements representing the total system. This is achieved as explained below.

4. As can be noted, the overall equilibrium has already been satisfied within each element as expressed by Eqns. (5.22) and (5.23). Hence by writing down the equilibrium equations at all the nodes of the total / global system / assembly and adding the contributions of the forces of corresponding elements connected at each node will result in the system / global equilibrium equations containing the nodal displacements as unknowns. These can be formulated as:

$$\{R\} = [K]\{\delta\} + \{F\}_p + \{F\}_{\varepsilon_0} \tag{5.24}$$

where $\{R\}$ = externally applied nodal forces of the system / assembly

    $[K]$ = $[k]_{\text{system}}$ = global stiffness matrix of the system obtained by superposition of stiffnesses of individual elements connected / contributing to each node of the system / assemblage

$\{\delta\}$ = nodal displacement matrix of the system

$\{F\}_p$ = equivalent nodal forces of the system due to inherent and (or) applied internal forces

$\{F\}_{\varepsilon_0}$ = equivalent nodal forces of the system due to initial strain, lack of fitness and other internal causes (if any)

5. Applying the boundary conditions specified on the boundary (boundaries) of the system by incorporating them in the above system equilibrium equations and solving the resulting equations, the unknown nodal displacements at all the nodes can be obtained. Once the nodal displacements are solved as above, the displacements, stresses and any other required parameters can be evaluated using the element characteristics as expressed in the above Eqns. (5.22) and (5.23).

The rest of the analysis and design procedure is the same as any other structural analysis problem.

5.10.2 FEM Analysis of Foundations using BEF Model

Assuming the soil structure system as BEF model i.e., footing is modelled as a beam in bending, totally bonded with soil which is modelled as Winkler's spring model, FEM analysis can be carried out as explained in the above section. This amounts to solve the beam bending problem while incorporating soil reaction / pressure as an additional equivalent externally applied nodal force (adding $-\{F\}_p$ to the RHS of Eq. (5.24)), or including it as an equivalent nodal force due to internal forces applied on the beam i.e., adding it as an additional internal nodal force to $\{F\}_p$ on the left hand side (LHS) of Eq. (5.24). Both these operations are the same whether it is added as $-\{F\}_p$ on the RHS or $+ \{F\}_p$ on the LHS of Eq. (5.24).

Thus, we can carry out FEM analysis of foundations as any other structural analysis problem using beam elements or plate elements or other shapes of structural elements for modelling the structure. The FEM analysis of foundations using Beam elements is discussed in the following sections.

5.10.3 Beam Elements

FEM analysis of beams is presented in Section 4.3 along with other structural components. Important details of the same are recalled below for carrying out BEF analysis. Beams being planar structures, beam convention used in Chapter 4 (Section 4.3) will be adopted for the following discussion. Accordingly, the beam is shown in xy – plane with v as vertical displacement along y axis in the following figures.

Beams are most commonly used structural components and carry transverse loads by bending following Euler-Bernoulli Bending theory (Crandall and Dahl, 1972). A typical planar beam element is shown in Figures 5.16 a and b, with two nodes i, j and positive forces, V_i, V_j and moments, M_i, M_j and vertical displacements, v_i, v_j and slopes,

θ_i, θ_j ,shown at the two node respectively The nodal coordinates are i ($x = 0$), j ($x = L$) where L is the length of the beam.

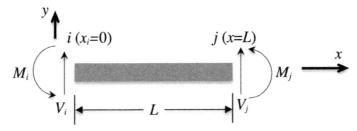

a) Nodal loads (positive as per solid mechanics)

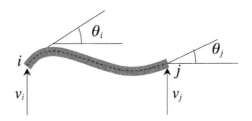

b) Nodal displacements and slopes
Figure 5.16

It may be noted that shear forces and bending moments are considered positive as per bending theory as shown below in Figure 5.17 and not as per mechanics of solids as shown in Figure 5.16a.

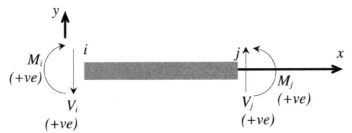

Figure 5.17 Positive Shear Force and Bending Moments as per Bending Theory

As usual x, y coordinate axes (without prime " ' ") refer to the global / system coordinates and if they are mentioned with prime, they refer to the local coordinates.

Following the standard FEM, direct stiffness analysis (Chapter 3), the beam element is associated with generalized deflections, i.e., vertical displacement v and slope $\theta = dv/dx$ at any point along the axis of the beam. Accordingly, the nodal displacement matrix can be expressed as:

$$\{\delta\} = \begin{Bmatrix} v_i \\ \theta_i \\ v_j \\ \theta_j \end{Bmatrix}$$

(5.25)

161

After carrying out the stiffness analysis (Section 4.3), the stiffness matrix is obtained as

$$[k]^e = \frac{EI}{L^3}\begin{bmatrix} 12 & 6L & -12 & 6L \\ 6L & 4L^2 & -6L & 2L^2 \\ -12 & -6L & 12 & -6L \\ 6L & 2L^2 & -6L & 4L^2 \end{bmatrix}$$ (5.26)

It can be noted that $[k]^e$ is a symmetric and positive definite matrix.

The element equilibrium equation (Section 4.3) of the beam is expanded to incorporate additional forces (external and internal, if any) and can be written from Eq. (5.22) as

$$\{R\}^e = \left\{\begin{matrix} F_i \\ F_j \end{matrix}\right\}^e = \left\{\begin{matrix} V_i \\ M_i \\ V_j \\ M_j \end{matrix}\right\}^e = [k]^e \left\{\begin{matrix} v_i \\ \theta_i \\ v_j \\ \theta_j \end{matrix}\right\}^e + \left\{\begin{matrix} F_i \\ M_i \\ F_j \\ M_j \end{matrix}\right\}^e_p + \{F\}^e_{\varepsilon_0}$$ (5.27)

where $\{R\}^e$ = the applied nodal forces (external) of the beam element = $\left\{\begin{matrix} V_i \\ M_i \\ V_j \\ M_j \end{matrix}\right\}^e$;

$[k]^e$ = stiffness matrix of the element; $\left\{\begin{matrix} v_i \\ \theta_i \\ v_j \\ \theta_j \end{matrix}\right\}^e$ = nodal displacement matrix of the element;

$\{F\}^e_p = \left\{\begin{matrix} F_i \\ M_i \\ F_j \\ M_j \end{matrix}\right\}^e_p$ = equivalent nodal forces (internal) of the element; and,

$\{F\}^e_{\varepsilon_0}$ = equivalent nodal forces (internal) of the element due to other causes, if any

It may be noted that θ, M and V can also be interpreted as slope, bending moments and shear forces at any cross section of the beam along its axis (x- axis) as per bending theory and are given by (Kameswara Rao, 2011),

$$\theta = v' = \frac{dv}{dx} = \text{slope of the beam at any point along its axis (x-axis)}$$

$$M = EI\frac{d^2v}{dx^2} = \text{bending moment (as per bending convention) of the beam}$$ (5.28)

$$V = -EI\frac{d^3v}{dx^3} = \text{shear force (as per bending convention)}$$

where v is the vertical displacement at that cross section.

These parameters can be calculated at any cross section of the beam from Equations 4.3(4) and 4.3 (6) (Section 4.3) and subsequent derivatives for calculating slope θ (involving first derivative of v), bending moment M (involving second derivative of v), shear force V (involving third derivative of v) as expressed in the above Equation 4.3 (13) (Section 4.3), once the deflections and the forces at the nodes are obtained from the FEM analysis. It may be observed that FEM analysis gives forces V, and moments M, as per mechanics convention which can be interpreted as shear forces and bending moments as per bending theory convention as may be required for design. Stresses at any point can be obtained using Eqn. (5.23).

5.10.4 System / Global Stiffness Matrix, Assembly and System Equilibrium Equations

Once the element stiffness matrices of all elements of the system are evaluated, the global stiffness matrix can be assembled by adding the contributions of individual elements connected at any node as explained in Chapter 4 and also in Sections 5.10.1, 5.10.2 and 5.10.3 Accordingly, after assembling the system, the system equations can be written by equating the internal equivalent nodal forces to the externally applied forces and moments as given in Equation 4.3 (12). Effects due to distributed forces, temperature, and initial lack of fitness and any peculiarities of individual element (if they are present) can all be included in the equilibrium equations, which are explained in the above sections. However, it may be noted that temperature will not affect the transverse forces though it produces axial forces. Thus the system equilibrium equations can be expressed in the standard form (Chapters 2 and 3) and expressed as in Eq. (5.24) as follows.

$$\{R\} = [K]\{\delta\} + \{F\}_p + \{F\}_{\varepsilon_0} \tag{5.29}$$

This can be written in the expanded form after assembling and superposing the stiffness characteristics, external and equivalent internal nodal forces contributed by each element connected at each node of the total assembly as:

$$\{R\}_{system} = [K]_{system}\{\delta\}_{system} + \{F\}_p + \{F\}_{\varepsilon_0} = [k]_{system}\begin{Bmatrix} v_1 \\ \theta_1 \\ v_2 \\ \theta_2 \\ \vdots \\ v_n \\ \theta_n \end{Bmatrix} + \sum \begin{Bmatrix} F_1 \\ F_2 \\ F_3 \\ \vdots \\ F_n \end{Bmatrix}_p + \sum \begin{Bmatrix} F_1 \\ F_2 \\ F_3 \\ \vdots \\ F_n \end{Bmatrix}_{\varepsilon_o} \tag{5.30}$$

where $\{R\}_{system}$ = total nodal force matrix of externally applied forces on the system. This

is obtained by adding contributions of all the externally applied forces on the elements connected at each node of the system;

$[K]_{system} = \sum_{i=1,2---n} [k]^i$ and elements of $[K]_{system}$ can be expressed as $K_{ij} = \sum k_{ij}$

with summation being carried out over all the elements of the system which have with i and j as common nodes, and $K_{ii} = \sum k_{ii}$ where the sum is carried out on all the elements connected at any node i of the assemblage; it may further be noted that $K_{ij} = K_{ji}$ in structural analysis problems indicating that the system stiffness matrix $[K]_{system}$ is symmetric.

$$\{\delta\}_{system} = \begin{Bmatrix} v_1 \\ \theta_1 \\ v_2 \\ \theta_2 \\ \vdots \\ v_n \\ \theta_n \end{Bmatrix} = \text{nodal displacement matrix of the system / assemblage;}$$

$$\{F\}_p = \sum \begin{Bmatrix} F_1 \\ F_2 \\ F_3 \\ \vdots \\ F_n \end{Bmatrix}_p = \text{total nodal force matrix due to internal forces } (p) \text{ in the elements}$$

This is obtained by adding all the equivalent forces due to internal forces in each of the element connected at each node of the system / assemblage;

$$\{F\}_{\varepsilon_o} = \sum \begin{Bmatrix} F_1 \\ F_2 \\ F_3 \\ \vdots \\ F_n \end{Bmatrix} = \text{total nodal force matrix due to other internal causes } (\varepsilon_0) \text{ in the}$$

elements;

This is obtained by adding all the equivalent nodal forces due to other internal causes in each of the element connected at each node of the system / assemblage;

The above equations can be assembled from the stiffness characteristics of individual elements as explained above in Section 5.10.3 and also in Section 4.3 as well as Chapter 3, in detail. In particular, element stiffness matrix of the beam element, $[k]^e$ is given by Eq. (5.26). Equivalent nodal forces due to internal forces, $\{F\}_p^e$ and equivalent nodal forces due to other internal causes in the element, $\{F\}_{\varepsilon_0}^e$ can be calculated from the stiffness analysis expressions given in Section 4.3 and in Chapter 3. $\{F\}_p^e$ can also be calculated using the simple lumped method or consistent method (Section 4.3). In practice, the lumped method is preferred wherein equivalent nodal forces are distributed between the nodes based on the proximity of the resultant load to the corresponding node and other common sense approaches.

Next step is, to apply the boundary conditions specified along the boundary (boundaries) of the system by incorporating them in the above system equilibrium equations. Solving these resulting equations, unknown nodal displacements at all the nodes can be obtained. Once the nodal displacements are solved as above, the displacements, stresses and any other required parameters can be evaluated using the element characteristics as expressed in the above Eqns. (5.22) and (5.23).

The rest of the analysis and design procedure is the same as any other structural analysis problem. This makes the analysis very modular since forces due various external and internal causes can be included in the analysis with the stiffness characteristics of the elements and the system remaining the same while analysis can be carried out with total flexibility for including the forces due to various causes affecting the elements and the system.

5.10.5 Incorporating Soil Contact Pressure in Stiffness Characteristics of Beam Elements

Now it is straightforward to incorporate soil pressure either as an external force / reaction on the element on the LHS of Eqns. (5.27), (5.29) and (5.30) with a –ve sign or as internal pressure / force on the beam element on the RHS of the same equations as $\{F\}_p^e$ with a +ve sign both resulting in the same equations (also explained in Section 5.10.2). Though the soil contact can be expressed using any of the discrete or continuum models as discussed in Sections 5.9 and 5.10, the Winkler's model is commonly preferred due to its simplicity and accessibility of reliable input soil parameters (Section 5.10.2). Accordingly, the same is adopted for incorporating the soil contact pressure for BEF analysis as presented below. Further, lumped method is used due to its simplicity, common sense / simple logic to evaluate equivalent nodal forces treating the soil contact pressure as an internal pressure acting on the beam element i.e., treating it as $\{F\}_p^e$ and retaining it on the LHS of Eqns. (5.27), (5.29) and (5.30).

The soil reaction using Winkler's model is expressed as (Sections 5.9.2, 5.9.3):

$$q(x) = kv \tag{5.31}$$

where $q(x)$ = soil contact pressure (in units of FL^{-1} compatible with BEF Eq. (5.18))

$\quad k$ = soil spring constant / foundation modulus for unit length of the beam i.e, area of $bx1$

$\quad\quad = k_s\, b1 = C_u\, b1$ (*1= unit length of the beam around the point at x*)

$\quad k_s = C_u$ = modulus of subgrade reaction with units of FL^{-3} (Sections 5.9.3, 5.9.4)

$\quad v$ = vertical displacement of the beam

$\quad b$ = width of the beam cross section

(F and L correspond to the force and length units respectively)

The corresponding equivalent soil contact pressures (considering them as internal forces acting on the beam element due to supporting soil) at the two nodes of the beam (Figure 5.16a) can be incorporated in the element equilibrium equation (Eq. 5.27) using simple lumped method as follows.

While the equivalent nodal forces due to internal forces and causes can be obtained using stiffness analysis of beams (Chapter 3 and Section 4.3), another simpler and alternate method to express the equivalent nodal forces due to internal forces applied on the beam, $\{F\}_p^e$ (or due concentrated loads applied in between the nodes), is to distribute the resultant force at both the nodes judiciously (depending on the proximity of the applied forces to the node under consideration) as mentioned in the following two alternatives.

1. Prorate loads to adjacent nodes using a simple beam model or
2. Prorate loads to adjacent nodes as if the element has fixed ends so that the values include fixed end moments and shears (vertical forces). This is more accurate but involves cumbersome computations.

Alternative 1 is preferable because of its simplicity which is referred to as lumped method / a judicious common-sense approach.

Though the soil reaction varies along the axis of the beam (x-axis) as per Eq. (5.31). using the lumped method, it is logical to assume that each node shares the soil pressure acting over half the contact area of the beam ($bL/2$) (half the length of the beam element which is assumed to be prismatic i.e., of uniform width, b and flexural rigidity, EI) since there are two nodes for the beam element (Bowles, 1996). Accordingly, the equivalent nodal force at the two nodes of the beam element to be included in matrix $\{F\}_p^e$ in the RHS of Eq. (5.27) are

$$F_{si} = k_i v_i = \frac{L}{2} b k_s v_i$$

$$F_{sj} = k_j v_j = \frac{L}{2} b k_s v_j$$

(5.32)

where b = width of the beam

k_s = modulus of subgrade reaction (FL^{-3})

L = length of the beam element

k_i, k_j = equivalent soil spring stiffnesses at nodes i, j

$$= \frac{L}{2} b k_s$$

(5.33)

v_i, v_j = vertical displacements at nodes $i.j$

F_{si}, F_{sj} = equivalent soil reactions at nodes

Accordingly,

$$\{F\}_{p-soil}^e = \begin{Bmatrix} F_{si} \\ 0 \\ F_{sj} \\ 0 \end{Bmatrix}_{soil}^e = \begin{Bmatrix} \frac{L}{2} b k_s v_i \\ 0 \\ \frac{L}{2} b k_s v_j \\ 0 \end{Bmatrix}_{soil}^e$$

(5.34)

Incorporating soil pressure in the element equilibrium equation, Eq. (5.27) can be rewritten while retaining all forces due to internal forces and other causes as:

$$\{R\}^e = \begin{Bmatrix} F_i \\ F_j \end{Bmatrix}^e = \begin{Bmatrix} V_i \\ M_i \\ V_j \\ M_j \end{Bmatrix}^e = [k]^e \begin{Bmatrix} v_i \\ \theta_i \\ v_j \\ \theta_j \end{Bmatrix}^e + \begin{Bmatrix} F_i \\ M_i \\ F_j \\ M_j \end{Bmatrix}^e_p + \{F\}^e_{\varepsilon_0} + \begin{Bmatrix} \dfrac{L}{2}bk_s v_i \\ 0 \\ \dfrac{L}{2}bk_s v_j \\ 0 \end{Bmatrix}^e_{soil} \tag{5.35}$$

$$= \frac{EI}{L^3}\begin{bmatrix} 12 & 6L & -12 & 6L \\ 6L & 4L^2 & -6L & 2L^2 \\ -12 & -6L & 12 & -6L \\ 6L & 2L^2 & -6L & 4L^2 \end{bmatrix}\begin{Bmatrix} v_i \\ \theta_i \\ v_j \\ \theta_j \end{Bmatrix}^e + \begin{Bmatrix} F_i \\ M_i \\ F_j \\ M_j \end{Bmatrix}^e_p + \{F\}^e_{\varepsilon_0} + \begin{Bmatrix} \dfrac{L}{2}bk_s v_i \\ 0 \\ \dfrac{L}{2}bk_s v_j \\ 0 \end{Bmatrix}^e_{soil} \tag{5.36}$$

Combining the first and last terms of the above equation (5.36), it can berewritten as:

$$\{R\}^e = \begin{Bmatrix} F_i \\ F_j \end{Bmatrix}^e = \begin{Bmatrix} V_i \\ M_i \\ V_j \\ M_j \end{Bmatrix} = \frac{EI}{L^3}\begin{bmatrix} 12+\dfrac{L^4}{2EI}bk_s & 6L & -12 & 6L \\ 6L & 4L^2 & -6L & 2L^2 \\ -12 & -6L & 12+\dfrac{L^4}{2EI}bk_s & -6L \\ 6L & 2L^2 & -6L & 4L^2 \end{bmatrix}\begin{Bmatrix} v_i \\ \theta_i \\ v_j \\ \theta_j \end{Bmatrix}^e + \begin{Bmatrix} F_i \\ M_i \\ F_j \\ M_j \end{Bmatrix}^e_p + \{F\}^e_{\varepsilon_0}$$

$$= [k]^e_{bef}\begin{Bmatrix} v_i \\ \theta_i \\ v_j \\ \theta_j \end{Bmatrix}^e + \begin{Bmatrix} F_i \\ M_i \\ F_j \\ M_j \end{Bmatrix}^e_p + \{F\}^e_{\varepsilon_0} = [k]^e_{bef}\{\delta\}^e + \{F\}^e_p + \{F\}^e_{\varepsilon_0} \tag{5.37}$$

where,

$$[k]^e_{bef} = \frac{EI}{L^3}\begin{bmatrix} 12+\dfrac{k_i L^4}{EI} & 6L & -12 & 6L \\ 6L & 4L^2 & -6L & 2L^2 \\ -12 & -6L & 12+\dfrac{k_j L^4}{EI} & -6L \\ 6L & 2L^2 & -6L & 4L^2 \end{bmatrix} \tag{5.38}$$

Thus, the modified stiffness matrix to be used for BEF of analysis (in place of $[k]^e$) can be expressed as

$$[k]^e_{bef} = \begin{bmatrix} 12+\dfrac{k_i L^3}{EI} & 6L & -12 & 6L \\ 6L & 4L^2 & -6L & 2L^2 \\ -12 & -6L & 12+\dfrac{k_j L^3}{EI} & -6L \\ 6L & 2L^2 & -6L & 4L^2 \end{bmatrix} \tag{5.39}$$

167

where $\left[k\right]^e_{bef}$ = stiffness matrix of beam on elastic foundatons (*bef*) element for BEF analysis (which includes stiffness due to soil reaction). It can be also noted that it is a symmetric matrix.

$k_i = k_j$ = equivalent soil spring stiffnesses at nodes of the beam element = $\dfrac{L}{2} b k_s$

(5.40)

$k_s = C_u$ = modulus of subgrade reaction

The rest of the analysis (such as assembling the system, formulating system equilibrium equations, applying boundary conditions and simplifying the resulting equilibrium equations of the system, solving for the unknown displacements and slopes, calculating forces and moments (or interpreting them as shear forces and bending moments as per bending convention, and other manipulations etc.) is the same as the standard structural analysis of beams as in Section 4.3 using the stiffness matrix of BEF element, $\left[k\right]^e_{bef}$ in place of $\left[k\right]^e$.

Examples of FEM analysis of foundations / BEF are presented in Section 5.12.

5.11 FEM Analysis of Foundations using PEF Model

Foundations which have comparable dimensions along both directions (length and breadth) need to be analyzed as plates on elastic foundations (PEF) in the rational approach. Few simple solutions and several numerical solutions for PEF are available in literature (Kameswara Rao, 2011 and others). The details for applying FEM to PEF are discussed in the following sections using classical bending theory of plates. The FEM analysis of PEF is essentially similar to that of BEF as presented in the above Section 5.10 except that the plate elements have to be used with plate bending in two directions (Section 5.8.3). Details of plate bending elements are available in several references (Zienkiewicz 1971, Desai and Abel 1971, Zienkiewicz and Taylor 1989, Rao 1982, Rajasekaran 1993, Kameswara Rao 2011 etc.) and software packages such as Ansys, Abaqus, Nastran etc.

5.11.1 FEM Analysis of PEF using Plate Bending Elements and Winkler's Soil Model

The state of deformation of a plate can be described entirely by the vertical displacement w of the middle plane of the plate. Continuity conditions between elements have to be imposed not only on this quantity but on its derivatives. This is to ensure that the plate remains continuous and does not 'kink'. At each node, therefore, three conditions of equilibrium and continuity need to be imposed.

Determination of suitable shape functions is now much more complex. Indeed, if complete slope continuity is required on the interfaces between various elements, the mathematical and computational difficulties often rise disproportionately fast. It is, however, relatively simple to obtain shape functions which, while preserving continuity of w, may violate the slope continuity between elements, though naturally not at the node where continuity is imposed. If such functions satisfy the 'constant strain' criterion, then convergence may still be found. Accordingly, non-conforming shape functions are usually adapted (Zienkiewicz, 1971). The solution with such shape functions will give bounds to the correct answer, but on many occasions, will yield an inferior accuracy. For practical applications, non-conforming shape functions are preferable.

The simplest type of element shape is now a rectangle with four corner nodes, while triangular and quadrilateral elements present some difficulties and are used for solutions of plates of arbitrary shape and shell problems (Zienkiewicz, 1971, Rao, 1982). Plate elements with higher degrees of freedom and performance such as 8 noded elements etc can also be used for higher accuracy (Chapter 2). Alternatively, finer discretization of the system (finer mesh) resulting in larger number of elements which are simple to analyses can also be adopted for achieving higher accuracy.

The general PEF analysis with Winkler's soil model is presented in Section 5.8.3 and general procedures for FEM analysis of soil-structure interaction (foundation analysis) are discussed in Section 5.10. Plate bending elements involve a large number of nodal degrees of freedom (nodal variables / generalized displacements at the nodes), and hence FEM analysis of PEF needs large computer resources though the analysis is same as that of BEF except for using plate bending elements instead of beam bending elements in the analysis. Similar as in BEF analysis (Section 5.10.5), the soil contact pressure is added as an internal force as $\{F\}^e_{p-soil}$. in the equilibrium equations of the plate element

Consequently, the plate element stiffness matrix gets modified as $[k]^e_{pef}$ by this addition of equivalent nodal forces due to soil contact pressure at the corresponding diagonal nodes of the plate element stiffness matrix $[k]^e$ of the plate element. The rest of the analysis follows the same steps as the BEF analysis and summarized below using the modified $[k]^e_{pef}$ in place of $[k]^e$ of the plate for formulating element and system equilibrium equations and solutions.

The element equilibrium of plate element with soil contact pressure can be expressed as:

$$\{R\}^e = [k]^e\{\delta\}^e + \{F\}^e_{p-soil} + \{F\}^e_p + \{F\}^e_{\varepsilon_0} \tag{5.41}$$

By combining the first two terms on the LHS of the above equation, the same can be rewritten as:

$$\{R\}^e = [k]^e_{pef}\{\delta\}^e + \{F\}^e_p + \{F\}^e_{\varepsilon_0}$$

where $[k]^e_{pef}$ is the modified stiffness matrix for the plate element with elastic foundation stiffness added to it i,e, referred to as *pef* element.

The system equilibrium equations of the assembly can be written accordingly as:

$$\{R\}_{system} = [K]_{pef-system}\{\delta\}_{system} + \{F\}_p + \{F\}_{\varepsilon_0} \tag{5.42}$$

$$=[K]_{pef-system}\{\delta\}_{system}+\sum\begin{Bmatrix}F_1\\F_2\\F_3\\\vdots\\F_n\end{Bmatrix}_p+\sum\begin{Bmatrix}F_1\\F_2\\F_3\\\vdots\\F_n\end{Bmatrix}_{\varepsilon_o} \tag{5.43}$$

i.e $\quad \{R\}_{system}=[K]_{pef-system}\{\delta\}_{system}+\{F\}_p+\{F\}_{\varepsilon_0}$ $\hspace{2cm}$ (5.44)

with $[K]_{pef-system}$ assembled using $[k]^e_{pef}$ (which includes the soil contact oressure in its stiffness characteristics) of all elements of the system for superposition from all elements connected at each node of the assembly as explained in BEF analysis in Section 5.10.

5.11.2 Additional Details of Incorporating Soil Reaction for PEF Analysis

The soil reaction for plates on elastic foundations is incorporated in FEM analysis by either of the following idealizations.

1. The soil is idealized as Winkler's model and the soil reaction $q(x)=kw$ is incorporated as an externally applied equivalent nodal force at all the nodes of the rectangular plate element or elements of other shapes. As in BEF analysis, lumped method is commonly used and preferred for calculating the equivalent external nodal forces due to soil reaction. The nodes of the plate are assumed to share the contact area of the plate element equally or on prorate basis for computing the soil reaction using Winkler's model. This can be done either at the time of formulation of the stiffness matrix of the plate element or while assembling the global stiffness matrix of the plate for applying equilibrium equations as explained in the above Section 5.11.1 (similar to the methods presented for FEM analysis of BEF).

2. Incorporating soil reaction in the element stiffness matrix for PEF (Bowles, 1966). Noting that the soil reaction has to be incorporated as equivalent nodal forces at the nodes of the plate element, the corresponding node soil "spring' (stiffness) has to be added to the corresponding diagonal elements of the plate element stiffness matrix as mentioned in the above Section 5.11.1, thus obtaining the modified stiffness matrix $[k]^e_{pef}$ for further FEM analysis. For example, for a typical four noded rectangular plate element of sides a and b (with nodes 1,2,3,4, and 3 degrees of freedom at each node), the equivalent nodal forces due to soil reaction (treated as external force) / soil contact pressure (treated as internal force) can be expressed using the approximate lumped method as follows (Kameswara Rao, 2011, Bowles, 1996).

$$F_{si}=k_{i-soil}w_i=k_s\frac{ab}{4}w_i \qquad i=1, 2, 3, 4 \tag{5.45}$$

where F_{si} = nodal force due to soil reaction at node i

$\qquad w_i$ = nodal displacement at the node i

a, b = dimensions of the rectangular plate element along x and y axes respectively.

k_s = modulus of subgrade reaction.

k_{i-soil} = equivalent spring stiffness at node i for the element under consideration

$$= k_s \frac{ab}{4}$$

Now by adding k_{i-soil} ($i = 1,2,3,4$) to the corresponding plate stiffness elements k_{11}^e k_{44}^e, k_{77}^e, k_{1010}^e, in the $[k]^e$ matrix for plate element (Kameswra Rao, 2011), the soil reactions gets incorporated in the PEF analysis. Thus, we have, the modified diagonal stiffness elements of the $[k]_{pef}^e$ as:

$$\left(k_{ii} \right)_{pef}^e = (k_{ii}^e)_{plate} + k_{i-soil}^e = (k_{ii}^e)_{plate} + \frac{ab}{4} k_s \qquad i = 1,4,7,10 \qquad (5.46)$$

It should be noted that the K_{ii} of the global stiffness matrix $[K]_{pef-system}$ gets contributions from all such *pef-* elements with i as the common node which includes the soil reaction in its stiffness coefficients. Similarly, all elements K_{ij} of the global stiffness matrix $[K]_{pef-system}$ get contributions from all the the *pef-* elements which have i and j as common nodes as per FEM procedure for assembling system equilibrium equations of the total assembly.

The rest of the FEM analysis is the same and follows the standard steps as in BEF analysis. Since FEM analysis of PEF involves large number of degrees of freedom, variables, equations for solutions, large computer resources are needed. Hence custom-made programs or software packages are used for the analysis such problems. FEM analysis of PEF is not included in the examples illustrated in Section 5.12 due to the above reasons though the procedure is explained in detail in the above sections.

5.11.3 Circular, ring shaped and plates of general shapes

The main difference in FEM for non-rectangular shapes including circular and ring-shaped foundations is in discretization using finite elements to fit the shape of the prototype plate as closely as possible and interpret displacements, strains, stresses and constitutive relationships appropriately. This can even be achieved using rectangular or triangular elements with finer mesh for discretization near the boundaries which may not be strictly fitting into rectangular or triangular geometry. One can also use quadrilateral or parallelogram shaped elements. For circular and ring-shaped elements, Bowles (1996) used beam elements for analysis, if the problems are axisymmetric. The formulation for other cases can be done using cylindrical polar coordinates (Zeinkiewicz, 1971, Rao, 1982).

The details of analysis for such plates can be referred to in standard books on FEM (Zienkiewicz, 1971, Zienkiewicz and Taylor, 1989, Rao 1982, Desai and Abel 1972). The rest of the FEM procedure is the same as for rectangular plates (Section 7.5.1 to 7.5.7) including the incorporation of soil reaction for PEF.

5.11.4 General Comments on FEM

From the details outlined in this chapter, following general comments can be made.

1. The FEM is very simple and very versatile to deal with complex problems.
2. The procedure involves too many unknowns to be solved from simultaneous equations and is practical only if programmed on computers or software packages are used. A large number of general-purpose software packages are available (eg.: SAP, STAAD, NASTRAN, ABAQUS, ANSYS, NISA etc.). Analysis is difficult to be carried out manually by hand calculators unless the unknowns are small in number (by using very few elements) and mathematical software like MATLAB is available.
3. A finite-element computer program should be somewhat self – checking. This is accomplished by checking the input and comparing sums of input versus output forces, even in commercial software packages.
4. One should not use a very short element next to a long element. Use more finite elements and effect a transition between short and long members. Try to keep the aspect ratio of rectangular elements within the following limits (L denotes length dimension):

$$\frac{L_{long}}{L_{short}} \leq 2 \text{ and not more than 3.}$$

5. The value of modulus of subgrade reaction, k_s directly affects the deflection but has very little effect on the computed bending moments at least for reasonable values of k_s.
6. Since k_s is usually estimated, the use of refined / advanced / higher order elements may give undeserved confidence in the computed results. Higher accuracy can also be achieved using more elements with smaller dimensions (refined mesh) and simple elements for easy computation.

5.12 Examples of FEM Analysis for Foundations using BEF Model

Several worked out examples of Foundation analysis using FEM are presented in this section using BEF model.

5.12.1 Example 1

Design a square footing / foundation as shown in Figure 5.12.1 to carry a load of 1200 kN transmitted through a 0.5 m x 0.5m square column reinforced with 20 mm (for steel) bars along its length. The footing is based at 2 m below ground level.

Input Data:
Soil Data:
Design soil pressure, $q = 90$ kN/m^2
Unit weight of overburden soil, $\gamma_s = 18$ kN/m^3
Modulus of subgrade reaction, $k_{s1} = 35.07 \times 10^5$ kN/m^3
(from load test on a square plate of 0.3 m $\times$ 0.3 m, i.e. $A_p = 0.09$ m^2)

Modulus of elasticity of soil, $E_s = k_{s1} \dfrac{\sqrt{A}(1-v^2)}{1.13} = \dfrac{35.01 \times \sqrt{0.09}}{1.13} = 9.31 \times 10^5 \, kN/m^2$

(assuming Poisson's ratio of soil, $v_s = 0$)

Concrete and steel data:
Use concrete of Grade 20
Use high tensil steel bars of Fe 415 grade
Hence, compressive strength of concrete, $\sigma_{ck} = 20$ N/mm^2
Hence, yield strength of steel bars, $\sigma_y = 415$ N/mm^2
Assume self-weight of the column and substructure as 10% of the column load. Load factor for design and over burden soil by Limit State Method (LSM) = 1.5. Use FEM for the analysis.

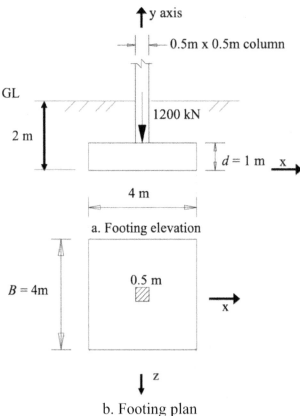

a. Footing elevation

b. Footing plan
Figure 5.12.1 Spread Footing

1. Foundation Analysis using FEM
a. Footing Dimension:
Total Column load + 20% (self-load of column footing and overburden soil)
= 1200 x 1.2 = 1440 kN

Area of footing needed $= \dfrac{1440}{90} = 16m^2$

Provide a square footing of 4 m x 4 m with 1 m depth for the FEM analysis. (The depth of the footing for carrying out FEM analysis can be estimated from conventional analysis).
Area of footing, $A = 16 \, m^2$

Net soil pressure for Limit State Design Method (LSD),

$$q_s = \frac{1.5 \times 1200}{16} = 112.5 \ kN/m^2 \tag{5.12.1}$$

Modulus of subgrade reaction corrected to footing area of 16 m^2,

$$k_{s4} = k_s = k_{s1}\sqrt{\frac{0.09}{16}} = 35.07 \times 10^5 \times \frac{0.3}{4} = 2.63 \times 10^5 kN/m^3 \tag{5.12.2}$$

Then soil reaction / contact pressure at any contact point $= k_s v$ (5.12.3)

where v = the displacement of the footing at that point

b. Flexural rigidity of foundation / footing:

Modulus of elasticity of concrete footing, $E_f = E = 5700\sqrt{\sigma_{ck}} = 5700\sqrt{20} = 2.5491 \times 10^7 kN/m^2$

Moment of Inertia of the footing,

$$I_{xx} = I_{zz} = I = \frac{Bd^3}{12} = 4 \times \frac{1^3}{12} = 0.33m^4$$

Flexural rigidity of footing, $EI = 2.5491 \times 10^7 \times 0.33 = 8.412 \times 10^6 kNm^2$ (5.12.4)

2. FEM analysis using beam elements incorporating soil reaction using Winkler's model

Maximum load from column including load factor (LF) = 1.5 x -1200 = -1800 kN

The footing is discretized into 2 elements as shown in Figures 5.12.2a and 5.12.2b.

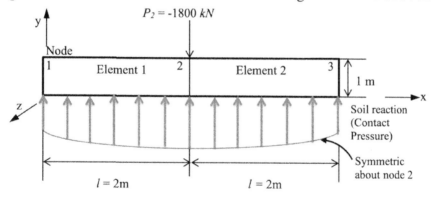

a) Footing on elastic foundation (soil)

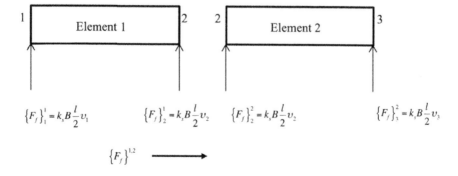

b) Equivalent nodal forces due to soil reaction, $\{F_f\}^{1,2,3}$

Figure 5.12.2

Plan dimensions of the footing $= L \times B$

For a square footing $L = B = 4$ m, element length $= l = L/2 = 2$ m

Stiffness matrix of beam elements:

from Eq. (5.26),

$$[k]_e = \frac{EI}{l^3} \begin{matrix} v_i & \theta_i & v_j & \theta_j \end{matrix} \begin{bmatrix} 12 & 6l & -12 & 6l \\ 6l & 4l^2 & -6l & 2l^2 \\ -12 & -6l & 12 & -6l \\ 6l & 2l^2 & -6l & 4l^2 \end{bmatrix}$$

(5.12.5)

Equilibrium equations of the footing elements incorporating soil reactions, as equivalent nodal forces from Figures 5.12.2(a) and 5.12.2 (b):

Element 1:

$$[k]^1 \{\delta\}^1 + \{F\}^1_f = \{R\}^1$$

$$\frac{EI}{l^3} \begin{matrix} v_1 & \theta_1 & v_2 & \theta_2 \end{matrix} \begin{bmatrix} 12 & 6l & -12 & 6l \\ 6l & 4l^2 & -6l & 2l^2 \\ -12 & -6l & 12 & -6l \\ 6l & 2l^2 & -6l & 4l^2 \end{bmatrix} \begin{Bmatrix} v_1 \\ \theta_1 \\ v_2 \\ \theta_2 \end{Bmatrix} + \begin{Bmatrix} k_s B \dfrac{l}{2} v_1 \\ 0 \\ k_s B \dfrac{l}{2} v_2 \\ 0 \end{Bmatrix} = \begin{Bmatrix} V_1 \\ M_1 \\ V_2 \\ M_2 \end{Bmatrix}$$

(5.12.6)

Element 2:

$$[k]^2 \{\delta\}^2 + \{F\}^2_f = \{R\}^2$$

$$\frac{EI}{l^3} \begin{matrix} v_2 & \theta_2 & v_3 & \theta_3 \end{matrix} \begin{bmatrix} 12 & 6l & -12 & 6l \\ 6l & 4l^2 & -6l & 2l^2 \\ -12 & -6l & 12 & -6l \\ 6l & 2l^2 & -6l & 4l^2 \end{bmatrix} \begin{Bmatrix} v_2 \\ \theta_2 \\ v_3 \\ \theta_3 \end{Bmatrix} + \begin{Bmatrix} k_s B \dfrac{l}{2} v_2 \\ 0 \\ k_s B \dfrac{l}{2} v_3 \\ 0 \end{Bmatrix} = \begin{Bmatrix} V_2 \\ M_2 \\ V_3 \\ M_3 \end{Bmatrix}$$

(5.12.7)

Introducing dimensionless relative stiffness constant of soil foundation,

$$C_1 = \left(k_s B \frac{l}{2} \right) \frac{l^3}{EI}$$

(5.12.8)

Equations (5.12.6) and (5.12.7) become:

Element 1

$$\frac{EI}{l^3} \begin{matrix} v_1 & \theta_1 & v_2 & \theta_2 \end{matrix} \begin{bmatrix} 12 + C_1 & 6l & -12 & 6l \\ 6l & 4l^2 & -6l & 2l^2 \\ -12 & -6l & 12 + C_1 & -6l \\ 6l & 2l^2 & -6l & 4l^2 \end{bmatrix} \begin{Bmatrix} v_1 \\ \theta_1 \\ v_2 \\ \theta_2 \end{Bmatrix} = \begin{Bmatrix} V_1 \\ M_1 \\ V_2 \\ M_2 \end{Bmatrix}$$

(5.12.9)

Element 2

$$\frac{EI}{l^3} \begin{bmatrix} 12+C_1 & 6l & -12 & 6l \\ 6l & 4l^2 & -6l & 2l^2 \\ -12 & -6l & 12+C_1 & -6l \\ 6l & 2l^2 & -6l & 4l^2 \end{bmatrix} \begin{Bmatrix} \upsilon_2 \\ \theta_2 \\ \upsilon_3 \\ \theta_3 \end{Bmatrix} = \begin{Bmatrix} V_2 \\ M_2 \\ V_3 \\ M_3 \end{Bmatrix} \qquad (5.12.10)$$

Global system equilibrium equations

Assembling the above elements to integrate the system, the systems / global stiffness matrix and system equilibrium equations can be obtained as below.

$$[k]_{system} = \frac{EI}{l^3} \begin{bmatrix} 12+C_1 & 6l & -12 & 6l & 0 & 0 \\ 6l & 4l^2 & -6l & 2l^2 & 0 & 0 \\ -12 & -6l & \begin{matrix}12+1C_1+12+C_1\\=24+2C_1\end{matrix} & -6l+6l=0 & -12 & 6l \\ 12 & 8 & 6l-6l=0 & 4l^2+4l^2=8l^2 & -6l & 2l^2 \\ 0 & 0 & -12 & -6l & 12+C_1 & -6l \\ 0 & 0 & 6l & 2l^2 & -6l & 4l^2 \end{bmatrix} \qquad (5.12.11)$$

$$= [K]_{Global} = [K]$$

and $[K]\{\delta\} = \{R\}$

$$\frac{EI}{l^3} \begin{bmatrix} 12+C_1 & 6l & -12 & 6l & 0 & 0 \\ 6l & 4l^2 & -6l & 2l^2 & 0 & 0 \\ -12 & -6l & 24+2C_1 & 0 & -12 & 6l \\ 6l & 2l^2 & 0 & 8l^2 & -6l & 2l^2 \\ 0 & 0 & -12 & -6l & 12+C_1 & -6l \\ 0 & 0 & 6l & 2l^2 & -6l & 4l^2 \end{bmatrix} \begin{Bmatrix} \upsilon_1 \\ \theta_1 \\ \upsilon_2 \\ \theta_2 \\ \upsilon_3 \\ \theta_3 \end{Bmatrix} = \begin{Bmatrix} V_1 \\ M_1 \\ V_2 \\ M_2 \\ V_3 \\ M_3 \end{Bmatrix} = \{R\} \qquad (5.12.12)$$

where $\{R\}$ = Matrix of externally applied Nodal forces.

3. Boundary Condition:

Noting that the beam is free / free at nodes 1 and 3, the boundary conditions are
$$V_1 = V_3 = M_1 = M_3 = 0 \qquad (5.12.13)$$

From symmetry of the footing with respect to node 2,
$$\upsilon_1 = \upsilon_3$$
$$\theta_1 = -\theta_3 \qquad (5.12.14)$$
$$\theta_2 = 0$$

Substituting boundary conditions Eq. (5.12.13) and symmetry conditions Eq. (5.12.14) and loads shown in Figure 5.12.2(a) into the system equilibrium equations Eq. (5.12.12), the resulting equations become

$$\frac{EI}{l^3}\begin{bmatrix} \overset{\upsilon_1}{12+C_1} & \overset{\theta_1}{6l} & \overset{\upsilon_2}{-12} & \overset{\theta_2}{6l} & \overset{\upsilon_3}{0} & \overset{\theta_3}{0} \\ 6l & 4l^2 & -6l & 2l^2 & 0 & 0 \\ -12 & -6l & 24+2C_1 & 0 & -12 & 6l \\ 6l & 2l^2 & 0 & 8l^2 & -6l & 2l^2 \\ 0 & 0 & -12 & -6l & 12+C_1 & -6l \\ 0 & 0 & 6l & 2l^2 & -6l & 4l^2 \end{bmatrix}\begin{Bmatrix} \upsilon_1=\upsilon_3 \\ \theta_1=-\theta_3 \\ \upsilon_2 \\ \theta_2=0 \\ \upsilon_3=\upsilon_1 \\ \theta_3=-\theta_1 \end{Bmatrix}=\begin{Bmatrix} 0 \\ 0 \\ P_2=-1800kN \\ 0 \\ 0 \\ 0 \end{Bmatrix} \qquad (5.12.15)$$

All the stiffness matrices $[k]^1,[k]^2,[k]_{system}=[K]$ can be noted to be symmetric and positive definite.

These can be rewritten after simplifications as:

$$(12+C_1)\upsilon_1+6l\theta_1-12\upsilon_2=0 \qquad\qquad (5.12.16)$$

$$6l\upsilon_1+4l^2\theta_1-6l\upsilon_2=0 \qquad\qquad (5.12.17)$$

$$-12\upsilon_1-6l\theta_1+(24+2C_1)\upsilon_2-12\upsilon_3+6l\theta_3=P_2\frac{l^3}{EI} \qquad\qquad (5.12.18)$$

$$6l\upsilon_1+2l^2\theta_1-6l\upsilon_3+2l^2\theta_3=0 \qquad\qquad (5.12.19)$$

$$-12\upsilon_2+(12+C_1)\upsilon_3-6l\theta_3=0 \qquad\qquad (5.12.20)$$

$$6l\upsilon_2-6l\upsilon_3+4l^2\theta_3=0 \qquad\qquad (5.12.21)$$

Noting that $\upsilon_1=\upsilon_3$, $\theta_2=0$, $\theta_1=-\theta_3$ from symmetry, Eq. (5.12.19) becomes $0 = 0$ (automatically satisfied).

Eqs. (5.12.20) and (5.12.21) can be noted to be same as Eqs. (5.12.16) and (5.12.17) respectively. Hence, solving Eqs. (5.12.16), (5.12.17) and (5.12.18),

$$\upsilon_1=\frac{3\upsilon_2}{3+C_1} \qquad\qquad (5.12.22)$$

Substituting for υ_2 from Eq. (5.12.24)

$$\upsilon_1=\frac{3}{2C_1(6+C_1)}\frac{P_2l^3}{EI}$$

$$\theta_1=\frac{3C_1}{6+2C_1}\frac{\upsilon_2}{l}=\frac{3}{4(6+C_1)}\frac{P_2l^2}{EI} \qquad\qquad (5.12.23)$$

$$\upsilon_2=\frac{3+C_1}{2C_1(6+C_1)}\frac{P_2l^3}{EI} \qquad\qquad (5.12.24)$$

Thus, the nodal displacement matrix $\{\delta\}$ is obtained as

$$\{\delta\} = \begin{Bmatrix} \upsilon_1 \\ \theta_1 \\ \upsilon_2 \\ \theta_2 \\ \upsilon_3 \\ \theta_3 \end{Bmatrix} = \begin{Bmatrix} \dfrac{3}{2C_1(6+C_1)} \\[3mm] \dfrac{3}{4(6+C_1)l} \\[3mm] \dfrac{3+C_1}{2C_1(6+C_1)} \\[2mm] 0 \\[2mm] \dfrac{3}{2C_1(6+C_1)} = \upsilon_1 \\[3mm] \dfrac{-3}{4(6+C_1)l} = -\theta_1 \end{Bmatrix} \dfrac{P_2 l^3}{EI} \qquad (5.12.25)$$

Substituting the values of dimensionless relative stiffness constant C_1 (Eq. 5.12.4), the load P_2 at node 2, the element length l, the flexural rigidity of the footing EI in Eqs. (5.12.25), the nodal displacements and slopes can be computed. Once the matrix $\{\delta\}$ is known, the forces, moments, soil reaction can be computed for each element respectively from Eqs. (5.12.5), (5.12.6), (5.12.7). The soil reaction should be less than the allowable soil pressure for LSD (Eq. 5.12.1). Knowing the forces and moments, the bending moment diagram (BMD) and shear force diagram (SFD) can be drawn as usual with mechanics of solids conventions. The structural design of the footing can now be carried out using the SFD and BMD after ensuring that the computed maximum soil reaction is within the allowable limits (Eq. 5.12.1).

Since the above solutions are obtained for general values of C_1, P_2, l, EI, these can be used for general applications of such individual footings with minor adjustments wherever necessary. However, the method is illustrated for the present problem of the analysis of square footing.

4. Results for the square footing (4 m x 4 m)
For the Example 1, the specific values of the parameters are as follows.
$B = L = 4$ m (square)
No. of beam elements = 2, length of beam element, $l = 2$ m
k_s (applicable to 4 m x 4 m footing, from Eq. 5.12.2) = $2.63 \times 10^5 \, \text{kN/m}^3$ (5.12.26)
EI (from Eq. 5.12.4) = $8.412 \times 10^6 \, \text{kNm}^2$
$P_2 = -1800$ kN (from Eq. 1.1)

C_1 (from Eqn. 5.12.8) $= 2.63 \times 10^5 \times 4 \times \dfrac{2}{2} \times \dfrac{2^3}{8.412 \times 10^6} = 1$

Substituting the above values in Eqs. 5.12.25, we get,

$$\{\delta\} = \begin{Bmatrix} \upsilon_1 \\ \theta_1 \\ \upsilon_2 \\ \theta_2 \\ \upsilon_3 \\ \theta_3 \end{Bmatrix} = \begin{Bmatrix} -367.5 \times 10^{-6} \, m \\ -91.88 \times 10^{-6} \\ -490.014 \times 10^{-6} \, m \\ 0 \\ -367.5 \times 10^{-6} \, m \\ 91.88 \times 10^{-6} \end{Bmatrix} \qquad (5.12.27)$$

178

5. Element wise Nodal forces, SFD, BMD and soil reactions
Element 1:
These can be obtained from the element stiffness characteristics given by Eq. (5.12.9) after substituting the values of parameters given in Eq. (5.12.26) and nodal displacement from Eq. (5.12.27). All units are in kN and meters.

$$\frac{8.412 \times 10^6}{8} \begin{bmatrix} 12+1=13 & 12 & -12 & 12 \\ 12 & 16 & -12 & 8 \\ -12 & -12 & 12+1=13 & -12 \\ 12 & 8 & -12 & 16 \end{bmatrix} \begin{Bmatrix} -367.5 \\ -91.88 \\ -490.014 \\ 0 \end{Bmatrix} \times 10^{-6} = \begin{Bmatrix} V_1 \\ M_1 \\ V_2 \\ M_2 \end{Bmatrix} \begin{Bmatrix} 0 \\ 0 \\ -900kN \\ 773kNm \end{Bmatrix} \quad (5.12.28)$$

q_{s1} = soil reaction/ contact pressure at node 1 $= k_s \cdot v_1 = 2.63 \times 10^5 \times 367.5 \times 10^{-6} = 96.65 kN / m^2$

q_{s2} = soil reaction/ contact pressure at node 2 $= k_s \cdot v_2 = 2.63 \times 10^5 \times 490 \times 10^{-6} = 128.808 kN / m^2$

It may be noted that the vertical deflection distribution function varies as a cubic function in x (Eqs. 7.27, 7.37). In between, the soil reaction / contact pressure at any point:

$$q_s \text{ (from Eq.1.3)} = k_s v = k_s [N]\{\delta\} = k_s [N] \begin{Bmatrix} v_1 \\ \theta_1 \\ v_2 \\ \theta_2 \end{Bmatrix}$$

where $N = \left\{ 1-3\left(\frac{x}{l}\right)^2 + 2\left(\frac{x}{l}\right)^3 \quad x - 2\left(\frac{x^2}{l}\right) + \frac{x^3}{l^2} \quad 3\left(\frac{x}{l}\right)^2 + 2\left(\frac{x}{l}\right)^3 \quad \frac{-x^2}{l} + \frac{x^3}{l^2} \right\}$ (5.12.29)

Therefore, SFD, BMD and contact pressure are shown in Figure 5.12.3.

Element 2
The nodal forces, SFD, BMD and soil reactions / contact pressure can be computed as above using the corresponding element characteristics (Eq. 5.12.10)

$$\frac{8.412 \times 10^6}{8} \begin{bmatrix} 12+1=13 & 12 & -12 & 12 \\ 12 & 16 & -12 & 8 \\ -12 & -12 & 12+1=13 & -12 \\ 12 & 8 & -12 & 16 \end{bmatrix} \begin{Bmatrix} -490.014 \\ 0 \\ -367.5 \\ 91.88 \end{Bmatrix} \times 10^{-6} = \begin{Bmatrix} -900kN \\ -773kNm \\ 0 \\ 0 \end{Bmatrix} \quad (5.12.30)$$

q_{s2} = soil reaction at node 2 $= k_s \cdot v_2 = 128.808 \ kN/m^2$

q_{s3} = soil reaction at node 3 $= k_s \cdot v_3 = q_{s1} = 96.65 \ kN/m^2$

The soil reactions can be seen to be symmetric about node 2 (since displacements are symmetric).

Soil reaction at mid-point of element 1 is same as at mid-point of element 2 due to symmetry: $q_s (x = 1) = q_s (x = 3) = k_s \ v_{x1} = k_s \ v_{x3}$

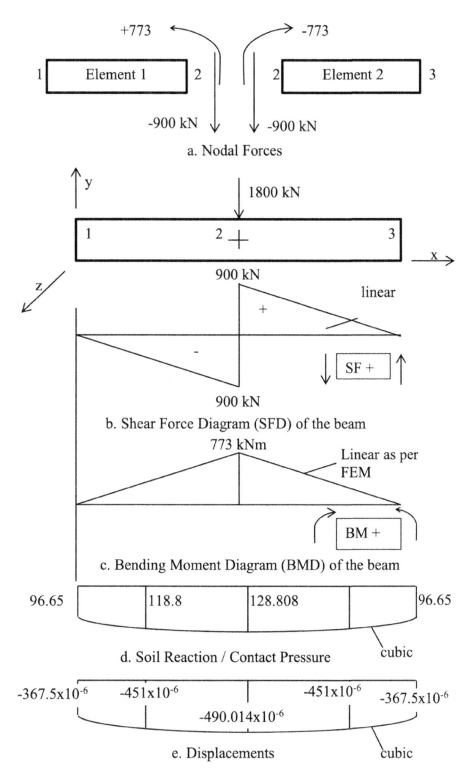

+773

−773

1 | Element 1 | 2

2 | Element 2 | 3

−900 kN

−900 kN

a. Nodal Forces

y

1800 kN

1 2 3

x

z

900 kN

linear

+

−

SF +

900 kN

b. Shear Force Diagram (SFD) of the beam

773 kNm

Linear as per FEM

BM +

c. Bending Moment Diagram (BMD) of the beam

96.65 | 118.8 | 128.808 | 96.65

cubic

d. Soil Reaction / Contact Pressure

-367.5×10^{-6} -451×10^{-6} -451×10^{-6} -367.5×10^{-6}

-490.014×10^{-6}

e. Displacements

cubic

Figure 5.12.3

From Eqs (5.12.29), (5.12.30) and (7.37),

$v(x=1) = v(x=3) =$

$$\left\{ 1-3\left(\frac{1}{2}\right)^2+2\left(\frac{1}{2}\right)^3 \quad 1-2\left(\frac{1}{2}\right)^2+\frac{1^3}{2^2} \quad 3\left(\frac{1}{2}\right)^2-2\left(\frac{1}{2}\right)^3 \quad \frac{1^2}{2}+\frac{1^3}{2^2} \right\} \begin{Bmatrix} -367.5 \\ -91.80 \\ -490.04 \\ 0 \end{Bmatrix} \times 10^{-6} = -451\times 10^{-6}$$

$q_s(x=1,3) = 2.63\times 10^5 \times 451 \times 10^{-6} = 118.8 kN/m^2$

In between the nodes, the displacements and the soil reactions / contact pressures vary as cubic functions. All units are in kN and meters.

Now the structural design can be carried out as per the SFD and BMD shown in Figure 5.12.3 using relevant codes and standards applicable for RCC design.

6. Results from Conventional Analysis

The results from conventional analysis are summarized below for comparison in Figure 5.12.4.

a. Soil reaction / contact pressure uniformly distributed

$$q_s = \frac{1800}{4\times 4} = 112.5 kN/m^2$$

- Uniformly distributed along the length of the beam as shown in Figure 5.12.4 a.

b. Displacement of footing

$$v_z = \frac{1800}{16\times k_s} = \frac{1800}{16\times 2.63\times 10^5} = -426.76\times 10^{-6}$$

- uniformly distributed as shown in Figure 5.12.4 b.

c. Shear force diagram (SFD)

Maximum at center $= \pm 112.5(4\times 2)$ at center

$= \pm 900$ - varying linearly as shown in Figure 5.12.4 c.

d. Bending Moment Diagram (BMD)

Bending Moment - Maximum at center $= 112.5\times 4\times \frac{2^2}{2} = 900 kNm$

- varying as a quadratic function as shown in Figure 5.12.4 d.

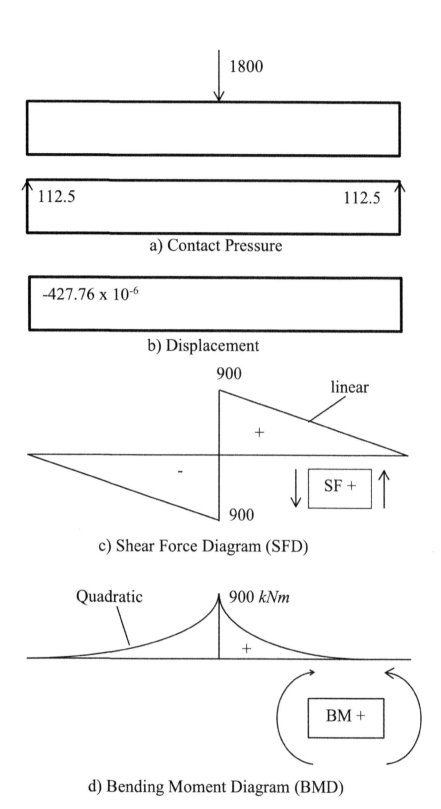

a) Contact Pressure

b) Displacement

c) Shear Force Diagram (SFD)

d) Bending Moment Diagram (BMD)

Figure 5.12.4

5.12.2 Further Illustration / Examples

Once the system matrix/global stiffness matrix is obtained, the same can be used for analysis of footing with different loads (including moments), different boundary conditions, parametric studies etc. Some of the many possible applications are illustrated below.

5.12.3 Footing for columns carrying loads and (or) moments (Figure 5.12.5)

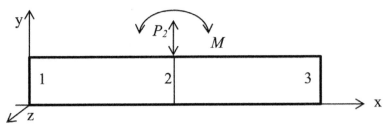

Figure 5.12.5 Footings Subjected to Loads and Moments

In this case of loading, the system equilibrium equation (5.12.15) gets modified as

$$[K]\{\delta\} = \{R\} \tag{5.12.31}$$

where $\{\delta\}^T = \left\{ \begin{array}{cccccc} v_1 & \theta_1 & v_2 & \theta_2 & v_3 & \theta_3 \end{array} \right\}$ $\qquad$ (5.12.32)

$\qquad \{R\}^T = \left\{ \begin{array}{cccccc} 0 & 0 & \pm P_2 & \pm M & v_3 & \theta_3 \end{array} \right\}$ $\qquad$ (5.12.33)

Since there may not be symmetry of the displacements, all the six components of displacement matrix $\{\delta\}$ have to be solved from the six-equilibrium equation (5.12.31) with $\{R\}$ matrix given by Eq. (5.12.33).

5.12.4 Footing with one end resting on rigid well / curtain well (Figure 5.12.6)

In this example, the unknown displacement and reactions have to be solved using Eq. (5.12.12) with the boundary conditions and unknown nodal forces as

$v_1 = 0$, V_1 = unknown reactions at node 1, $V_3 = 0$, $M_3 = 0$

Hence also, there are five unknowns in the displacements (Since $v_1 = 0$ is known) and one unknown reaction at node 1. All the six unknowns can be solved from resulting system equilibrium equation (5.12.12).

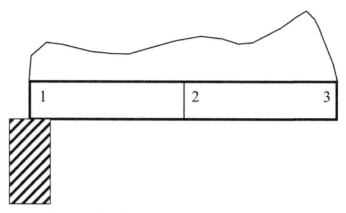

Figure 5.12.6 Footing with one end simply supported

5.12.5 Different sizes of foundations (Figure 5.12.7)

Maintaining two elements, the lengths can be different based on discretization (may be due to eccentric loads, foundation rehabilitation, non-homogeneous soils, varying loads and other possible field situations).

In this case, use $l = l_1$ for element 1, and $l = l_2$ for element 2 and then assemble $[k]$ matrix (Eq. 5.12.11) and system equilibrium equations (5.12.12). The resulting equations can then be solved for the unknowns after inserting the boundary conditions and nodal forces.

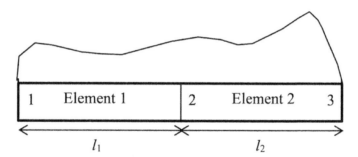

Figure 5.12.7 Footing discretized with different element lengths

5.12.6 Non-homogeneous and non-uniform soils (Figure 5.12.8)

In such cases, the relative soil stiffness contents C_1 and C_2 for the soils will be different. Hence, $[k]^1$ and $[k]^2$ can be calculated from Eqs. (5.12.9) and (5.12.10) with appropriate values of C_1, C_2 and l_1 and l_2 and even $[EI]_1$ and $[EI]_2$ if the footing has non uniform sizes as shown in Figure 5d. After getting $[k]^1$ and $[k]^2$, then the system matrices can be assembled as shown in Eqs. (5.12.11) and (5.12.12). The unknowns can then be solved from the resulting six equations.

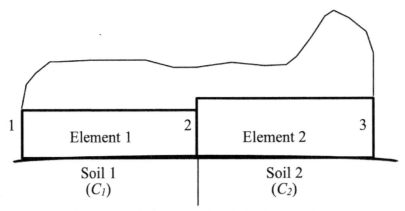

Figure 5.12.8 Foundations on soils with Non uniform properties

5.12.7 Combined footings, footings with more than one load/moment at different points

In such situations as illustrated in Figure 5.12.9, we have to introduce more elements. In such cases $[k]^1$, $[k]^2$, $[k]^3$ ... etc. have to be calculated using Eq. (5.12.5) and then assemble the system matrices like Eqs. (5.12.11) and (5.12.12). However, the number of degrees of freedom will increase at the rate of two per element and so also the nodal forces. It may be difficult to solve the higher order simultaneous equations manually.

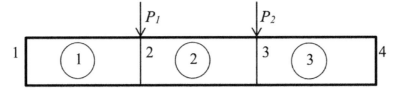

Figure 5.12.9 Combined footing

5.12.8 Distributed loads (Figure 5.12.10)

While consistent methods are more accurate to deal with distributed loads, it is computationally simple to lump the uniformly distributed loads equally as equivalent nodal forces as shown in Figure 5.12.10. These equivalent loads can be added to the nodal force matrix to solve for the unknowns using the system matrices Eqs. (5.12.11) and (5.12.12). If the loads are distributed differently (i.e., linearly varying on other patterns), the loads are judiciously lumped at both ends of each element and included in the nodal force matrix for solutions.

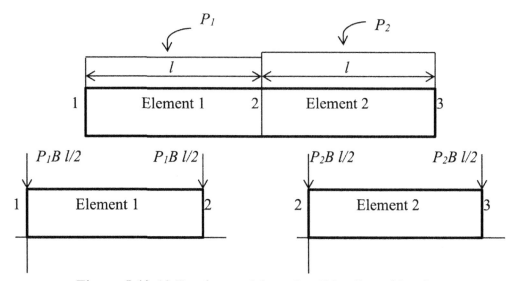

Figure 5.12.10 Footings subjected to Distributed loads

5.12.9 Examples of beams without foundation soil

(only 2 elements of equal length are used for illustration, Figure 5.12.11)

A few examples of beams without soil support below (beams without foundations or beams with negligible soil support). These examples can be helpful in extending the applications to beams with partial support / tensionless soil support / tensionless foundations /and other variations of the soil structure interaction problems.

Noting that the beam (shown in Figure 5.12.11) does not have the soil support (i.e. $k_s = 0$) and hence $(C_1 = 0)$. System matrices can be obtained from Eqs. (5.12.11) and (5.12.12) by substituting $C_1 = 0$ in these matrices.

Accordingly, Eq. (5.12.12) becomes

$$[K]\{\delta\} = \frac{EI}{l^3} \begin{bmatrix} \overset{v_1}{12} & \overset{\theta_1}{6l} & \overset{v_2}{-12} & \overset{\theta_2}{6l} & \overset{v_3}{0} & \overset{\theta_3}{0} \\ 6l & 4l^2 & -6l & 2l^2 & 0 & 0 \\ -12 & -6l & 24 & 0 & -12 & 6l \\ 6l & 2l^2 & 0 & 8l^2 & -6l & 2l^2 \\ 0 & 0 & -12 & -6l & 12 & -6l \\ 0 & 0 & 6l & 2l^2 & -6l & 4l^2 \end{bmatrix} \begin{Bmatrix} v_1 \\ \theta_1 \\ v_2 \\ \theta_2 \\ v_3 \\ \theta_3 \end{Bmatrix} = \begin{Bmatrix} V_1 \\ M_1 \\ V_2 \\ M_2 \\ V_3 \\ M_3 \end{Bmatrix} = \{R\} \qquad (5.12.34)$$

Once the system equilibrium equations are assembles as in Eq. (5.12.34), many problems involving beams can be solved taking into account all practical situations. These may include non-uniform beams, multiple and distributed loads, beams partially embedded in soils/solids, different combination of classical and non classical boundary conditions including spring supports/ sinking supports / guided supports / anchor supports etc. However, the beam with simply supported ends is illustrated below.

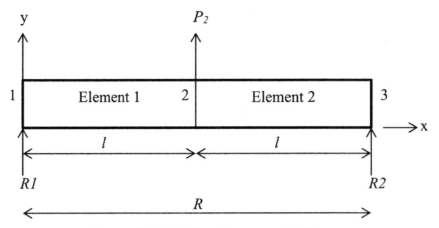

Figure 5.12.11 Simply supported beam

5.12.10 Simply supported beams (SSB)

For the simply supported (SS) beam shown in Figure 5.12.11, the system equilibrium equations (5.12.34) are rewritten for beams with 2 elements of equal length l as:

$$[K] \cdot \{\delta\} = R \tag{5.12.35}$$

Bounding conditions: For simply supported ends of nodes 1 and 3,

$$\upsilon_1 = 0 = \upsilon_3$$

$$V_1 = R_1, \quad M_1 = 0 \tag{5.12.36}$$

$$V_2 = R_2, \quad M_2 = 0$$

Symmetry conditions:
Noting that the beam is symmetric about node 2, these symmetry conditions are:

$$R_1 = R_3$$

$$\theta_1 = -\theta_3 \tag{5.12.37}$$

$$\theta_2 = 0$$

Taking P_2 to be a negative load $= -P$, acting in the opposite direction of y, and substituting these conditions in Eqs. (5.12.34) / (5.12.35), we get

$$6l\theta_1 - 12\upsilon_2 = R_1 \frac{l^3}{EI} \qquad \text{(a)}$$

$$4l\theta_1 - 6\upsilon_2 = 0 \qquad \text{(b)}$$

$$-6l\theta_1 + 24\upsilon_2 + 6l\theta_3 = -P\frac{l^3}{EI} \qquad \text{(c)} \tag{5.12.38}$$

$$\theta_1 + \theta_3 = 0 \qquad \text{(d)}$$

$$-12\upsilon_2 - 6l\theta_3 = R_3 \frac{l^3}{EI} \qquad \text{(e)}$$

$$6\upsilon_2 + 4l\theta_3 = 0 \qquad \text{(f)}$$

It may be noted that Eq. (5.12.38(d)) is the same symmetry condition as in Eq. (5.12.37) and Eq. (5.12.38(f)) is the same as Eq. (5.12.38(b)) taking $\theta_1 = -\theta_3$, Eqs. (5.12.38(a)) and (5.12.38(e)) are expressions for the reactions R_1 and R_3 ($R_1 = R_3$). Hence the unknowns θ_1 and v_2 can be solved from Eqs. (5.12.38(a)) and (5.12.38(c)) as

$$v_2 = -\frac{Pl^3}{6EI}, \quad \theta_1 = -\frac{Pl^2}{4EI}, \quad R_1 = R_3 = \frac{P}{2} \tag{5.12.39}$$

Equations (5.12.39) can be noted to be the same expression obtained using Euler-Bernoulli's bending theory of beams, i.e.

$$v_2 = -\frac{Pl^3}{6EI} = \frac{-PL^3}{48EI} \text{ (since } l = \frac{L}{2} \text{)}$$

$$\theta_1 = -\theta_3 = \frac{-Pl^2}{4EI} = \frac{-PL^2}{16EI} \tag{5.12.40}$$

5.12.11 Element Nodal forces, shear force diagram (SFD) and bending moment diagram (BMD)

Element 1:
From Eqn. (2.5) and taking $C_1 = 0$ (as there is no soil foundation below),

$$\frac{EI}{l^3} \begin{bmatrix} 12 & 6l & -12 & 6l \\ 6l & 4l^2 & -6l & 2l^2 \\ -12 & -6l & 12 & -6l \\ 6l & 2l^2 & -6l & 4l^2 \end{bmatrix} \begin{Bmatrix} v_1 = 0 \\ \theta_1 = \dfrac{-Pl^2}{4EI} \\ v_2 = \dfrac{-Pl^3}{6EI} \\ \theta_2 = 0 \end{Bmatrix} = \begin{Bmatrix} V_1 \\ M_1 \\ V_2 \\ M_2 \end{Bmatrix} \tag{5.12.41}$$

From the above we can simplify,

i.e. $V_1 = \dfrac{P}{2}, \quad M_1 = 0, \quad V_2 = \dfrac{-P}{2}, \quad M_2 = \dfrac{PL}{2}$

Then nodal forces are shown in Figure 5.12.12.

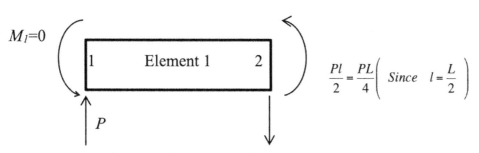

Figure 5.12.12 Nodal forces for Element 1

Element 2:

From Eq. (5.12.9) and noting $C_1 = 0$ (as there is no soil foundation below), we get

$$\frac{EI}{l^3}\begin{bmatrix} 12 & 6l & -12 & 6l \\ 6l & 4l^2 & -6l & 2l^2 \\ -12 & -6l & 12 & -6l \\ 6l & 2l^2 & -6l & 4l^2 \end{bmatrix}\begin{Bmatrix} v_2 = \dfrac{-Pl^3}{6EI} \\ \theta_2 = 0 \\ v_3 = 0 \\ \theta_3 = \dfrac{P_2 l^2}{4EI} = -\theta_1 \end{Bmatrix} = \begin{Bmatrix} V_2 \\ M_2 \\ V_3 \\ M_3 \end{Bmatrix} \qquad (5.12.42)$$

After simplifications of Eq. (5.12.42), we get

$$V_2 = -\frac{P}{2}, \quad V_3 = \frac{P}{2}, \quad M_2 = \frac{-Pl}{2} \qquad (5.12.43)$$

The fourth row of Eq. (5.12.42) is automatically satisfied noting that from symmetry $\theta_1 = -\theta_3$.

The nodal forces of element 2 are shown in Figure 5.12.13.

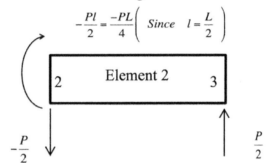

Figure 5.12.13 Nodal forces for Element 2

The SFD and BMD for the entire beam are shown in Figure 5.12.14.

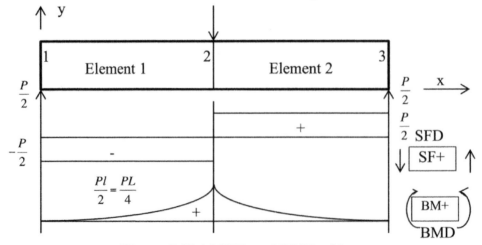

Figure 5.12.14 SFD and BMD of beam

BM and SF can also be obtained all along the beam using the corresponding FEM expressions for beam elements given by Eqs. (7.37), (7.39) and (7.44).

Accordingly,

$$M_b = EI\frac{d^2v}{dx^2} = EI[B]\{\delta\} = EI\begin{bmatrix} B_1 & B_2 & B_3 & B_4 \end{bmatrix}\begin{Bmatrix} v_i \\ \theta_i \\ v_j \\ \theta_j \end{Bmatrix} \tag{5.12.44}$$

in which

$$B_1 = -\frac{6}{l^2} + \frac{12x}{l^3}; \quad B_2 = -\frac{4}{l} + \frac{6x}{l^2}; \quad B_3 = \frac{6}{l^2} - \frac{12x}{l^3}; \quad B_4 = -\frac{2}{l} + \frac{6x}{l^2} \tag{5.12.45}$$

and shear force $V_F = -EI\frac{d^3v}{dx^3} = -EI\frac{d}{dx}\{B\}\{\delta\}$ \hfill (5.12.46)

$$= -EI\begin{Bmatrix} \dfrac{12}{l^3} & \dfrac{6}{l^2} & -\dfrac{12}{l^3} & \dfrac{6}{l^2} \end{Bmatrix}\begin{Bmatrix} v_i \\ \theta_i \\ v_j \\ \theta_j \end{Bmatrix} \tag{5.12.47}$$

These values for elements (1) and (2) can be obtained from Eqs. (5.12.44) to (5.12.47) as:

Element 1: $0 \leq x \leq l$

$$M_{b1} = +\frac{Px}{2}$$
$$V_{F1} = -\frac{P}{2} \tag{5.12.48}$$

Element 2: $0 \leq x \leq l$

$$M_{b2} = \frac{P}{2}(l - x)$$
$$V_{F2} = \frac{P}{2} \tag{5.12.49}$$

It may be noted from Eqs. (5.12.48) and (5.12.49), M_b varies linearly with x and V_F is constant in each element.

5.13   Assignment Problems

Note:
1.   All the problems are based on the methods explained in Chapter 5 and worked out examples given in Section 5.12.
2.   Analysis the assignment problems using FEM analysis using BEF model and by conventional analysis for comparison of all the results. Use a maximum of 2 to elements for carrying out calculations manually / calculator.
3.   Solve for all the responses of the foundation and compare the results from both the methods. Analyze the problems using any software package if possible and compare the results wherever possible.
4.   Assume the following data for foundation and soil unless specified otherwise in the problem.
    Depth of foundation, $d = 0.8$ m
    Modulus of elasticity of concrete, $E_c = 20 \times 10^6$ kN/m^2
    Modulus of elasticity of soil, $E_s = 4 \times 10^5$ kN/m^2
    Poisson's ratio of soil, $v_s = 0.3$
    Unit weight of soil, $\gamma_s = 18$ kN/m^3
    Allowable soil pressure / bearing capacity of soil, $q_{allowable} = 150$ kN/m^2
5.   Assume the foundation to be a rigid plate for evaluation of modulus of subgrade reaction k_s.
6.   Assume the loads and moments acting on the foundation / transmitted thorough column (s) or other super structures to be net design loads including load factors etc.
7.   Assume any other missing data needed for the FEM analysis and conventional analysis. Typical convention sketches of individual and combined footings are shown below for reference if necessary.

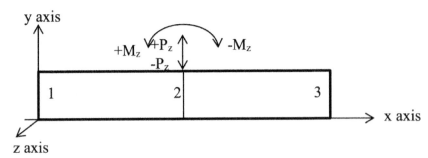

Figure 5.13.1 Convention sketch of spread / individual footing with concentrated loads and moments

Square footings:
1.   Analyze a square footing / foundation of size 4m x 4m to carry a central concentrated   vertically downward load of $P_z = P_2 = -1200$ kN transmitted through a column.

2.   Redo the above problem if the load is eccentrically applied with eccentricity $e_x = +0.3$m.

3. Redo the above problem 1 with an additional moment of $M_z = M_2 = +200$ kNm transmitted the column in addition to the concentrated load of 1200 kN.

4. Redo the problem 3 when the column is eccentrically placed with eccentricity of $e_x = +0.3$m. Compare the results of problems 3 and 4 and comment on the r.

5. Analyze problem 4 with the right-hand end of the foundation ($x = 4$m) is resting on a rigid support like a curtain wall.

6. Analyze problem 5 if the right hand support ($x = 4$m) has settled by 0.5m.

7. Analyze problem 4 if the beam depth is increased to 1.2m from $x = 2.3$m to 4m while it is 0.8m from $x = 0$ to 2.3m.

8. Redo the problem 7 with the modulus of subgrade reaction k_s varies along x axis as:
   i. $k_s = k_{s1}$ (as in problem 1 with total area of the foundation being 16 m^2) from $x = 0$ to 2.3m, $k_s = 1.5\ k_{s1}$
   ii. k_s varies linearly from k_{s1} at $x = 0$ to 1.5 k_{s1} at x = 4m.

9 to 32. Redo problems with the addition of a uniformly distributed downward load on the as follows.
   i. -20 kN / m from $x = 0$m to $x = 2$m and -40 kN / m from $x = 2$m to $x = 4$m
   ii. -50 kN / m from $x = 2$m to $x = 4$m
   iii -60 kN / m from $x = 0$m to $x = 3$m

Rectangular footings, combination of square and rectangular footings, Trapezoidal footings:

17 to 64 . Redo the problems 1 to 32 using a rectangular footing (instead of square footing) of size 4 m x 3 m along x and y directions respectively.

65 to 96. Redo the problems 1 to 32 with the footing with varying sizes as below 2 m x 2 m from $x = 0$ to $x = 2$m and 2m x 4m along x and y directions respectively.

97 to 128. Redo the problems 1 to 32 using a trapezoidal footing (instead of a square footing) with width of the footing varying uniformly from 3 m to 4 m from $x = 0$ to $x = 4$m.

Combined footings (Figure 5.13.2)

129. Analyze a combined footing / foundation of size 12 m x4 m (along x and y axes respectively) to carry central concentrated vertically downward loads of $P_1 = -1600$ kN at $x = 4$m and $P_2 = -2400$ kN at $x = 8$m transmitted through two columns.

130. Redo the above problem 129 if the loads are applied at $x = 3$m and $x = 9$m.

131. Redo the above problem 129 with an additional moment of +1000 kNm and -1600 kNm along with the loads P_1 and P_2.

132. Redo the problem 130 with an additional moment of +1200 kNm and -1400 kNm along with the loads P_1 and P_2.

133.	Analyze problem 132 with the left-hand end of the foundation ($x = 0$ m) is resting on a rigid support like a curtain wall.
134.	Analyze problem 133 if the left-hand support ($x = 0$m) has settled by 0.6m
135.	Analyze problem 132 if the beam depth is increased to 1.2m from x = 3m to 8 while it is 0.8m from x = 0 to 3m and x = 9m to 12m.
136.	Redo the problem 135 with the modulus of subgrade reaction k_s varies along x axix as:

 i. $k_s = k_{s1}$ (as in problem 65 with total area of the foundation being 48 m²) from $x = 0$ to 3m, $k_s = 1.5\ k_{s1}$ from $x = 3$m to 9m and $k_s = 2\ k_{s1}$ for $x = 9$m to 12m.

 ii. k_s varies linearly from k_{s1} at $x = 0$ to $2\ k_{s1}$ at $x = 12$m.

137 to160. Redo problems 129 to 136 with the addition of a uniformly distributed downward load on the as follows.

 i.   -20 kN/m from $x = 0$ m to $x = 4$ m and -40 kN/m from $x = 4$ m to $x = 8$ m and -60 KN from $x = 8$ m to 12 m

 ii.   -50 kN / m from x = 8m to x = 12m

 iii.   -60 kN / m from x = 0m to x = 12m

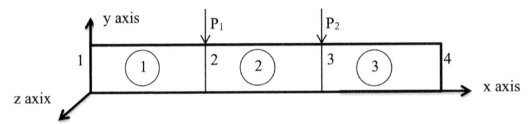

Figure 5.13.2 Sketch of a combined footing

Appendix 5A

Evaluation of Modulus of Subgrade Reaction, k_s and Elastic Properties of Soil, E_s, v_s using Plate Load Test

5A.1 Evaluation of Modulus of Subgrade Reaction of the Soil, k_s – Plate Load Test

The idea of modeling soil as an elastic medium was first introduced by Winkler (1888) and this representation of soil is referred to as the Winkler's soil model. The subgrade reaction at any point along the structure is assumed to be directly proportional to its vertical displacement at that point. In other words, the soil is assumed to be elastic and obeys Hooke's Law. Hence, the modulus of subgrade reaction (k_s) for the soil is given by

$$k_s = \frac{q}{w} = C_u \qquad (5A.1)$$

where, q= bearing / contact pressure / soil reaction at any point of structure –soil interface / contact area and w = the vertical deflection at that point of interface

Then, Winkler's spring constant of the soil, $k = C_u A = k_s A$

where q is the (vertical) bearing pressure at a point along the structure, and w is the vertical displacement of the structure at that point. k_s is also referred to as coefficient of elastic uniform compression C_u. A is the contact area of the plate /foundation.

The main difficulty in applying the Winkler's soil model is that of quantifying the modulus of subgrade reaction (k_s) to be used in the analysis, as soil is a very variable material. In practical terms, k_s can be evaluated either by carrying out in-situ plate load tests or appropriate laboratory tests or relating it in some way to elastic characteristics of the soil. The plate load test is widely used in practice and is described in BS 1377: Part 9: 1990, IS: 1888, 1982 (Kameswara Rao, 2011 etc). Plate load test is described in detail in the above references. A few important aspects are summarized below. The test set up is also shown in Figure 5A.1.

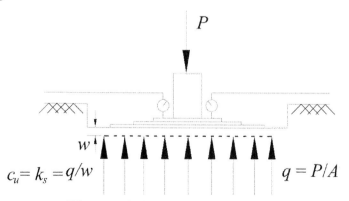

Figure 5A.1 Plate load test setup

The plate should obviously be as large as possible, consistent with the requirements and being able to exert the vertical forces required. The standard plate is either of circular shape of 760mm diameter or 760mm x 760mm square shape, 16mm thick and requires stiffening by means of other circular / square plates placed concentrically above it. Invariably, a large plate does not settle uniformly. The settlement must, therefore, be monitored by means of three or four dial gauges equally spaced around the perimeter in order to determine the mean settlement. Supports for these dial gauges should be sited well outside the zone of influence of the jacking load which is measured by a proving ring. When choosing a diameter of plate to use for the test, due consideration should also be given to the limited zone of influence of the loaded plate. Typically, the soil will only be effectively stressed to a depth of 1.25 to 1.50 times the diameter of the plate. This limitation can be overcome to some extent by carrying out the plate test at depth in pit, rather than on the surface. Small diameter plates are often used to overcome the practical difficulties of providing the requisite reaction/vertical forces. Terzaghi (1955) used 305mm square plate for evaluation.

Figure 5A.2 shows a typical plot of q against w that would be obtained from a plate bearing test. In Foundation design, as distinct from pavement design, the value of k_s is the secant modulus of the graph over the estimated working range of bearing pressure as indicated in Figure 5A.2. The value of the modulus of subgrade reaction (k_s) obtained from the test varies according to the size of plate used. Figure 5A.3 shows the variation of k_s with plate diameter based on experimental evidence. It is apparent, therefore, that k_s depends not only on the deformation characteristics of the soil but also on the size of contact area between plate and subgrade. The variation of k_s with plate size creates an obvious difficulty in deciding which plate size should be used as the standard or reference for defining values of k_s for analysis.

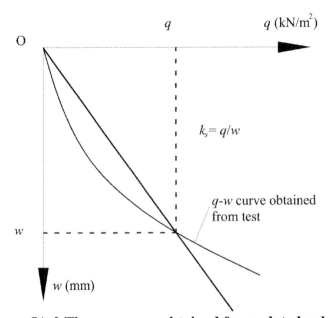

Figure 5A.2 The q-w curve obtained from plate load test

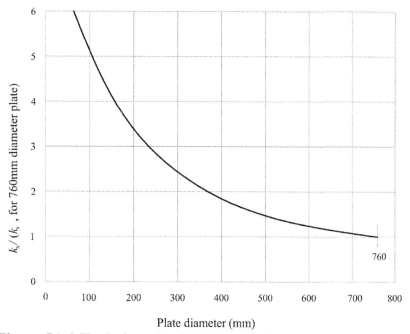

Figure 5A.3 Variation of subgrade reaction with plate diameter

Furthermore, due account must also be taken of the size and geometry of the loaded area. Terzaghi (1955) made several useful recommendations to overcome these difficulties. Basically, he first proposed reference values of k_s for sands and clays based on plate bearing tests carried out using a 305mm square plate. He then advocated methods of conditioning these values to allow for the geometry of the base. His recommendations are presented in Section 5A.2. As most plate bearing tests are carried out using circular plates, it is necessary to relate the performances of circular and square plates in order to follow Terzaghi's recommendations. The theoretical relationship between values of k_s obtained from plate bearing tests using circular and square plates can be derived as follows.

Square Plate

The mean settlement (w_{sp}) is given by

$$w_{sp} = \frac{0.95(1 - v_s^2)qB}{E_s} \qquad (5A.2)$$

where v_s is the Poisson's ratio of the soil, q is the average bearing pressure under plate, B is the side of square plate, and E_s is the modulus of elasticity of the soil.

Circular Plate

The mean settlement (w_{cp}) for the same value of q is

$$w_{cp} = \frac{0.85(1 - v_s^2)qB}{E_s} \qquad (5A.3)$$

where B is the diameter of the plate.

196

It can be noted from Eqns. (5A.2) and (5A.3) that

$$k_{sp} = 0.895 \, k_{cp} \tag{5A.4}$$

where the suffixes *sp* and *cp* refer to square plate and circular plate respectively.

Horizontal plate load tests can also be carried out in trial pits to obtain corresponding values of the horizontal modulus of subgrade reaction (k_h) which is relevant to the analysis of laterally loaded piles, pile groups and sheet piling etc.. More information about k_h is given in Kameswara Rao (2011) and other references.

Values of k_s for long beams may also be assessed by relating them to the intrinsic parameters of the soil such as elastic modulus (E_s), Poisson's ratio (v_s) and California Bearing Ratio (CBR). E_s and v_s can both be derived from the results of triaxial tests. Salvadurai (1979) developed the following expressions for k_s in terms of E_s and v_s for beams having a L/B ratio $\geq$ 10, where L is the length and B is the breadth of beam.

$$k_s = \frac{0.65 \, E_s}{B(1-v_s^2)} \tag{5A.5}$$

$$k_s = \frac{\pi \, E_s}{2B(1-v_s^2)\log_e\left(\dfrac{L}{B}\right)} \tag{5A.6}$$

The two expressions are in close agreement for values of *L/B* in the range 10 to 13. An approximate, empirical relationship between the modulus of subgrade reaction (k_s) obtained using the standard 760mm diameter plate and the California Bearing Ratio (CBR) for soils that are uniform in depth is plotted in Figure 5A.4.

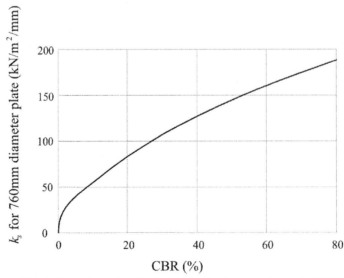

Figure 5A.4 Empirical relationship between k_s and CBR value

5A.2 Size of Contact Area

It is evident from Figure 5A.3 that the value of modulus of subgrade reaction (k_s) varies according to the size of the plate used in the plate bearing test. Similarly, k_s varies with the breadth B of a continuous beam resting on an elastic subgrade. This fact was first reported by Engesser (Jones, 1997) when he confirmed that the value of k_s decreases with increasing width (B) of the beam. Terzaghi (Taylor, 1964) also investigated this phenomenon and derived expressions for beams supported by both cohesionless and cohesive soils (Kameswara Rao, 2011).

Vesic, 1961 (Bowles, 1996) proposed an expression for k_s in terms of E_s and ν_s (of the soil) as

$$k_s = \frac{1}{B}\left[0.65 \sqrt[12]{\frac{E_s B^4}{E_f I_f}}\right] \frac{E_s}{1-\nu_s^2} \approx \frac{E_s}{B(1-\nu_s^2)} \tag{5A.7}$$

where B, I_f, E_f = width, moment of inertia of the cross section and modulus of elasticity
of the footing respectively.

E_s, ν_s = modulus of elasticity and Poisson's ratio of the soil.

One can also adopt the expression of k (neglecting t_l) from Vlasov's elasticity model (Eq. 5.17a – Chapter 5) with appropriate choice of $\psi(z)$ (Vlasov and Leontev, 1966).

5A.3 Winkler's Soil Medium with or without Tension

The common assumption made in conventional methods of analyzing loaded continuous beams resting on a horizontal subgrade is that tension is not allowed to develop between the beam and the underlying subgrade. It is therefore, necessary when using beams on elastic foundation approach to condition the Winkler soil medium to detach springs which are not in compression under the action of the applied loading under consideration. However, in most other applications, tension in the springs is allowed. Comparisons between the results of analyses carried out using these two conditions for the same foundation beam problem usually differ by only small amounts (Jones, 1997).

5A.4 Sensitivity of Responses on k_s

When using the beam on elastic foundation concept to analyze foundation problems, it is imperative that a range of values of the modulus of subgrade reaction (k_s) are tested to ascertain the sensitivity of soil parameter in the analysis. Usually, the resulting bending moments and shear forces are not sensitive to changes in the value of k_s.

5A.5 Modulus of Subgrade Reaction for different Plate Sizes and Shapes

The same plate load test results described in Section 5A.1 can be used for obtaining k_s of different plate sizes and shapes (some of these have been discussed in the above sections) using theory of elasticity solutions (Kameswara Rao, 2011) as follows.

For a rigid circular plate of area A on an elastic half space subjected to vertical load,

$$k_s = C_u = 1.13 \frac{E_s}{1-v_s^2} \cdot \frac{1}{\sqrt{A}} \qquad (5A.8)$$

where E_s and v_s are modulus of elasticity and Poisson's ratio of the soil.

C_u = coefficient of elastic uniform compression / modulus of subgrade reaction.

For a rectangular plate of sides a and b on an elastic half space, subjected to vertical load, k_s can be expressed as

$$k_s = c_f \frac{E_s}{(1-v_s^2)} \frac{1}{\sqrt{A}} \quad \text{for flexible plate} \qquad (5A.9)$$

$$k_s = c_r \frac{E_s}{(1-v_s^2)} \frac{1}{\sqrt{A}} \quad \text{for rigid plate} \qquad (5A.10)$$

where A = area of the plate = ab,

c_f, c_r = shape constants depending on the flexibility or rigidity of the test plate used.

E_s, v_s = modulus of elasticity and Poisson's ratio of the soil medium respectively. The values of c_f and c_r are given in Table 5A.1 for ready reference.

It can be seen from Table 5A.1 and Eqs. (5A.9) and (5A.10), that the value of k_s is not much dependent on the flexibility or rigidity of the plate. It can be further observed from Eqns. (5A.8), (5A.9) and (5A.10) that k_s is proportional to $(A)^{-1/2}$, i.e.

$$\frac{k_{s1}}{k_{s2}} = \sqrt{\frac{A_2}{A_1}} \qquad (5A.11)$$

where k_{s1} is the value corresponding to a bearing plate of area A_1 and k_{s2} is the value corresponding to a bearing plate of area A_2. From Eq. (5A.11), it is evident that if the value of k_{s1} corresponding to a plate area A_1 is known, k_{s2} corresponding to any other plate area A_2 can be easily evaluated. It is convenient to convert all the test results and present the values of corresponding to a bearing plate area of 10 m² (Kameswara Rao, 2011). The average values of k_s (corresponding to plate area of 10 m²) for different soils are given in Table 5A.2.

Table 5A.1 Coefficients c_f and c_r

Shape of the plate	$\frac{a}{b}$	c_f	c_r
Circular	---	---	1.13
Square	1.0	1.06	1.08
Rectangular	1.5	1.07	---
	2.0	1.09	1.10
	3.0	1.13	1.15
	5.0	1.22	1.24
	10.0	1.41	1.41

Table 5A.2 Average Values of k_s for Different Soils
(Corresponding to a plate area of 10 m²)

Soil Description	Permissible Pressure on Soil, kN/m²	k_s kN/m³
Gray plastic silty clay with sand & organic salt	98	137.34 x 10²
Brown saturated silty clay with sand	147.15	196.20 x 10²
Dense silty clay with some sand (above ground water level)	Up to 490.5	1049.67 x 10²
Medium moist sand	196.2	196.20 x 10²
Dry sand with gravel	196.2	196.20 x 10²
Fine saturated sand	245.25	294.30 to 343.35 x 10²
Medium sand	245.25	304.11 x 10²
Gray fine dense, saturated sand	245.25	333.54 x 10²
Loess with natural moisture content	294.3	441.45 x 10²
Moist loess	196.2	461.07 x 10²

These results can also be summarized as presented in Table 5A.3 (Kameswara Rao, 2011).

5A.6 Poisson's Ratio ν_s of the Soil Medium

It is possible to evaluate Poisson's ratio, ν_s using dynamic test results such as wave propagation tests etc. (Kameswara Rao, 2011). Also, Poisson's ratio can be evaluated using some simple soil tests which may result in large errors. It has been observed that in general Poisson's ratio varies from about 0.25 to 0.35 for cohesionless soils and form about 0.35 to 0.45 for cohesive soils which are capable of supporting foundation blocks. Hence, in the absence of any test data Poisson's ratio can be assumed as 0.3 for cohesionless soils and 0.4 for cohesive soils without causing any appreciable error in the analysis and design of foundations (Kameswara Rao, 2011).

This is further justifiable from the fact that the responses to foundation soil systems have been found to be not much sensitive to the variations in the value of Poisson's ratio, ν_s.

5A.7 Evaluation of Young's Modulus of Elasticity of Soil, E_s

While it is ideal to evaluate Young's modulus of elasticity, E_s (or shear modulus of soil, μ_s) from dynamic tests and other appropriate field and laboratory tests, often it may not be possible to do so due to physical limitations and costs.

Table 5A.3 Values of k_s and μ_s for different soil categories (assuming $v_s = 0.3$)

Soil Category	Soil Description	Permissible Pressure on soil, kN/m^2	k_s for A=10 m^2 kN/m^3	Shear Modulus, μ_s (kN/m^2)
I	Weak soils (clay and silty clays with sand in a plastic state)	upto 147.15	98.1 x 10^2 196.2 x 10^2 294.3 x 10^2	100.06 x 10^2 201.11 x 10^2 301.17 x 10^2
II	Soils of medium strength (clays and silty clays with sand close to plastic limit)	147.15 to 343.35	294.3 x 10^2 392.4 x 10^2 490.5 x 10^2	301.17 x 10^2 401.23 x 10^2 502.27 x 10^2
III	Strong soils (clays and silty clays with sand of hard consistency; Gravels and gravelly Sand; loess and loessial soils)	343.35 to 490.5	490.5 x 10^2 583.6 x 10^2 686.77 x 10^2 784.8 x 10^2 882.9x 10^2 981 x 10^2	502.27 x 10^2 602.33 x 10^2 702.40 x 10^2 803.44 x 10^2 903.50 x 10^2 1003.56 x 10^2
IV	Rocks	> 490.5	> 981 x 10^2	>1003.56x 10^2

However, Modulus of Elasticity, E_s can also be evaluated from the results of cyclic plate load test. It is obvious that E_s can be calculated (and hence shear modulus $\mu_s = E_s / (2(1 + v_s))$ from Eqns. (5A.8), (5A.9), and (5A.10), knowing the value of k_s, and the shape and size of the plate used (flexible or rigid plate solutions do not make much difference as can be seen from Table (5A.1). Poisson's ratio can be assumed to be either 0.3 or 0.4 as mentioned above (Section 5A.6), or evaluated form appropriate field and (or) laboratory tests.

For a square plate of 30 cm side (fairly rigid), the values of shear modulus, μ_s for different soil categories listed in Table 5A.3, can be computed using Eq. (5A.10) with c_r = 1.08 (from Table 5A.1) and $v_s = 0.3$. The same are presented in Table 5A.3.

5A.8   Evaluation of k_s for Foundations subjected to Dynamic Loads

For the design of foundations subjected to dynamic loads such as in machine foundations, modulus of subgrade reaction k_s (also referred to as coefficient of elastic uniform compression, C_u) has to be obtained from a cyclic plate load test (instead of a plate load test) or wave propagation tests (Kameswara Rao, 2011) and extensive literature is available on this topic. Some important details of cyclic plate load test are summarized below.

5A.9 Cyclic Plate Load Test

As the name itself indicates, this is a modified version of the standard plate load test as shown in Figure 5A.1. The test is conducted using either standard plate (square of circular shape of 0.305 m to 0.760 m size or any other size) and involves several loading cycles (loading, unloading and reloading schedule), as per standard practice (Kameswara Rao, 2011). This facilitates in separating the elastic (recoverable) part of the deformation from the plastic part (irrecoverable part or permanent set) which in turn can be related to the Young's modulus of elasticity, E_s of the soil.

Some Salient Features of Cyclic Plate Load Test:

The bearing pressure-settlement curve obtained from a typical cyclic plate load test is shown in Figure 5A.4. After each load application, sufficient time is allowed to ensure that the settlement has attained a final value for all practical purposes at that load level. In Figure 5A.4, the recoverable part of the settlement (during unloading) represents the elastic part and the non-recoverable part signifies the plastic settlement (permanent set). The elastic part of the settlement is plotted as a function of average contact pressure (bearing pressure) in Figure 5A.6 and the relationship is observed to be generally linear. The slope of this curve is referred to as dynamic modulus of subgrade reaction k_s or dynamic coefficient of elastic uniform compression, C_u.

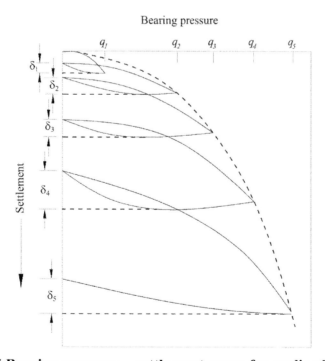

Figure 5A.5 Bearing pressure – settlement curve for cyclic plate load test

Dynamic Modulus of Subgrade Reaction, k_s and Spring Constant, k:
From Figure 5A.5, the slope of the curve is referred to as coefficient of elastic uniform compression, C_u (or k_s) and can be expressed as

$$k_s = C_u = \frac{q}{w} \tag{5A.12}$$

where q is the bearing pressure (load per unit area), and w is the elastic vertical settlement / displacement. Then the spring constant for vertical deformation is given by

$$k = C_u A = k_s A \tag{5A.13}$$

where A = contact area of plate, i.e., bearing area.

All other expressions discussed in Sections. 5A.5, 5A.6, 5A.7 remain the same except that in dynamic situations k_s ($=C_u$) obtained from cyclic plate load test discussed above is recommended to be used instead of k_s from normal plate load test.

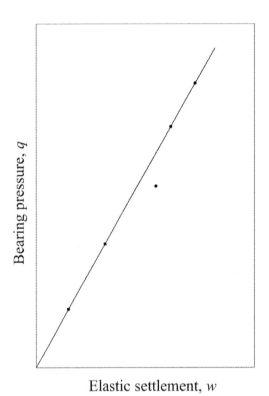

Elastic settlement, w

Figure 5A.6 Determination of k_s (C_u) from cyclic plate load test data

5A.10 Summary

Plate load test is commonly used for determination of modulus of subgrade reaction of soil, k_s and elastic parameters of the soil, E_s and v_s (usually assumed). Following points may be noted for applications in analysis.

1. If k_s of one plate size is calculated using plate load test or other tests, it can be calculated for any other plate size / area of actual foundation being analysed using

Eqns. (5A.8) to (5A.11). Only the k_s applicable to the actual foundation contact area is to be calculated using either the soil parameters E_s and v_s, or from plate load test results of known plate area and recalculating the value as per equation (5A.11).

2. Knowing k_s, elastic properties of the soil can be calculated from Eqns. (5A.8) to (5A.10). Usually Poisson's ratio, v_s is assumed between 0 to 0.5 (though it can be estimated from field and laboratory experiments), for applying Eqns. (5A.8) to (5A.10) to calculate modulus of elasticity of soil, E_s. Knowing k_s the spring constant, k to be used in the foundation analysis can be calculate using Eq. (5A.1).

3. If E_s and v_s of the soil are known from other tests, k_s can be evaluated using Eqns. (5A.8) to (5A.10).

4. There are various empirical, semi empirical, analytical expressions available for evaluating k_s, E_s, v_s and other related parameters in literature besides the methods discussed in the above sections. The choice of input parameters has to be judiciously made based on the available options, guidelines, reliability of data, compatible with the method of analysis used for optimum accuracy of results.

Chapter 6

FEM for the Analysis of Field Problems - Seepage / Flow through Porous Media

6.1 Introduction

The water / fluid which flows through the pores of any soil / porous media under external stresses is an important area in soil mechanics / Civil Engineering, fluid mechanics and other related areas. Such a flow in soil media is commonly referred to as seepage in soils or, flow through porous media for broader applications in the above areas. While the terminology and the flow phenomena are mostly similar in theory, governing equations, methods of solution, parameters for evaluation, and analysis and design, it is briefly explained below to familiarize the same using seepage in soil media.

Water pressures and flow quantities under or inside dams, around foundations, cofferdams, or retaining walls, or in compressible soil layers, ground water problems, flow through pipes etc. are important factors in the design of such hydraulic structures and related areas. The theories and equations which represent the flow of water through porous media in general and soil media in particular are presented in Appendix 6A. The solution of the appropriate equations for particular cases of flow through soils gives estimates of pore-water pressures and seepage quantities in practical circumstances. Various methods of solution are available for solving such problems in literature as summarized in Appendix 6A. In particular Finite Element Method (FEM) of Analysis of such problems is discussed in the following sections of this chapter.

6.2 FEM for the Analysis of Seepage / Flow through Porous Media

The concept of the finite element method of analysis to problems in elasticity as a process in which the total potential energy is minimized with respect to nodal displacements can be extended to a variety of physical problems where the extremum principle is applicable. If the problems are of a type in which the functional to be minimized is an integral of terms involving the unknown functions and its derivatives in powers of zero, one or two only, then the equations resulting from the minimizing process will be of the same type as those involved in linear stiffness analysis of structures.

Field problems which are governed by a general Quasi-Harmonic differential equation will be discussed in this chapter. The main interest in this category is due to the fact that the general Quasi-harmonic differential equation includes, as particular cases, the well-known Laplace and Poisson's equations which are encountered in the analysis of many physical phenomena such as (1) Heat conduction (2) Distribution of electric or magnetic potential (3) Seepage through porous media (4) Irrotational flow of ideal fluids (5) Bending of prismatic beams etc. (Zienkiewicz, 1971; Desai and Abel; 1972, Rao S.S., 1982). Isotropic or anisotropic regions can be included in the analysis.

6.3 General Formulation using Extremum Principle

A general Quasi-harmonic equation in Cartesian coordinates can be written as

$$\frac{\partial}{\partial x}\left(k_x \frac{\partial \phi}{\partial x}\right) + \frac{\partial}{\partial y}\left(k_y \frac{\partial \phi}{\partial y}\right) + \frac{\partial}{\partial z}\left(k_z \frac{\partial \phi}{\partial z}\right) + Q = 0 \qquad (6.1)$$

where ϕ is the only unknown physical parameter assumed to be single valued in the region.

In the case of seepage / flow through porous media,

ϕ = fluid potential = total head = pressure head + elevation head (neglecting velocity head as the velocity is very small in seepage problems). At any point in the domain / continuum / region, it has the same value in all directions i.e., it is not a vector and is a unique parameter at that point.

k_x, k_y, k_z are the coefficients of permeabilities of the flow region along x, y and z directions respectively.

Q = the externally applied flow/seepage flux i.e., rate of flow generated.

The definitions of the above quantities will be different for other applications such as heat conduction, electrical applications etc., though the mathematical form Eq. (6.1) remains the same.

The boundary conditions normally specified for solving the above Eq. (6.1) are

(a) The value of ϕ is specified on the boundary $\qquad$ (6.2)

or (b) $\quad k_x \frac{\partial \phi}{\partial x} l_x + k_y \frac{\partial \phi}{\partial y} l_y + k_z \frac{\partial \phi}{\partial z} l_z + q + \alpha \phi = 0 \qquad (6.3)$

on the boundary in which l_x, l_y and l_z are the direction cosines of the outward normal to the boundary surface.

If $k_x = k_y = k_z = k$ (for isotropic regions with same coefficient of permeability in all directions) and if $q = \alpha = 0$, then boundary condition (6.3) reduces to a non-conducting boundary i.e., $\frac{\partial \phi}{\partial n} = 0$.

For a two-dimensional formulation in x-y plane, the above phenomenon for example the equation (6.1) becomes:

$$\frac{\partial}{\partial x}\left(k_x \frac{\partial \phi}{\partial x}\right) + \frac{\partial}{\partial y}\left(k_y \frac{\partial \phi}{\partial y}\right) + Q = 0 \qquad (6.4)$$

The above Eqns. (6.1) and (6.4) are the governing equations for seepage in three and two dimensions and solutions can be obtained for simple problems using theory of partial differential equations or numerical methods etc. as summarized in Appendix 6A. They also are noted to be amenable for solution using extremum principles using Euler's theory as explained below.

Euler's theorem in calculus of variations (Crandall, 1956) can be used to solve Eq. (6.1) as an alternative approach in addition to the standard methods available to solve partial differential equations. The theorem states that if the functional given by the following Eq. (6.5)

$$\chi(\varphi) = \iiint f\left(x, y, z, \phi, \frac{\partial \phi}{\partial x}, \frac{\partial \phi}{\partial y}, \frac{\partial \phi}{\partial z}\right) dx\,dy\,dz \qquad (6.5)$$

is to be minimized over a bounded domain, the necessary and sufficient condition for this minimum to be attained, is that the unknown function has to satisfy the equations

$$\frac{\partial}{\partial x}\left(\frac{\partial f}{\partial\left(\partial \phi / \partial x\right)}\right) + \frac{\partial}{\partial y}\left(\frac{\partial f}{\partial\left(\partial \phi / \partial y\right)}\right) + \frac{\partial}{\partial z}\left(\frac{\partial f}{\partial\left(\partial \phi / \partial z\right)}\right) - \frac{\partial f}{\partial \phi} = 0 \qquad (6.6)$$

in the same region with the variable ϕ satisfying the same prescribed boundary conditions.

This alternative formulation can be tried for solving Eqn. (6.1) by applying the above stated principle of calculus of variations using Eqn. (6.6). Accordingly, minimization of the following integral

$$X = \iiint \left\{ \frac{1}{2}\left\{ k_x\left(\frac{\partial \phi}{\partial x}\right)^2 + k_y\left(\frac{\partial \phi}{\partial y}\right)^2 + k_z\left(\frac{\partial \phi}{\partial z}\right)^2 \right\} - Q\phi \right\} dx\,dy\,dz \qquad (6.7)$$

where the integral is taken over the entire region and ϕ satisfying boundary conditions as specified will result in solution of Eq. (6.1).

It is noted that the simultaneous implementation of boundary conditions (a) and (b) as expressed in Equations. (6.2) and (6.3) are difficult to implement. While the boundary condition (a) can be implemented easily, the boundary condition (b) poses practical difficulties in implementation. This can be overcome by not constraining the boundary values on those parts of the boundary where boundary condition (b) needs to be satisfied but to add another integral pertaining to that boundary surface to the functional expressed as in either Eq. (6.5) or Eq. (6.7) which, on minimization as per Euler's theory satisfies the required boundary condition (b) automatically. Accordingly, the integral to be added is:

$$\int_S \left(q\phi + 1/2\alpha\phi^2\right) dS \qquad (6.7a)$$

in which S is the surface of the boundary where boundary condition (b) needs to be satisfied. Thus, the integral given by Eq. (6.7a) needs to be added to the functional given by either Eq. (6.5) or Eq. (6.7) which on minimization satisfies the boundary condition (b) expressed by Eq. (6.3) automatically. The details can be referred from Zienkiewicz (1971). The minimization process can then be carried out using FEM for general regions though other methods may also be used for simple regions.

These aspects are explained using FEM for the solution of flow in two dimensional regions / continua (x-y plane) of general geometry and boundary conditions for example, in the following section.

207

6.3.1 Discretization of the Region / Continuum

The governing equation for seepage in two dimensional region is given in Eq. (6.4) above as:

$$\frac{\partial}{\partial x}\left(k_x\frac{\partial \phi}{\partial x}\right)+\frac{\partial}{\partial y}\left(k_y\frac{\partial \phi}{\partial y}\right)+Q=0 \tag{6.4a}$$

The corresponding functional to be minimized for solving the above equation in two dimensions using extremum principle (as per Euler's theorem) can be written from Eq. (6.7) as:

$$X=\iiint\left\{\frac{1}{2}\left\{k_x\left(\frac{\partial \phi}{\partial x}\right)^2+k_y\left(\frac{\partial \phi}{\partial y}\right)^2\right\}-Q\phi\right\}dxdy \tag{6.4b}$$

For pursuing the above minimization process for the analysis of flow through porous media using FEM, consider for example, a general region / continuum in two dimensions (*x-y* plane) is as shown in Figure 6.1. The region is discretized into an assembly of triangular elements following the steps of FEM (Chapter 2 and 3). In particular, consider one of these triangular elements '*e*' with *i, j, m* as nodes marked in an anti-clockwise direction as shown in the figure for evaluating element characteristics.

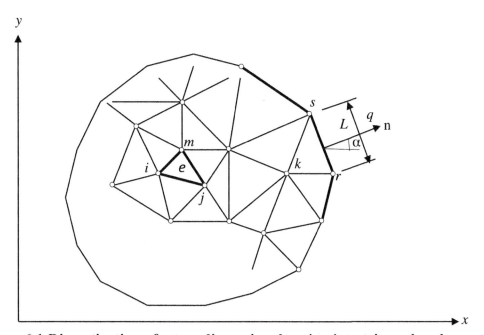

Figure 6.1 Discretization of a two-dimensional region into triangular elements

6.3.2 Element Characteristics

Following similar procedure as in solid / structural mechanics as explained in Chapters 2 and 3, element characteristics can be derived as follows. The element *e* of plane continuum is shown magnified in Figure 6.2 (with *i, j, m* as nodes numbered in an anticlockwise direction along with their x and y coordinates) for evaluating element characteristics in a similar manner as elastic solid as explained in Section 4.5.

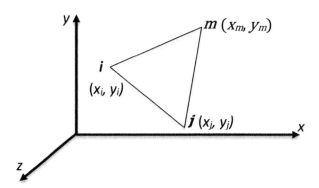

Figure 6.2 Plane triangular element of the continuum

Proceeding on similar lines as in Chapters 2 and 3 and Section 4.5, let nodal values of ϕ of a typical triangular element define the function within each element as

$$\phi = \{N\}\{\phi\}^e = \{N_i, N_j, N_m\}\{\phi\}^e \tag{6.8}$$

where

$$\{\phi\}^e = \begin{Bmatrix} \phi_i \\ \phi_j \\ \phi_m \end{Bmatrix}^e \tag{6.9}$$

$\phi_i \cdot \phi_j$, ϕ_m are the nodal values of ϕ at the nodes i, j, m respectively of the element e and are single valued components as they have the same value in all directions at any point in the context of seepage problems. (However, it may be noted that for solid mechanics problems and other applications, they are vectors which may have one, two or three (or more) components one (or more) each along the coordinate directions depending on the dimensionality of the region and the problem under consideration).

N_i, N_j, N_m are the distribution functions of ϕ (which are also single components as ϕ_i, ϕ_j, ϕ_m are single components) within the element as explained in Chapter 2, Appendix 2B.

The distribution functions N_i, N_j, N_m can be expressed (from Chapter2) as:

$$N_i = (a_i + b_i x + c_i y)/2\Delta$$
$$N_j = (a_j + b_j x + c_j y)/2\Delta \tag{6.10}$$
$$N_m = (a_m + b_m x + c_m y)/2\Delta$$

where

$$2\Delta = a_i + a_j + a_m = \begin{vmatrix} 1 & x_i & y_i \\ 1 & x_j & y_j \\ 1 & x_m & y_m \end{vmatrix} = 2 \times (\text{area of the triangular element } i, j, m)$$

$$\tag{6.11}$$

and

$$a_i = x_j y_m - x_m y_j \qquad b_i = y_j - y_m \qquad c_i = x_m - x_j$$
$$a_j = x_m y_i - x_i y_m \qquad b_j = y_m - y_i \qquad c_j = x_i - x_m$$
$$a_m = x_i y_j - x_j y_i \qquad b_m = y_i - y_j \qquad c_m = x_j - x_i$$

$$(6.12)$$

It can be noted that the coefficients in Equation (6.12) can be written by cyclic permutation of the subscripts i, j, m.

Having known the nodal values of ϕ which define the function throughout the region, the functional χ can be minimized with respect to ϕ which is now expressed in terms of nodal values of each element as expressed in Eq. (6.4b). Accordingly, χ has to be minimized with respect to all these nodal values of ϕ i.e., ϕ_i ($i = 1, 2, \ldots,$ n). To achieve this, the contributions to each differential, such as $\partial \chi / \partial \phi_i$ from a typical element are first evaluated and then, the sum of all these contributions coming from all elements connected at each node is equated to zero, as the functional χ is to be minimized (χ being expressed as in Eq. (6.4b).

If the value of χ associated with the element is called χ^e, by differentiating Eq. (6.4b) with respect to ϕ_i for example, we get,

$$\frac{\partial \chi^e}{\partial \phi_i} = \iint \left\{ k_x \frac{\partial \phi}{\partial x} \frac{\partial}{\partial \phi_i}\left(\frac{\partial \phi}{\partial x}\right) + k_y \frac{\partial \phi}{\partial y} \frac{\partial}{\partial \phi_i}\left(\frac{\partial \phi}{\partial y}\right) - Q \frac{\partial \phi}{\partial \phi_i} \right\} dx\, dy$$

$$(6.13)$$

From eqns. (6.8) and (6.9), Eq. (6.11) can be written as

$$\frac{\partial X^e}{\partial \phi_i} = \frac{1}{(2\Delta)^2} \iint \left\{ k_x \left(b_i, b_j, b_m\right)(\phi)^e \, b_i + k_y \left(c_i, c_j, c_m\right)(\phi)^e \, c_i \right\} dx\, dy - \frac{1}{2\Delta} \iint Q\left(a_i + b_i x + c_i y\right) dx\, dy$$

$$(6.14)$$

The other partial derivatives of X^e with respect to the other nodal values ϕ_j, ϕ_m associated with element e can be similarly written as in the above equation.

Any element contributes only three of the differentials associated with its nodes (since discretization of the region is done using three noded triangular elements in the present formulation) and can be written as

$$\left\{ \frac{\partial \chi^e}{\partial \varphi^e} \right\} = \left\{ \begin{array}{c} \dfrac{\partial \chi^e}{\partial \varphi_i} \\[2mm] \dfrac{\partial \chi^e}{\partial \varphi_j} \\[2mm] \dfrac{\partial \chi^e}{\partial \varphi_m} \end{array} \right\}$$

$$(6.15)$$

From Eq. (6.12), using similar expressions for the other two components, Eq. (6.13) can be expressed as

$$\left\{\frac{\partial \chi^e}{\partial \phi^e}\right\} = [h]^e \{\phi^e\} - \{F\}^e \tag{6.16}$$

which can be identified to be similar to element equilibrium equation derived in Chapters 2, 3, 4 and 5 and in which only the stiffness and distributed force contributions are included. If k_x and k_y are taken as constant within the element, and noting that over the area of the element $\iint dx\,dy = \Delta$, we can obtain the matrix $[h]^e$ using Eqs. (6.13), (6.14) and (6.15) as:

$$[h]^e = \frac{k_x}{4\Delta}\begin{bmatrix} b_i b_i & b_j b_i & b_m b_i \\ b_i b_j & b_j b_j & b_m b_j \\ b_i b_m & b_j b_m & b_m b_m \end{bmatrix} + \frac{k_y}{4\Delta}\begin{bmatrix} c_i c_i & c_j c_i & c_m c_i \\ c_i c_j & c_j c_j & c_m c_j \\ c_i c_m & c_j c_m & c_m c_m \end{bmatrix} \tag{6.17}$$

or $h_{rs} = \left(k_x b_r b_s + k_y c_r c_s\right)/4\Delta$ for all values of r and $s = i, j, m$.

It may be noted that $[h]^e$ matrix is symmetric as in elasticity problems. If Q is assumed to be constant, the nodal values F_i^e, F_j^e, F_m^e of the matrix value $\{F\}^e$ in Eq. (6.16) can be expressed as:

$$F_i^e = F_j^e = F_m^e = Q\iint \left(a_i + b_i x + c_i y\right) dx\,dy \Big/ 2\Delta = Q\left(a_i + b_i \bar{x} + c_i \bar{y}\right)/2 \tag{6.18}$$

where $\bar{x}$ and $\bar{y}$ are the coordinates of the C.G. (Center of Gravity) of the triangle as given below.

$$\bar{x} = \left(x_i + x_j + x_m\right)/3 \quad \text{and} \quad \bar{y} = \left(y_i + y_j + y_m\right)/3 \tag{6.19}$$

In fluid flow problems, $[h]^e$ is called element characteristic permeability matrix and $\{F\}^e$ is called the nodal vector matrix for applied flow.

Substituting, Eq. (6.19) in Eq. (6.18) the integral becomes equal to $(1/3)\Delta$ and

hence $\{F\}_Q^e = \dfrac{Q\Delta}{3}\begin{Bmatrix} 1 \\ 1 \\ 1 \end{Bmatrix}$ \hfill (6.20)

In flow problems the gradient of ϕ which is proportional to velocity of flow is of interest. Hence from Eq. (6.8) we can write the gradients as

$$\{\text{Grad } \phi^e\} = \{G\}^e = \begin{Bmatrix} \dfrac{\partial \phi}{\partial x} \\[2mm] \dfrac{\partial \phi}{\partial y} \end{Bmatrix}^e = \frac{1}{2\Delta}\begin{bmatrix} b_i & b_j & b_m \\ c_i & c_j & c_m \end{bmatrix}\begin{Bmatrix} \phi_i \\ \phi_j \\ \phi_m \end{Bmatrix} = [B]^e \{\phi\}^e \tag{6.21}$$

The matrix $[G]^e$ is similar to the strain matrix, $\{\varepsilon\}^e$ in solid / structural mechanics problems (Chapter 2, Appendix 2B).

Then velocity of flow υ as per Darcy's law (Scott. 1963) is,

$\upsilon = -ki$ (- sign is added since flow will be in the decreasing direction of head, i.e., ϕ), where k = coefficient of permeability of the porous medium,

$$i = \text{hydraulic gradient at that point} = \frac{\partial \varphi}{\partial n}$$

in which φ = total head at that point, and n = direction of flow (direction of outer normal to the surface across which flow is taking place, as shown in Figure 6.1 for example). Then the quantity of flow Q is:

$$Q = \upsilon A = -kiA \tag{6.22}$$

where A = area of cross section through which flow is taking place.

Accordingly, the velocity in any element e can be expressed using Equations (6.21) and (6.22) as:

$$\upsilon = \begin{Bmatrix} \upsilon_x \\ \upsilon_y \end{Bmatrix} = -\begin{bmatrix} k_x & 0 \\ 0 & k_y \end{bmatrix} \begin{Bmatrix} \dfrac{\partial \varphi}{\partial x} \\ \dfrac{\partial \varphi}{\partial y} \end{Bmatrix}^e = -[R][B]^e \{\varphi\}^e \text{ (m/sec)} \tag{6.23}$$

where $[R] = \begin{bmatrix} k_x & 0 \\ 0 & k_y \end{bmatrix}$ is the matrix of coefficients of permeability of the porous medium / region.

Also, flow across a unit area (i.e., $A = 1$) in any element e, $\{Q\}$ can be written as

$$\{Q\} = -[R][B]^e \{\phi\}^e$$

$$= -\begin{bmatrix} k_x & 0 \\ 0 & k_y \end{bmatrix} \frac{1}{2\Delta} \begin{bmatrix} b_i & b_j & b_m \\ c_i & c_j & c_m \end{bmatrix} \begin{Bmatrix} \phi_i \\ \phi_j \\ \phi_m \end{Bmatrix}^e = -\frac{1}{2\Delta} \begin{bmatrix} k_x & 0 \\ 0 & k_y \end{bmatrix} \begin{Bmatrix} b_i\phi_i + b_j\phi_j + b_m\phi_m \\ c_i\phi_i + c_j\phi_j + c_m\phi_m \end{Bmatrix}$$

$$= -\frac{1}{2\Delta} \begin{Bmatrix} k_x(b_i\phi_i + b_j\phi_j + b_m\phi_m) \\ k_y(c_i\phi_i + c_j\phi_j + c_m\phi_m) \end{Bmatrix} = \begin{Bmatrix} Q_x \\ Q_y \end{Bmatrix} \text{ (m}^3 \text{ per unit area)} \tag{6.24}$$

If $k_x = k_y = k$

$$\{Q\}^e = \begin{Bmatrix} Q_x \\ Q_y \end{Bmatrix} = -\frac{k}{2\Delta} \begin{Bmatrix} b_i\phi_i + b_j\phi_j + b_m\phi_m \\ c_i\phi_i + c_j\phi_j + c_m\phi_m \end{Bmatrix}^e \tag{6.25}$$

It may be noted that if Q^e is positive, the flow direction is along the positive directions of x and y axes and if Q^e is negative, then the flow direction is along the negative direction of x and y axes. Flow in any other direction 'n' can be obtained as:

$$Q_n = Q_x \cdot l_x + Q_y \cdot l_y = Q \cdot \frac{\partial \phi}{\partial n} \tag{6.26}$$

when l_x and l_y are direction cosines of 'n' with respect to x and y as shown in Figure 6.1.

The net flows in x and y directions at the nodes can also be evaluated as the uniform flow tributary to node (Desai and Abel, 1972) as:

$$\{Q\}^e = \begin{Bmatrix} Q_{xi} \\ Q_{xj} \\ Q_{xm} \\ Q_{yi} \\ Q_{yj} \\ Q_{ym} \end{Bmatrix} = -\frac{1}{2}\begin{bmatrix} b_i & 0 \\ b_j & 0 \\ b_m & 0 \\ 0 & c_i \\ 0 & c_j \\ 0 & c_m \end{bmatrix}\{v\} \qquad (6.27)$$

6.3.3 Assembling System Equations governing Flow in the Region / Continuum

The next step is to apply the above expressions of minimization process of χ for the whole region / porous medium which is now a function of $\{\phi\}^{1,2,3...n}$ after discretization of the region into several elements. This can be achieved by equating its first partial derivatives with respect to the variables $\{\phi\}^{1,2,3...n}$ to zero as per extremum principle of Euler and then solve for these unknown variables from the resulting equations from minimization process of the reassembled region. Accordingly, contributions to first partial derivatives $\partial\chi^e / \partial\phi_{i(i=1,2,...,n)}$ coming from all elements connected at each node are added and equated to zero to satisfy the extremum principle as per Euler's theory mentioned above (Crandall, 1956). Thus the system of assembled equations are obtained which is similar to assembling the system equilibrium equations in solid / structural mechanics wherein the system equilibrium equations are assembled by adding contribution of internal forces coming from each of the elements connected at each node and equating them to the external nodal forces at the corresponding node to satisfy the equilibrium conditions of the system (Chapters 2, 3, 4 and 5).

Only the elements connected at the node i will contribute to $\partial\chi / \partial\phi_i$ just as only forces acting on such elements contributed in elasticity to the equilibrium equations at each node. The final equations of the assembly can be identified to be similar to the system equilibrium equations as explained in Chapters 2, 3, 4 and 5.

The final equations of the minimization process of χ requires the sum of all contributions of the first partial differentials with respect to each of the nodal variables φ_r ($r = 1, 2, ..., n$) to be zero, i.e.

$$\frac{\partial\chi}{\partial\phi_r} = \sum\frac{\partial\chi^e}{\partial\phi_r} = 0 \text{ for all values of } r \text{ (i.e. } r = 1,2,...,n) \qquad (6.28)$$

the summation being taken over all the elements, From Eqns. (6.16) to (6.20), the resulting expressions for the above conditions (Eq. 6.28) can be written as:

$$\frac{\partial\chi}{\partial\phi_r} = \sum\sum[h_{rs}]^e\phi_r - \sum F_r = 0 \text{ for all values of } r \text{ and } s \ (1,2...,n). \qquad (6.29)$$

Eqns. (6.29) can also be expressed in matrix form as:

$$[H]\{\phi\} - \{F\} = 0$$

i.e. $\qquad\qquad [H]\{\phi\} = \{F\}$ $\qquad\qquad\qquad\qquad$ (6.30)

where, $[H]$ is called the global permeability matrix and $\{F\}$ is called the matrix of nodal vectors of applied flow.

In Eqns. (6.29), the summation $\sum\sum[h_{rs}]^e$ is being carried out over all the elements e and nodes r and s. This is similar to the structural formulation of the assembly as explained in Chapters 2, 3, 4 and 5. The summation $\sum\sum[h_{rs}]^e$ will have contributions only from those elements which have r and s as common nodes. For all other elements for which r and s are not common nodes $h_{rs} = 0$.

The rest of the solution procedure is the same as is applicable to solid mechanics problems i.e. applying boundary conditions along the boundary of the region / continuum / porous media as specified, simplifying the system equations after incorporating the boundary conditions (as explained in the following section), solving the resulting simultaneous equation for unknown nodal values of $\{\phi\}$ and the evaluation of required flow parameters such as gradients, flow quantities within each element or at any node(s) of the element etc.

6.3.4 Boundary Conditions

The way to deal with the two types of boundary conditions normally specified in seepage problems as given by Eqns. (6.2) and (6.3) is explained in Section 6.3 above. For boundary conditions of type (a) as given by Eq. (6.2) where ϕ is specified on the boundaries, the nodal values are substituted as prescribed on the boundaries for the boundary nodes while assembling the system of equations and then the resulting equations ae simplified for further solution and processing as explained in Chapters 2 to 5. Thus, the problem becomes identical with the equivalent structural problem where displacements are prescribed on the boundaries and the problem can be analysed following the usual steps as explained in the above chapters.

However, the boundary conditions of type (b) as given by Eq. (6.3), present some difficulty and the way to deal with it is explained above in Section 6.3. Accordingly, in such a case it is advantageous to modify the variational approach so that the values of ϕ on the boundary can take up any value without any constraint. This can be easily achieved by adding appropriate additional terms to the functional which is to be minimized as given in Eq. (6.7a). Accordingly, the type (b) boundary condition given by Eq. (6.3) in two dimensions becomes:

$$k_x \frac{\partial\phi}{\partial X} l_x + k_y \frac{\partial\phi}{\partial Y} l_y + q + \alpha\phi = 0 \qquad\qquad (6.31)$$

By variational principle it can be shown that if a portion 'S' of the boundary is subjected to such a condition then the functional given by Eq. (6.7) (also explained in Eq. .7a) is to be modified as:

214

$$\chi = \iint \left[\frac{1}{2} \left\{ k_x \left(\frac{\partial \phi}{\partial x} \right)^2 + k_y \left(\frac{\partial \phi}{\partial y} \right)^2 \right\} - Q\phi \right] dxdy + \int_S q\phi ds + \int_S \alpha\phi^2 ds \qquad (6.32)$$

The last two terms are integrated along the boundary subject to boundary condition (b) as given in Eq. (6.3) and ϕ is not constrained now. Also, the contributions of partial derivatives with respect to nodal values of ϕ have to be added to Eq. (6.29). Such partial differentials will exist only for elements forming the boundary specified with type (b) boundary condition.

A condition like $\alpha = 0$, and $q = 0$ represents a special case of an unloaded boundary. Symmetry lines as well as line at impermeable boundaries fall into this category.

The net flows at each node $\{F\}$ i.e., applied flow matrix can be computed using Eq. (6.29). For example, the net flow at any node 'i' can be computed as F_i using Eq. (6.29) as:

$$F_i = \left[H_i \right] \{\phi\} \qquad (6.33)$$

where $\left[H_i \right]$ is the i row of the global permeability matrix and $\{\phi\}$ is the total head / potential matrix i.e., listing of the values of the total head at all the nodes of the system.

Similarly, the net gradients at each of the nodes can be obtained by adding the contributions to gradients coming from all the elements connected at each node using Eq (6.21). For example, the gradients at a typical node i and can be expressed as:

$$\{Grad\ \phi\}_i = \sum \{G\}^e = \sum \left\{ \begin{array}{c} \frac{\partial \phi}{\partial x} \\ \frac{\partial \phi}{\partial y} \end{array} \right\}^e = \sum \frac{1}{2\Delta^e} \begin{bmatrix} b_i & b_j & b_m \\ c_i & c_j & c_m \end{bmatrix}^e \left\{ \begin{array}{c} \phi_i \\ \phi_j \\ \phi_m \end{array} \right\}^e = \sum [B]^e \{\phi\}^e$$

$$(6.34)$$

the summation being carried out over all the elements connected at each node. The same procedure can be repeated for all the nodes where the gradients need to be evaluated.

Only the steady state seepage / flow analysis is presented in this chapter for easy understanding. The cases of non-homogeneity and (or) any other arbitrary properties as well as transient flow problems can be formulated and analyzed similarly and may be referred from other references (Desai and Abel, 1972, Zienkiewicz, 1971 etc,). As the formulation is analogous to elasticity formulation, similar computer programs can also be used for the analysis, by suitably redefining the input parameters.

6.4 Examples of Seepage / Flow through Porous Media

Examples illustrating FEM analysis of steady state seepage / flow through porous media problems are presented in this section.

Example 1

Obtain the values of the equivalent fluid flow (seepage) at nodes A, B and C for the seepage problem shown in Figure 6.4.1, using Finite Element Approach (use one element ABD). ϕ is the total head in meters. Also find the seepage Q in the element.

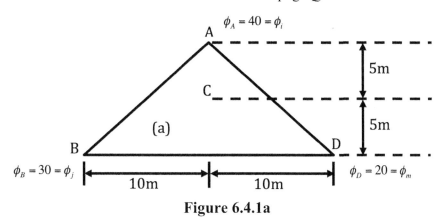

Figure 6.4.1a

Solution:

All units are in meters. The coordinates of A, B, D are shown in brackets.
Let:

$$i = A(0, 5) \qquad j = B(-10, -5) \qquad m = D(10, -5)$$

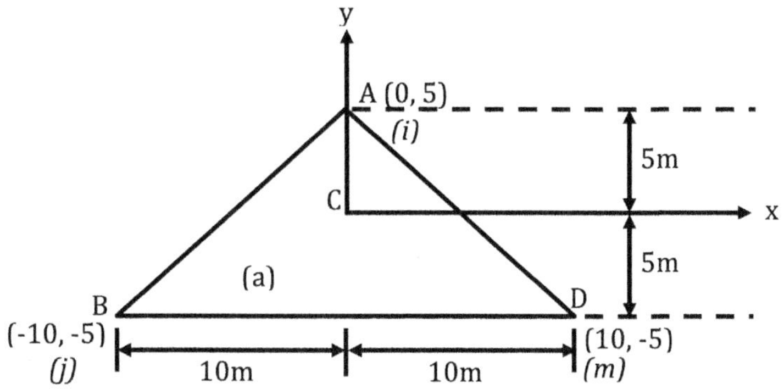

Figure 6.4.1b

Following the steps as explained in Section 6.3.2, the characteristics of the above element can be evaluated as follows.

$$\Delta = 10 \times 10 = 100$$

$$a_A = a_i = x_j y_m - x_m y_j = (-10)(-5) - (10)(-5) = 100$$

$$b_A = b_i = y_j - y_m = -5 - (-5) = 0$$

$$c_A = c_i = x_m - x_j = 10 - (-10) = 20$$

$$a_B = a_j = x_m y_i - x_i y_m = (10)(5) - (0)(-5) = 50$$

216

$$b_B = b_j = y_m - y_i = -5 - 5 = -10$$
$$c_B = c_j = x_i - x_m = 0 - 10 = -10$$

$$a_D = a_m = x_i y_j - x_j y_i = 0 - (-10)5 = 50$$
$$b_D = b_m = y_i - y_j = 5 + 5 = 10$$
$$c_D = c_m = x_j - x_i = -10$$

As explained in Section 6.3.2 we can write the characteristic permeability matrix of the element, (which is analogous to stiffness matrix $[k]$ in structural analysis), $[h]^e$ as

$$[h]^e = \frac{k_x}{4\Delta}\begin{bmatrix} 0 & 0 & 0 \\ 0 & 100 & -100 \\ 0 & -100 & 100 \end{bmatrix} + \frac{k_y}{4\Delta}\begin{bmatrix} 400 & -200 & -200 \\ -200 & 100 & 100 \\ -200 & 100 & 100 \end{bmatrix}$$

Noting $k_x = k_y = k$,

$$[h]^e = \frac{k}{4}\begin{bmatrix} 4 & -2 & -2 \\ -2 & 2 & 0 \\ -2 & 0 & 2 \end{bmatrix} = \frac{k}{2}\begin{bmatrix} 2 & -1 & -1 \\ -1 & 1 & 0 \\ -1 & 0 & 1 \end{bmatrix}$$

Then the flow equilibrium equation of the element 'a' can be written as (Section 6.3.2):

$$[H]\{\varphi\} = [h]^e \begin{Bmatrix} \varphi_i \\ \varphi_j \\ \varphi_m \end{Bmatrix} = \begin{Bmatrix} F_i \\ F_j \\ F_m \end{Bmatrix}$$

On simplification, the above equation becomes

$$\begin{bmatrix} 2 & -1 & -1 \\ -1 & 1 & 0 \\ -1 & 0 & 1 \end{bmatrix}\begin{Bmatrix} \phi_A = 40 \\ \phi_B = 30 \\ \phi_D = 20 \end{Bmatrix} = \begin{Bmatrix} F_A \\ F_B \\ F_D \end{Bmatrix}$$

$\therefore F_A = 30$ (storage)

$\quad F_B = -10$ (loss of storage)

$\quad F_C = -20$ (loss of storage)

The total head ϕ at any point of the element is given by

$$\phi = \begin{Bmatrix} N_i & N_j & N_m \end{Bmatrix}\begin{Bmatrix} \phi_i \\ \phi_j \\ \phi_m \end{Bmatrix}$$

where ϕ = total head = fluid potential

Accordingly, ϕ at point C with coordinates $x = 0$, $y = 0$, is

$$\phi = \frac{1}{2\Delta} \begin{bmatrix} 100 + 0x + 20y & 50 - 10x - 10y & 50 + 10x - 10y \end{bmatrix} \begin{Bmatrix} 40 \\ 30 \\ 20 \end{Bmatrix}$$

$$\phi_c = \frac{1}{2\Delta} \begin{bmatrix} 40 \times (100 + 50 + 50) \end{bmatrix} = \frac{1}{2(100)} \times 40 \times 200 = 40$$

Gradient of flow in the element

$$\text{Grad}, \; \phi = \begin{bmatrix} \dfrac{\partial \phi}{\partial x} \\[2mm] \dfrac{\partial \phi}{\partial y} \end{bmatrix} = \frac{1}{2\Delta} \begin{bmatrix} b_i & b_j & b_m \\ c_i & c_j & c_m \end{bmatrix} \begin{Bmatrix} \phi_i \\ \phi_j \\ \phi_m \end{Bmatrix} = \{G\}\{\phi\}^e$$

$$= \frac{1}{2\Delta} \begin{Bmatrix} 0 & -10 & 10 \\ 20 & -10 & -10 \end{Bmatrix} \begin{Bmatrix} 40 \\ 30 \\ 20 \end{Bmatrix} = \text{constant}$$

$\dfrac{\partial \phi}{\partial x} = -\dfrac{1}{2}$ (flow along positive x direction)

$\dfrac{\partial \phi}{\partial y} = \dfrac{3}{2}$ (flow along negative y direction)

It may be noted that Q is constant in the element as gradients are constant (i.e., Grad ϕ = constant)

Then the flow $Q = -RG\phi = \begin{Bmatrix} Q_x \\ Q_y \end{Bmatrix} = -k \begin{Bmatrix} -\dfrac{1}{2} \\[2mm] \dfrac{3}{2} \end{Bmatrix} = \begin{Bmatrix} \dfrac{k}{2} \\[2mm] -\dfrac{3k}{2} \end{Bmatrix}$

Flow across AD (per unit area) – Figure 6.4.1b

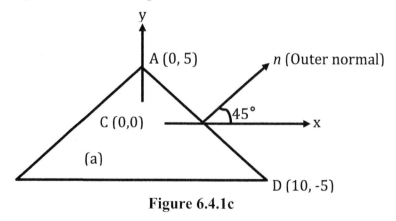

Figure 6.4.1c

$$Q_n = Q_x \cdot \cos 45 + Q_y \cdot \sin 45 = \frac{k}{2}\sqrt{2} - \frac{3k}{2} \cdot \sqrt{2} = -\sqrt{2}k$$

(negative flow, i.e., flow inward)

where n in the direction of outer normal of AD as shown in Figure 6.4.1c.

Example 2:

For the two-dimensional soil profile shown (take the thickness as unity) evaluate the value of the head at point A, for steady state flow through the medium using Finite Element Method. The soil is homogeneous and isotropic with constant permeability coefficient, k. Compare the result with the one using Finite Difference Method. All units are in meters.

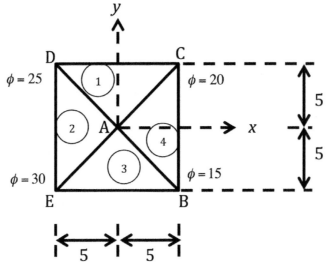

Figure 6.4.2a

Solution:

Discretize the medium into 4 triangular elements as shown in Figure 6.4.2a.

Element characteristics:

Area of each element $= \Delta = \dfrac{1}{2} \times 10 \times 5 = 25$

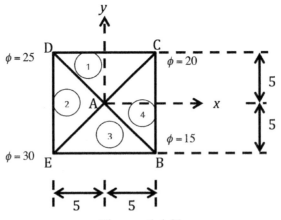

Figure 6.4.2b

Coordinates of the nodes with A as origin (Global Coordinates)

A $(0, 0)$

B $(5, -5)$

C $(5, 5)$

E $(-5, -5)$

Area of each element using the above coordinates, element 1 (triangle 1 - A, C, D for example)

$$2\Delta = \begin{vmatrix} 1 & 1 & 1 \\ 0 & 5 & -5 \\ 0 & 5 & 5 \end{vmatrix} = 50 \qquad \Delta = 25$$

Element 1:

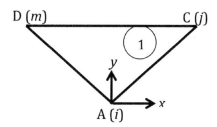

$a_i = a_A = x_C y_D - x_D y_C = 5 \times 5 - (-5 \times 5) = 25 + 25 = 50$

$b_i = b_A = y_C - y_D = 0$

$c_i = c_A = x_D - x_C = -5 - 5 = -10$

$a_j = a_C = x_D y_A - x_A y_D = -5 \times 0 - 0 \times 5 = 0$

$b_j = b_C = y_D - y_A = 5$

$c_j = c_C = x_A - x_D = 5$

$a_m = a_D = x_A y_C - x_C y_A = 0 \times 5 - 5 \times 0 = 0$

$b_m = b_D = y_A - y_C = -5$

$c_m = c_D = x_C - x_A = 5$

$$\begin{bmatrix} a_i & b_i & c_i \\ a_j & b_j & c_j \\ a_m & b_m & c_m \end{bmatrix} = \begin{bmatrix} a_A & b_A & c_A \\ a_C & b_C & c_C \\ a_D & b_D & c_D \end{bmatrix} = \begin{bmatrix} 50 & 0 & -10 \\ 0 & 5 & 5 \\ 0 & -5 & 5 \end{bmatrix}$$

Element 2:

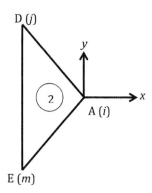

$a_i = a_A = x_D y_E - x_E y_D = (-5 \times -5) - (-5 \times 5) = 50$

$b_i = b_A = y_D - y_E = 5 - (-5) = 10$

$c_i = c_A = x_E - x_D = -5 - (-5) = 0$

$$a_j = a_D = x_E y_A - x_A y_E = 0$$
$$b_j = b_D = y_E - y_A = -5$$
$$c_j = c_D = x_A - x_E = 5$$

$$a_m = a_E = x_A y_D - x_D y_A = 0$$
$$b_m = b_E = y_A - y_D = -5$$
$$c_m = c_E = x_D - x_A = -5$$

$$\begin{bmatrix} a_i & b_i & c_i \\ a_j & b_j & c_j \\ a_m & b_m & c_m \end{bmatrix} = \begin{bmatrix} a_A & b_A & c_A \\ a_C & b_C & c_C \\ a_D & b_D & c_D \end{bmatrix} = \begin{bmatrix} 50 & 10 & 0 \\ 0 & -5 & 5 \\ 0 & -5 & -5 \end{bmatrix}$$

Element 3:

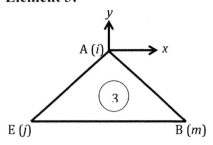

$$a_i = a_A = x_E y_B - x_B y_E = (-5 \times -5) - 5(-5) = 50 \qquad a_j = a_E = x_B y_A - x_A y_B = 0$$
$$b_i = b_A = y_E - y_B = -5 - (5) = 0 \qquad\qquad\qquad b_j = b_E = y_B - y_A = -5$$
$$c_i = c_A = x_B - x_E = 5 - (-5) = 10 \qquad\qquad\qquad c_j = c_E = x_A - x_B = -5$$

$$a_m = a_B = x_A y_E - x_E y_A = 0$$
$$b_m = b_B = y_A - y_E = 5$$
$$c_m = c_B = x_E - x_A = -5$$

$$\begin{bmatrix} a_i & b_i & c_i \\ a_j & b_j & c_j \\ a_m & b_m & c_m \end{bmatrix} = \begin{bmatrix} a_A & b_A & c_A \\ a_E & b_E & c_E \\ a_B & b_B & c_B \end{bmatrix} = \begin{bmatrix} 50 & 0 & 10 \\ 0 & -5 & -5 \\ 0 & 5 & -5 \end{bmatrix}$$

Element 4:

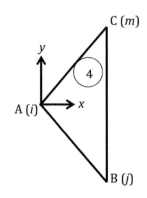

$a_i = a_A = x_B y_C - x_C y_B = (5\times 5) - 5(-5) = 50$

$b_i = b_A = y_B - y_C = -5 - 5 = -10$

$c_i = c_A = x_C - x_B = 0$

$a_j = a_B = x_C y_A - x_A y_C = 0$

$b_j = b_B = y_C - y_A = 5$

$c_j = c_B = x_A - x_C = -5$

$a_m = a_C = x_A y_B - x_B y_A = 0$

$b_m = b_C = y_A - y_B = 5$

$c_m = c_C = x_B - x_A = 5$

$$\begin{bmatrix} a_i & b_i & c_i \\ a_j & b_j & c_j \\ a_m & b_m & c_m \end{bmatrix} = \begin{bmatrix} a_A & b_A & c_A \\ a_B & b_B & c_B \\ a_C & b_C & c_C \end{bmatrix} = \begin{bmatrix} 50 & -10 & 0 \\ 0 & 5 & -5 \\ 0 & 5 & 5 \end{bmatrix}$$

Element characteristic matrices (or permeability matrices), i.e. [h] matrices:
Area of all elements, $\Delta = 25m^2$

Element 1:

$$[h]^1 = \frac{k}{4\Delta}\left[\begin{bmatrix} b_ib_i & b_ib_j & b_ib_m \\ b_ib_j & b_jb_j & b_jb_m \\ b_ib_m & b_jb_m & b_mb_m \end{bmatrix} + \begin{bmatrix} c_ic_i & c_ic_j & c_ic_m \\ c_ic_j & c_jc_j & c_jc_m \\ c_ic_m & c_jc_m & c_mc_m \end{bmatrix}\right] \begin{matrix} A \\ C \\ D \end{matrix} \quad (\text{columns } A, C, D)$$

$$= \frac{k}{4\Delta}\left[\begin{bmatrix} 0 & 0 & 0 \\ 0 & 25 & -25 \\ 0 & -25 & 25 \end{bmatrix} + \begin{bmatrix} 100 & -50 & -50 \\ -50 & 25 & 25 \\ -50 & 25 & 25 \end{bmatrix}\right] = \frac{k}{4\Delta}\begin{bmatrix} 100 & -50 & -50 \\ -50 & 50 & 0 \\ -50 & 0 & 50 \end{bmatrix}$$

$$= \frac{k}{4}\begin{bmatrix} 4 & -2 & -2 \\ -2 & 2 & 0 \\ -2 & 0 & 2 \end{bmatrix} \begin{matrix} A \\ C \\ D \end{matrix} \quad (\text{columns } A, C, D)$$

Element 2:

$$[h]^2 = \frac{k}{4\Delta}\left[\begin{bmatrix} 100 & -50 & -50 \\ -50 & 25 & 25 \\ -50 & 25 & 25 \end{bmatrix} + \begin{bmatrix} 0 & 0 & 0 \\ 0 & 25 & -25 \\ 0 & -25 & 25 \end{bmatrix}\right] \begin{matrix} A \\ D \\ E \end{matrix} \quad (\text{columns } A, D, E) = \frac{k}{4\Delta}\begin{bmatrix} 100 & -50 & -50 \\ -50 & 50 & 0 \\ -50 & 0 & 50 \end{bmatrix}$$

$$= \frac{k}{4}\begin{bmatrix} 4 & -2 & -2 \\ -2 & 2 & 0 \\ -2 & 0 & 2 \end{bmatrix} \begin{matrix} A \\ D \\ E \end{matrix} \quad (\text{columns } A, D, E)$$

Element 3:

$$[h]^3 = \frac{k}{4\Delta}\left[\begin{array}{ccc} & {\scriptstyle A} & {\scriptstyle E} & {\scriptstyle B} \\ 0 & 0 & 0 \\ 0 & 25 & -25 \\ 0 & -25 & 25 \end{array}\right] + \left[\begin{array}{ccc} {\scriptstyle A} & {\scriptstyle E} & {\scriptstyle B} \\ 100 & -50 & -50 \\ -50 & 25 & 25 \\ -50 & 25 & 25 \end{array}\right]\begin{array}{c} A \\ E \\ B \end{array} = \frac{k}{4\Delta}\left[\begin{array}{ccc} 100 & -50 & -50 \\ -50 & 50 & 0 \\ -50 & 0 & 50 \end{array}\right]$$

$$= \frac{k}{4}\left[\begin{array}{ccc} {\scriptstyle A} & {\scriptstyle E} & {\scriptstyle B} \\ 4 & -2 & -2 \\ -2 & 2 & 0 \\ -2 & 0 & 2 \end{array}\right]\begin{array}{c} A \\ E \\ B \end{array}$$

Element 4:

$$[h]^4 = \frac{k}{4\Delta}\left[\begin{array}{ccc} {\scriptstyle A} & {\scriptstyle B} & {\scriptstyle C} \\ 100 & -50 & -50 \\ -50 & 25 & 25 \\ -50 & 25 & 25 \end{array}\right] + \left[\begin{array}{ccc} {\scriptstyle A} & {\scriptstyle B} & {\scriptstyle C} \\ 0 & 0 & 0 \\ 0 & 25 & -25 \\ 0 & -25 & 25 \end{array}\right]\begin{array}{c} A \\ B \\ C \end{array} = \frac{k}{4\Delta}\left[\begin{array}{ccc} 100 & -50 & -50 \\ -50 & 50 & 0 \\ -50 & 0 & 50 \end{array}\right]$$

$$= \frac{k}{4}\left[\begin{array}{ccc} {\scriptstyle A} & {\scriptstyle B} & {\scriptstyle C} \\ 4 & -2 & -2 \\ -2 & 2 & 0 \\ -2 & 0 & 2 \end{array}\right]\begin{array}{c} A \\ B \\ C \end{array}$$

Total assemblage permeability matrix:

Global system matrix [H] (analogous of global stiffness matrix in solids)

$$[H]_{system} = \frac{k}{4}\times \left[\begin{array}{ccccc} {\scriptstyle A} & {\scriptstyle B} & {\scriptstyle C} & {\scriptstyle D} & {\scriptstyle E} \\ 4+4+4+4=16 & -2-2=-4 & -2-2=-4 & -2-2=-4 & -2-2=-4 \\ -2-2=-4 & 2+2=4 & 0 & 0 & 0 \\ -2-2=-4 & 0 & 2+2=4 & 0 & 0 \\ -2-2=-4 & 0 & 0 & 2+2=4 & 0 \\ -2-2=-4 & 0 & 0 & 0 & 2+2=4 \end{array}\right]\begin{array}{c} A \\ B \\ C \\ D \\ E \end{array}$$

$$= \frac{k}{4}\times \left[\begin{array}{ccccc} {\scriptstyle A} & {\scriptstyle B} & {\scriptstyle C} & {\scriptstyle D} & {\scriptstyle E} \\ 16 & -4 & -4 & -4 & -4 \\ -4 & 4 & 0 & 0 & 0 \\ -4 & 0 & 4 & 0 & 0 \\ -4 & 0 & 0 & 4 & 0 \\ -4 & 0 & 0 & 0 & 4 \end{array}\right]\begin{array}{c} A \\ B \\ C \\ D \\ E \end{array}$$

For example, the components of the first row of the global permeability matrix can be obtained as

$H_{AA} = \sum h_{AA}^e$ of all elements with A as common mode $= \dfrac{k}{4}(4+4+4+4) = 4k$

$H_{AB} = \sum h_{AB}^e$ of all elements with A and B as common mode $= h_{AB}^3 + h_{AB}^4 = \dfrac{k}{4}(-2-2) = -k$

$H_{AC} = \sum h_{AC}^e$ of all elements with A and C as common mode $= h_{AC}^1 + h_{AC}^4 = \dfrac{k}{4}(-2-2) = -k$

$H_{AD} = \sum h_{AD}^e$ of all elements with A and D as common mode $= h_{AD}^1 + h_{AD}^2 = \dfrac{k}{4}(-2-2) = -k$

$H_{AE} = \sum h_{AE}^e$ of all elements with A and E as common mode $= h_{AE}^2 + h_{AE}^3 = \dfrac{k}{4}(-2-2) = -k$

The flow equilibrium equations can be written as

$$k \begin{array}{cc} & \begin{array}{ccccc} A & B & \ C & D & E \end{array} \\ & \begin{bmatrix} 4 & -1 & -1 & -1 & -1 \\ -1 & 1 & 0 & 0 & 0 \\ -1 & 0 & 1 & 0 & 0 \\ -1 & 0 & 0 & 1 & 0 \\ -1 & 0 & 0 & 0 & 1 \end{bmatrix} \end{array} \begin{Bmatrix} \phi_A \\ \phi_B = 15 \\ \phi_C = 20 \\ \phi_D = 25 \\ \phi_E = 30 \end{Bmatrix} = \begin{Bmatrix} F_A = 0 \\ F_B \\ F_C \\ F_D \\ F_E \end{Bmatrix}$$

Assuming that no flow is added at node A (F_A=0)

From the first equation: $4\phi_A - 90 = 0$, $\rightarrow$ $\therefore$ $\phi_A = 22.5$

Further, the flow characteristics in each element and the seepage quantities Q can be calculated as explained in Section 6.3.2.

Example 3:
Consider the region 1354 shown in **Figure 6.4.3** for the analysis of seepage.
$\phi_1 = 20$, $\phi_2 = 16$, $\phi_3 = 12$. Calculate ϕ_4 and ϕ_5.

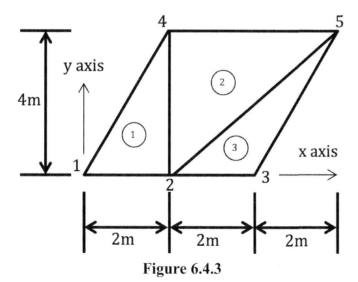

Figure 6.4.3

Solution:
The region is divided into 3 triangular elements for carrying out FE analysis. The coordinates of the nodes are listed below.

Element	Coordinates of nodes (meters)			Area (per meter length along z axis)
	i / Node	j / Node	m / Node	
1	0, 0 (1)	2, 0 (2)	2, 4 (4)	4
2	2, 0 (2)	6, 4 (5)	2, 4 (4)	8
3	2, 0 (2)	4, 0 (3)	6, 4 (5)	4

The FE characteristics for these elements are tabulated below:

	Element 1	Element 2	Element 3
$a_i = x_j y_m - x_m y_j$	8	16	16
$b_i = y_j - y_m$	-4	0	-4
$c_i = x_m - x_j$	0	-4	2
$a_j = x_m y_i - x_i y_m$	0	-8	-8
$b_j = y_m - y_i$	4	4	4
$c_j = x_i - x_m$	-2	0	-4
$a_m = x_i y_j - x_j y_i$	0	8	0
$b_m = y_i - y_j$	0	-4	0
$c_m = x_j - x_i$	2	4	2

Permeability matrices of all elements:

$$[h]^a = \frac{k_x}{4\Delta}\left[\begin{bmatrix} b_i b_i & b_j b_i & b_m b_i \\ b_i b_j & b_j b_j & b_m b_j \\ b_i b_m & b_j b_m & b_m b_m \end{bmatrix} + \frac{k_y}{4\Delta}\begin{bmatrix} c_i c_i & c_j c_i & c_m c_i \\ c_i c_j & c_j c_j & c_m c_j \\ c_i c_m & c_j c_m & c_m c_m \end{bmatrix}\right]$$

$k_x = k_y = k$

Element 1

$$[h]^1 = \frac{k}{4\Delta}\begin{array}{ccc} 1 & 2 & 4 \\ \begin{bmatrix} 16+0 & -16+0 & 0+0 \\ -16+0 & 16+4 & 0-4 \\ 0+0 & 0-4 & 0+4 \end{bmatrix}\end{array} = \frac{k}{16}\begin{array}{ccc} 1 & 2 & 4 \\ \begin{bmatrix} 16 & -16 & 0 \\ -16 & 20 & -4 \\ 0 & -4 & 4 \end{bmatrix}\end{array}\begin{array}{c} 1 \\ 2 \\ 4 \end{array}$$

Element 2

$$[h]^2 = \frac{k}{4\Delta}\begin{bmatrix} 0+16 & 0+0 & 0-16 \\ 0+0 & 16+0 & 0-4 \\ 0-16 & -16+0 & 16+16 \end{bmatrix} = \frac{k}{16}\begin{array}{ccc} 2 & 5 & 4 \\ \begin{bmatrix} 8 & 0 & -8 \\ 0 & 8 & -8 \\ -8 & -8 & 16 \end{bmatrix}\end{array}\begin{array}{c} 2 \\ 5 \\ 4 \end{array}$$

Element 3

$$[h]^3 = \frac{k}{4\Delta}\begin{bmatrix} 16+4 & -16-8 & 0+4 \\ -16-8 & 16+16 & 0-8 \\ 0+4 & 0-8 & 0+4 \end{bmatrix} = \frac{k}{16}\begin{array}{ccc} 2 & 3 & 5 \\ \begin{bmatrix} 20 & -24 & 4 \\ -24 & 32 & -8 \\ 4 & -8 & 4 \end{bmatrix}\end{array}\begin{array}{c} 2 \\ 3 \\ 5 \end{array}$$

225

Global permeability matrix [H] on superposition (assembling) [H]$_{system}$ can be obtained as

$$[H] = \frac{k}{16} \begin{bmatrix} 16 & -16 & 0 & 0 & 0 \\ -16 & 48 & -24 & -12 & 4 \\ 0 & -24 & 32 & 0 & -8 \\ 0 & -12 & 0 & 20 & -8 \\ 0 & 4 & -8 & -8 & 12 \end{bmatrix} \begin{matrix} 1 \\ 2 \\ 3 \\ 4 \\ 5 \end{matrix} = k \begin{bmatrix} 1 & -1 & 0 & 0 & 0 \\ -1 & 3 & -1.5 & -0.75 & 0.25 \\ 0 & -1.5 & 2 & 0 & -0.5 \\ 0 & 0.75 & 0 & 1.25 & -0.5 \\ 0 & 0.25 & -0.5 & -0.5 & 0.75 \end{bmatrix} \begin{matrix} 1 \\ 2 \\ 3 \\ 4 \\ 5 \end{matrix}$$

Boundary Conditions

Let $\phi_1 = 20$, $\phi_2 = 16$, $\phi_3 = 12$

Let us evaluate the total heads at nodes 4 and 5 i.e., ϕ_4 and ϕ_5. Taking the assembly equations corresponding to the nodes of 4 and 5 (assuming F$_4$, F$_5$ nodal vectors of applied flow to be zero) i.e., $F_4 = F_5 = 0$.

$-12\phi_2 + 20\phi_4 - 8\phi_5 = 0$

$5\phi_4 - 2\phi_5 = 48$

$4\phi_2 - 8\phi_3 - 8\phi_4 + 12\phi_5 = 0$

$-2\phi_4 + 3\phi_5 = 8$

$\phi_4 = 14.55$, $\phi_5 = 12.37$

The quantity of flow in Element 3 can be calculated as follows (Section 6.3.2).

$$\text{Grad, } \phi = \frac{1}{2\Delta} \begin{bmatrix} b_i & b_j & b_m \\ c_i & c_j & c_m \end{bmatrix} \begin{Bmatrix} \phi_i \\ \phi_j \\ \phi_m \end{Bmatrix} = \frac{1}{8} \begin{bmatrix} -4 & 4 & 0 \\ 0 & -2 & 2 \end{bmatrix} \begin{Bmatrix} 20 \\ 16 \\ 14.55 \end{Bmatrix} = \frac{1}{8} \begin{Bmatrix} -16 \\ -2.9 \end{Bmatrix}$$

$$\{Q\} = \begin{Bmatrix} Q_x \\ Q_y \end{Bmatrix} = -k \cdot \frac{1}{8} \begin{Bmatrix} -16 \\ -2.9 \end{Bmatrix} = -k \cdot \frac{1}{8} \begin{Bmatrix} -16 \\ -2.9 \end{Bmatrix} = k \begin{Bmatrix} 2 \\ 0.3625 \end{Bmatrix}$$

The Q_x and Q_y are constant throughout the element. If Q is positive, the flow takes place along the positive direction of the axes.

Example 4:

Permeameter Analysis:

Consider the soil mass in Figure 6.4.4 (shown below), which is enclosed between two impervious surfaces which represents a typical permeameter used to determine soil permeability in the laboratory. The soil mass is subjected to the fluid pressure, p (distribution shown in the figure). From the definition of the total head, i.e.

$$\phi = p/\gamma + H$$

where ϕ = the total head, γ = unit density of fluid and H = elevation head at that point, the total head distribution at each end of the soil sample is uniform (Figure 6.4.4). No flow occurs across the impervious top and bottom boundaries, and the sample is assumed to have a unit thickness with no flow normal to the x-y plane.

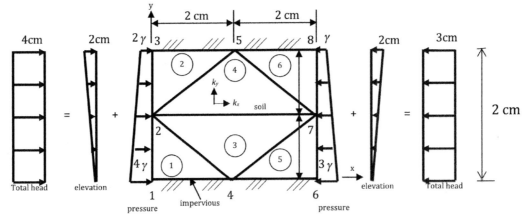

Figure 6.4.4 Permeameter

Solution:

Total head on the left $= \dfrac{p}{\gamma} + H = 4cm$

Total head on the right $= \dfrac{p}{\gamma} + H = 3cm$

Applied nodal heads:

$$\phi_1 = \phi_2 = \phi_3 = 4cm$$

$$\phi_6 = \phi_7 = \phi_8 = 3cm$$

ϕ_4 and ϕ_5 have to be determined

γ = density of fluid = 1gm/cm³

$k_x = k_y = 1$cm/sec = horizontal and vertical permeabilities

Discretize the region into 6 triangular elements as shown in Figure 6.4.4.
Following the steps explained in Section 6.3.2, the element characteristics can be calculated as follows.

Δ is the area of the element of unit thickness. By using the values of k_x and k_y and the dimensions of the elements as shown in Figure 6.4.4, we can compute element characteristic (permeability) matrices as:

$$[h]^1 = \frac{1}{4}\begin{array}{ccc} 1 & 4 & 2 \\ \begin{bmatrix} 5 & -1 & -4 \\ -1 & 1 & 0 \\ -4 & 0 & 4 \end{bmatrix} \end{array} \qquad [h]^2 = \frac{1}{4}\begin{array}{ccc} 2 & 5 & 3 \\ \begin{bmatrix} 4 & 0 & -4 \\ 0 & 1 & -1 \\ -4 & -1 & 5 \end{bmatrix} \end{array}$$

$$[h]^3 = \frac{1}{4}\begin{array}{ccc} 2 & 4 & 7 \\ \begin{bmatrix} \dfrac{5}{2} & -4 & \dfrac{3}{2} \\[2mm] -4 & 8 & -4 \\[2mm] \dfrac{3}{2} & -4 & \dfrac{5}{2} \end{bmatrix} \end{array} \qquad [h]^4 = \frac{1}{4}\begin{bmatrix} \dfrac{5}{2} & \dfrac{3}{2} & -4 \\[2mm] \dfrac{3}{2} & \dfrac{5}{2} & -4 \\[2mm] -4 & -4 & 8 \end{bmatrix}$$

227

$$[h]^5 = \frac{1}{4}\begin{bmatrix} 1 & -1 & 0 \\ -1 & 5 & -4 \\ 0 & -4 & 4 \end{bmatrix}, \quad [h]^6 = \frac{1}{4}\begin{bmatrix} 1 & 0 & -1 \\ 0 & 4 & -4 \\ -1 & -4 & 5 \end{bmatrix}$$

Now assemble the global permeability matrix and obtain the assemblage equations as

$$[H] = \begin{matrix} & \begin{matrix} 1 & 2 & & 3 & 4 & & 5 & 6 & 7 & & 8 \end{matrix} \\ \begin{bmatrix} 5 & -4 & 0 & -1 & 0 & 0 & 0 & 0 \\ & 13 & -4 & -4 & -4 & 0 & 3 & 0 \\ & & 5 & 0 & -1 & 0 & 0 & 0 \\ & & & 10 & 0 & -1 & -4 & 0 \\ & & & & 10 & 0 & -4 & -1 \\ & \text{Symmetrical} & & & & 5 & -4 & 0 \\ & & & & & & 13 & -4 \\ & & & & & & & 5 \end{bmatrix} \end{matrix} \begin{Bmatrix} 4 \\ 4 \\ 4 \\ \phi_4 \\ \phi_5 \\ 3 \\ 3 \\ 3 \end{Bmatrix} = \begin{Bmatrix} F_1 \\ F_2 \\ F_3 \\ F_4 = 0 \\ F_5 = 0 \\ F_6 \\ F_7 \\ F_8 \end{Bmatrix}$$

In this simple example, all nodal potentials are known except those at nodes 4 and 5. Since we know that the flows added at nodes 4 and 5, F_4 and F_5, are zero, we can solve for ϕ_4 and ϕ_5 directly as:

$$-4 - 16 + 10\phi_4 - 3 - 12 = 0, \quad \phi_4 = 3.5$$

and

$$-16 - 4 + 10\phi_5 - 12 - 3 = 0, \quad \phi_5 = 3.5$$

Hence, pressures at points 4 and 5 are

$$p_4 = (3.5 - 0)\gamma = 3.5\gamma$$
$$p_5 = (3.5 - 2)\gamma = 1.5\gamma$$

which indicates that the pressure variation in the x direction is linear.

Using Darcy's law, we can compute the element velocities and flow quantities as follows.

For example, in element 3 ($i=2, j=4, m=7$):
$\phi_2 = 4$, $\phi_4 = 3.5$, $\phi_7 = 3$, $b_i = -1$, $b_j = 0$, $b_m = 1$, $c_i = 2$, $c_j = -4$, $c_m = 2$, and the area $\Delta = 2cm^2$. Hence,

$$\begin{Bmatrix} \upsilon_x \\ \upsilon_y \end{Bmatrix} = -\frac{1}{4}\begin{bmatrix} 1 & 0 \\ 0 & 1 \end{bmatrix}\begin{bmatrix} -1 & 0 & 1 \\ 2 & -4 & 2 \end{bmatrix}\begin{Bmatrix} 4 \\ 3.5 \\ 3 \end{Bmatrix}$$

$$\begin{Bmatrix} \upsilon_x \\ \upsilon_y \end{Bmatrix} = \begin{Bmatrix} 0.25 \\ 0 \end{Bmatrix} cm/sec$$

The net flows from element (3) can be computed using element flow equations

228

$$[h]^3\{\phi\}^3 = \{F\}^3 = \frac{1}{4} \begin{array}{ccc} 2 & 4 & 7 \end{array} \begin{bmatrix} \frac{5}{2} & -4 & \frac{3}{2} \\ -4 & 8 & -4 \\ \frac{3}{2} & -4 & \frac{5}{2} \end{bmatrix} \begin{Bmatrix} \phi_2 = 4 \\ \phi_4 = 3.5 \\ \phi_7 = 3 \end{Bmatrix} = \begin{Bmatrix} F_2 \\ F_4 \\ F_7 \end{Bmatrix}^3 = \begin{Bmatrix} \frac{1}{8} \\ 0 \\ -\frac{1}{8} \end{Bmatrix}$$

The equivalent fluid flows at the nodes of element 3 can also be calculated as:

$$\begin{Bmatrix} Q_{x1} \\ Q_{x2} \\ Q_{x3} \\ Q_{y1} \\ Q_{y2} \\ Q_{y3} \end{Bmatrix} = -\frac{1}{2} \begin{bmatrix} -1 & 0 \\ 0 & 0 \\ 1 & 0 \\ 0 & 2 \\ 0 & -4 \\ 0 & 2 \end{bmatrix} \begin{Bmatrix} 0.25 \\ 0 \end{Bmatrix} = \begin{Bmatrix} \frac{1}{8} \\ 0 \\ -\frac{1}{8} \\ 0 \\ 0 \\ 0 \end{Bmatrix}$$

The assembled/systems net flows at nodes can be computed as:
$F_2 = -4 + 4 + 13 \times 4 - 4 \times 4 - 4 \times 3.5 - 4 \times 3.5 + 0 + 3 \times 3 + 0 = 1$

6.5   Assignment Problems

Note:
1. Use triangular elements for discretization of the domain / region / continuum for FEM analysis.
2. Use only small number of elements for discretization for making manual computations using calculator.
3. Use the value of coefficient permeability, $k = 4 \times 10^{-7}$ m/sec unless specified otherwise in the problem.
4. All the dimensions are in meters unless specified otherwise in the problem.
5. Compare the FEM results with results using any other methods including results from Software packages, wherever possible
6. Assume the thickness of the region to be 1 m in the z direction.

1. Obtain the values of the equivalent fluid flow (seepage) at nodes A, B and C for the seepage problem shown in Figure 6.5.1, using Finite Element Approach (use one element ABC). ϕ is the total head in meters. Also find the seepage Q in the element. All dimensions are in meters. Nodal coordinates: A (15,10); B (0,0); C (20,0); Total head values at nodes: $\phi_A = 40$; $\phi_B = 30$; $\phi_C = 50$
   Calculate the head at
   i.   D (10,5)
   ii.   C.G. of the element.

iii.   What should be the total head at A in case the head at the C.G. is 35? iv. What is the total flow cross AC?

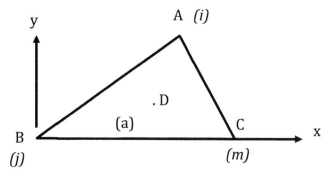

Figure 6.5.1

2. Redo the Problem 1 with $k_x - 2 k_y = 4 \times 10^{-7}$ m/sec.

3. Redo problem with AC as a non-conducting boundary

4. Redo problem 2 with AB as a non-conducting boundary.

5. For the two-dimensional soil domain shown in Figure 6.5.2 (take the thickness as unity) evaluate the value of the head, gradient and quantity of flow at point A, using Finite Element Method. The soil is homogeneous and isotropic with constant permeability coefficient, k. THE applied total heads at the corner nodes is indicated in the Figure below. Calculate the flow quantities at all the nodes and boundaries BC and CD.

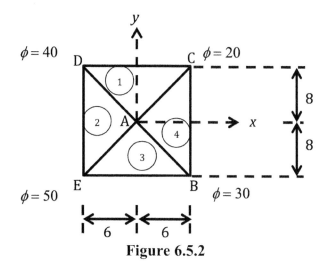

Figure 6.5.2

6. Redo Problem with $k_x = 4 k_y = 4 \times 10^{-7}$ m /sec.

7. Redo Problem 5 with DC and EB as non-conducting boundaries.

8. Redo Problem 6 with DC and EB as non-conducting boundaries.

230

9. Carry out seepage analysis for the two-dimensional region ABCD shown in Figure 6.5.3. The values of the total head at the nodes A, E, B are as follows.

$$\phi_A = 30, \ \phi_E = 24, \ \phi_B = 16.$$

i. Calculate ϕ_C and ϕ_D

ii. Calculate total flow in all the elements

iii. Calculate the flow across the surface BC

iv. Calculate gradients of flow at all the nodes

v. Assuming AB and CD to be non-conducting boundaries, calculate the flow across DE and BC

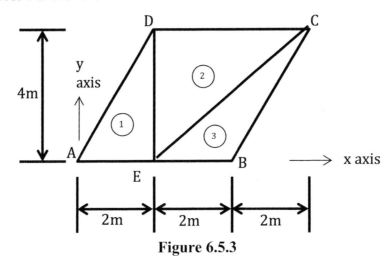

Figure 6.5.3

10. Redo Problem 9 with $k_x = 2 \ k_y = 4 \times 10^{-7}$ m /sec.

11. Redo Problem 9 using only two elements ABD and BDC for discretization and compare the results.

12. Redo problem 11 with $k_x = 2 \ k_y = 4 \times 10^{-7}$ m /sec. and compare the results with those from Problem 10.

13. Analyze the seepage characteristics of the laboratory permeameter shown in Figure 6.5.4. The soil mass is enclosed between two impervious surfaces, 1-6 and 3-7 and is subjected to the fluid pressure, p (distribution shown in the figure). Neglecting the velocity head, the total head at any point is,

$$\phi = p/\gamma + H$$

where ϕ = the total head, γ = unit density of fluid and H = elevation head at that point. The total head distribution at each end of the soil sample is uniform as shown in Figure 6.5.4. No flow occurs across the impervious top and bottom boundaries, and the sample is assumed to have a unit thickness with no flow normal to the x-y plane.

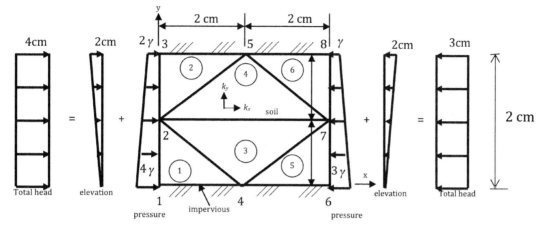

Figure 6.5.4

14. Redo the Problem 13 with anisotropic permeabilities i.e. $k_x = 2\,k_y = 4 \times 10^{-7}$ m /sec.

15. Analyse for seepage characteristics of the pyramidal permeameter (symmetric about its axis along x axis) shown below in Figure 6.5.5 using three elements as marked. Assume AB and CD to be non-conducting boundaries, The values of the total head at the nodes are: $\phi_A = 50,\ \phi_B = 50,\ \phi_C = 36, \phi_D = 36$. Calculate

   i. ϕ_E and ϕ_D

  ii. Calculate total flow in all the elements

 iii. Calculate the flow across the surface BC

 iv. Calculate gradients of flow at all the nodes. calculate the flow across CE and BC.

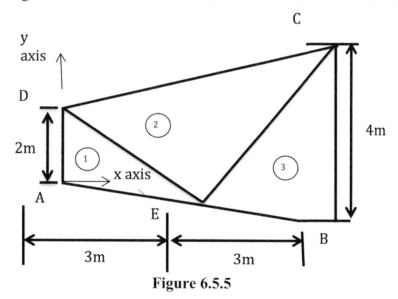

Figure 6.5.5

16. Redo Problem 15 with isotropic permeabilities i.e., $k_x = 2,\ k_y = 4 \times 10^{-7}$ m /sec.

Appendix 6 A
Seepage / Flow through Porous Media

6A.1   Introduction

The presence of water in soils affects the behavior whether it be granular or cohesionless, or cohesive. Generally, in cohesionless soils, only the effective unit weight changes when water fills the voids, while in cohesive soils, the conditions of their deposition/formation, and response of the soil to applied stresses are all dependent on the water around the soil particles.

Water exists in cohesive soils in chemically combined, adsorbed on the surface, and (or) free, unbonded forms. In cohesionless soils, it exists in significant quantity only in the form of free water. No differentiation may be made between the adsorbed surface water on the finer particles and the free pore water. However, there is a gradual transition from the ordered distribution of water molecules nearest the surface to the randomly oriented molecules slightly away from the surface.

Under such circumstances, it is somewhat difficult to explain the term "free" water in the pores of a clay soil because of the gradual change in the degree of freedom. For the purposes of discussion, free water will be considered to be that water in the pores of a soil which can flow through the pores by pressures applied to the soil structure or pore water. In cohesive soils, the water which moves under these conditions will have to be determined by physicochemical conditions, the imposed stress levels, and the system temperature etc.

The water which flows through the pores of any soil under external stresses is an important area in Soil Mechanics / Civil Engineering. Water pressures and flow quantities under or inside dams, around foundations, cofferdams, or retaining walls, or in compressible soil layers are important parameters in the design of such structures. In the following section, the theories and equations which have been developed to represent the flow of water through porous media in general and soil media in particular are presented. The solution of the appropriate equations for particular cases of flow through soils makes it possible to arrive at estimates of pore-water pressures and seepage quantities for practical applications. Various methods of solution including Finite Element Method (FEM) are available to analyze a wide range of problems in this area.

6A.2   Flow Equations

The flow of fluids under general circumstances can be formulated completely and analyzed by specifying the following seven conditions.

(1) Fluid

(a) Condition of continuity - The net mass or weight flow of fluid into or out of a geometrically circumscribed volume in a given time interval must be equal to the storage or loss of storage of fluid in the volume in the interval. If there is no storage in the volume, the net flow must be zero.

(b) *Equation of state of the fluid* - If the flow of a fluid is under varying conditions of temperature and pressure, the pressure-density-temperature relation of the fluid is to be specified. Such a relationship is referred to as the equation of state of the fluid.

(c) *Dynamic condition* - The response of the fluid to applied pressures must be specified by a relationship analogous to Newton's second law of motion for solids relating proportionality between force and acceleration.

(2) Soil solids

(a) *Condition of continuity* - The net mass or weight flow of soil solids into or out of a geometrically circumscribed volume in a given time interval must be equal to the storage or loss of storage of the soil solids in the volume in the interval.

(b) *Equation of state of soil solids and structure* - Since the soil solids and structure may deform or change in volume under varying stresses, the response of these components to changing stress must be characterized in a manner analogous to condition 1(b) for the fluid phase.

(c) *Dynamic condition* - In general, the soil structure will be subjected to stresses which will vary in time, and its reaction to such dynamic stresses requires description. However, in all seepage problems, the forces and stresses vary very slowly in time so that the accelerations of the soil skeleton are negligibly small thus representing static behavior.

(3) *Fluid-solid interrelationship* - Since the soil-fluid system will be subjected to external stresses, these should be balanced by the stresses in the soil and fluid at all points in the system. Thus the state of the soil or fluid can be completely described at any point and at any time by specifying a relationship between applied, fluid, and soil stresses.

The first three conditions, 1(a) through (c), relate to the fluid phase only, and it is convenient to discuss them for the analysis of flow through porous media. The last four, 2(a) to (c) and 3 relate to transient flow problems and can be studied when necessary.

Water flows in the soil through pores of greatly varying cross sections. Across each section of a pore the distribution of water velocity cannot be determined because of the random nature of the pore spaces. Hence, the fluid flow / velocity can be analyzed in terms of average quantities, based either on the gross cross-sectional area of an element of the soil mass or on that proportion of the gross area which is occupied by voids. If the rate of discharge of water through a section of soil of gross area A is q, then the *superficial* velocity of flow, v, is defined as

$$v = \frac{q}{A}$$

(6A.1a)

234

If the porosity of an element of soil of cross-sectional area A and height H is n, it can be assumed that an average area nA will be shared by void space in each section in the height H, since

$$\int_0^H nA\,dH = nAH \qquad (6A.2)$$

Considering that water flow takes place through the pores only, we can calculate the average *seepage* velocity v_s as

$$v_s = \frac{q}{nA}, \qquad (6A.1b)$$

Hence, from Eqns. (6A.1 a, b) we get

$$v = nv_s \qquad (6A.3)$$

Because of the convenience of using gross cross-sectional areas, most calculation involving the velocity of fluid flow in soil make use of the superficial velocity v, which is commonly used for such studies.

Another complexity may arise, if the soil is not fully saturated since there is gas / air present in the soil structure, part of which may be immobile and some of it may move with the pore fluid as it passes through the soil. It is difficult to assess the proportions of the gas /air which may be included in either process. Hence it may be logical to hypothesize that if the bubbles do not move, they can be considered to be part of the compressible solid component, taking part in the process of change of fluid storage in the element but nor in the flow mechanism. Thus for further analysis, it is convenient to assume that the major part of the gas in unsaturated soils performs this function as this is quite possible. If gas moves with the pore fluid, the equation of state of the fluid needs to be modified. In either case, the mass quantity of static or mobile gas/air must be assumed to be constant. Accordingly, the gas/air, if present, is commonly assumed to be attached to the soil structure for analysis.

Continuity condition - An element of a mass of soil ABCDEFGH through which fluid flow is taking place is shown in Figure 6A.1 with dimensions of dx, dy, and dz along the Cartesian coordinate axes x, y, and z respectively. Fluid / Water may flow into or out of the element through each of its faces, and fluid / water may be stored (positively or negatively including zero storage) within the element during a time interval, for various reasons such as compressibility of the water, compressibility of any gas bubbles present in the soil, compressibility of the soil solids and of the soil structure.

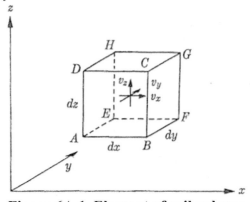

Figure 6A.1 Element of soil volume

As per the conditions of continuity, the quantity of water flowing into the element less the quantity of water flowing out equals the amount of water stored within the element, in any time interval.

Then the weight flux of water per unit cross-sectional area passing through the center in the x-direction is given by $\gamma_w v_x$ where γ_w and v_x are the unit weight of water and superficial velocity at the center (at the water pressure existing at the element center). The rate of change of weight flux per unit area with distance in the x-direction is $(\partial/\partial x)(\gamma_w v_x)$, and therefore the weight flux of water through the face $ADHE$ is given by

$$\left[\gamma_w v_x - \frac{\partial}{\partial x}(\gamma_w v_x)\frac{dx}{2}\right]dzdy \qquad (6A.4a)$$

Similarly, the weight flux through face $BCGF$ is

$$\left[\gamma_w v_x + \frac{\partial}{\partial x}(\gamma_w v_x)\frac{dx}{2}\right]dzdy \qquad (6A.4b)$$

The net weight of water entering or leaving the element through faces $ADHE$ and $BCGF$ is given by the difference between the above two fluxes:

$$(\text{Net weight flux})_x = \frac{\partial}{\partial x}(\gamma_w v_x)dxdzdy \qquad (6A.4c)$$

Flow will also take place in the z-direction with velocity v_z at the center of the element, and the net weight flux parallel to the z-axis is

$$(\text{Net weight flux})_z = \frac{\partial}{\partial z}(\gamma_w v_z)dzdxdy \qquad (6A.5)$$

Similarly, along the direction y (perpendicular to the paper), the net weight flux is

$$(\text{Net weight flux})_y = \frac{\partial}{\partial y}(\gamma_w v_y)dydxdz \qquad (6A.6)$$

The rate of storage or loss of fluid in the element is given by the sum of Eqs. (6A.4c) (6A.5) and (6A.6). Therefore

Rate of change of storage of fluid / water in the element $= \dfrac{\partial W}{\partial t}$

$$= \left[\frac{\partial}{\partial x}(\gamma_w v_x) + \frac{\partial}{\partial y}(\gamma_w v_y) + \frac{\partial}{\partial z}(\gamma_w v_z)\right]dxdydz \qquad (6A.7a)$$

where W is the weight of fluid stored in the volume $dx\ dy\ dz$. This is the equation of continuity representing the conservation of water during the flow process. In other branches of study, similar expressions may be written, for example, in terms of charge in electrical systems or in terms of heat in thermal flow.

The left-hand side of Eq. (6A.7a) accounts for the storage change due to either all or any of the factors i.e., compression of the soil, or water, and (or) air / gas constituents of the soil. The relative amounts of the different components of storage may vary widely for a given soil, and only significant terms may be retained for practical applications. For example, if the soil is fully or almost saturated, the change in the gas volume may be neglected. Since this is usually the case in nature, the change of storage in an elemental soil volume is primarily due to the compressibility of the soil structure and the pore water. If soil layers hundreds or thousands of feet thick are being studied (for example, in a problem involving a deep water well or pumping from an oil field), both the soil and fluid compressibility need to be considered. However, in Civil engineering (with the possible exception of high earth dams and embankments) only relatively shallow depths of soil of the order of a few of meters are analyzed. For such depths the fluid compressibility is usually negligible. Further, the soil itself may be relatively incompressible if it consists of a dense granular material.

Thus, if the soil structure and pore fluid may reasonably be assumed to be incompressible, and no gas exists in the pores, there will be no change in fluid storage in any soil element during the flow process, and Eq. (6A.7a) becomes

$$\frac{\partial}{\partial x}(\gamma_w \upsilon_x) + \frac{\partial}{\partial y}(\gamma_w \upsilon_y) + \frac{\partial}{\partial z}(\gamma_w \upsilon_z) = 0 \qquad (6A.7b)$$

This is the general equation of continuity of flow when the there is no change in storage. It therefore describes flow conditions which do not change in time and is referred to as a *steady-state* equation.

Equations of state - The effect of the behaviors of the solid, fluid, and gaseous components of the soil as a function of temperature and pressure on the equation of storage i.e. Eq. (6A.7) may need to be studied further. These relations are called *equations of state*, in which the densities of the various substances are usually expressed as functions of pressure and temperature. Leaving the discussion of the equation of state of the soil alone for the present, we can examine the equations for water and gas.

Muskat [Scott, 1963] uses the general expression for all fluids,

$$\gamma = \gamma_0 (p/p_a)^m e^{\beta p} \qquad (6A.8)$$

where,

Liquids: $m=0$;
    Incompressible: $\beta = 0$;
    Compressible: $\beta \neq 0$

Gases: $\beta = 0$;
    Isothermal expansion: $m = 1$;
    Adiabatic expansion: m = constant volume specific heat / constant
        pressure specific heat

Since Eq. (6A.8), does not match well with experimental data a simpler expression (Scott, 1963) for liquids is used (especially water) as follows.

$$\gamma_w = \gamma_{w0} (1 + \beta p) \qquad (6A.9)$$

where p, the pressure, is measured above the standard atmospheric pressure at which the unit weight of the liquid (water) is γ_{w0}; β is the compressibility of the fluid. Its value for water is of the order of

$$1/\beta = 300,000\,psi$$

The equation of state for gas under isothermal volume changes is expressed as

$$\gamma_g = \gamma_{g0}\left(p_g + p_a\right)/p_a \tag{6A.10}$$

where γ_{g0} is the unit weight of the gas at atmosphere pressure p_a and the appropriate temperature, and p_g is the pressure in the gas measured above atmospheric pressure. The equations of state of the soil constituents which occur in the left side of Eq. (6A.7a) require more detailed studies which may be relevant for transient flow and consolidation.

Fluid dynamics - Eq. (6A.7) is formulated in terms of space derivatives of weight flux / flow, which involve the superficial velocity of the fluid as a function of coordinates (the fluid velocity being a vector). To describe the motion of the fluid completely, the effect of applied forces / pressures on the movement of the fluid is to be known. Then only Eq. (6A.7), together with the equations of state can be analyzed. For example, for a compressible fluid and completely saturated soil, we have, in the case of isothermal flow, seven unknowns. They are the velocities of the fluid, v_x, v_y, v_z, the pressure in the fluid, p, the pressure in the soil structure, $\bar{\sigma}$, the unit weight of fluid, γ_w, and that of the soil structure, γ_d, at every point in the medium. Combining Eqns. (6A.7) and (6A.9) with another expression for the soil compressibility yields only a total of three relationships among the quantities. Therefore, we need four more expressions to solve a particular case for which the applied forces and boundaries are known. One of these equations arises from condition (3), and the other three necessary equations are obtained from the expressions relating the response of the fluid to applied stress. The relationship between stresses in the fluid and those in the solid components (condition 3) become relevant for transient flow. Since the present discussion is focused on response of fluids to applied pressures the dynamics of fluid flow is presented below.

In general, the fluid will move under the action of the internal pressure gradients (which develop from the initial pressure conditions in the fluid and along the boundaries of the region under consideration) and *body forces*, which, in fluids in soil mechanics problems, are almost entirely due to the force of gravity on the mass of the fluid. The motion of the fluid will be resisted by internal friction due to its viscosity.

Combining the effects of the three forces, pressure gradients, body forces, and viscous friction resistance, results in the well-known Navier-Stokes equations of motion in fluid mechanics. These three equations, give, the total time derivative of the velocity (that is, the acceleration) in the x-, y-, and z-directions respectively as a function of the pressure gradients, body forces, and viscous resistance forces acting in respective directions. Very few analytical solutions of these equations are available in literature.

Solutions to simple problems, such as the flow between parallel flat plates or through a tube of constant circular cross section (Hagen-Poiseuille equation) are available (Taylor, 1967). Under these circumstances, the solution of the equations for flow through soils is extremely difficult.

Hence the objective now is to formulate a dynamical equation for the flow of fluids through porous media in particular, analogous to the general Navier-Strokes relations. Accordingly, Darcy, presented the results of his investigations in the form of an equation since known as Darcy's law for flow in one direction. It may be generalized to describe three-dimensional flow in conjunction with the continuity relation and the equations of state for the constituents which can lead to equations completely describing flow through a given region of porous medium. However, it has some limitations as discussed below.

6A.3   Darcy's Law

Darcy carried out experiments on the flow of water through porous sand filter beds, and concluded that the superficial velocity of flow is directly proportional to the pressure gradient and the constant of proportionality which included both the soil and water properties (Taylor,1967). This constant is called the coefficient of permeability, k of the porous medium / soil medium.

Experimental results - In most hydraulics and soil mechanics investigations, it is conventional to represent fluid pressure in terms of *heads*, that is, in terms of the static height of a column of fluid which would result in a given pressure at a point. Thus, instead of referring to a *fluid pressure, p,* where p is expressed in, say pascals, it is referred to as *fluid pressure head* denoted by the symbol h_p. Pressure and pressure head are related through the unit weight of the fluid,

$$h_p = \frac{p}{\gamma_w} \qquad (6A.11a)$$

if the fluid can be considered incompressible, and

$$h_p = \int \frac{dp}{\gamma_w} \qquad (6A.11b)$$

if compressibility is to be considered. then, the equations for superficial velocity may in general be written as:

$$v_x = -k_x \frac{\partial h_p}{\partial x} \qquad (6A.12a)$$

$$v_y = -k_y \frac{\partial h_p}{\partial y} \qquad (6A.12b)$$

$$v_z = -k_z \frac{\partial h_p}{\partial z} \qquad (6A.12c)$$

In these equations the minus sign indicates that velocity is positive in the direction of decreasing pressure.

In soil mechanics terminology, the constants of proportionality k_x, k_y, k_z, are called the *coefficients of permeability* or sometimes the *permeability* of a soil in the x, y-, and z- directions. It can be understood that such a term implicitly involves the unit weight and viscosity of the pore fluid, besides the size and geometry of the pores of the soil through which flow is taking place. In other areas of study (petroleum engineering) where fluid flow through porous media is of interest, discrimination is made between the fluid and medium properties, so that, for example

$$k_x = \frac{\gamma_w K_x}{\mu} \tag{6A.13}$$

where K_x, is a function of the porous medium alone and μ, is the viscosity of the pore fluid.

Eqns. (6A.12) express the velocity in terms the internal pressure gradients and the fluid viscosity, to obtain the Navier-Stokes equations, but do not take into account possible body forces. In soil mechanics, as mentioned above, the principal body force of interest is gravity, which can be included in the equations as follows:

$$v_x = -k_x \left(\frac{\partial h_p}{\partial x} - \frac{F_x}{\gamma_w} \right) \tag{6A.14a}$$

$$v_y = -k_y \left(\frac{\partial h_p}{\partial y} - \frac{F_y}{\gamma_w} \right) \tag{6A.14b}$$

$$v_z = -k_z \left(\frac{\partial h_p}{\partial z} - \frac{F_z}{\gamma_w} \right) \tag{6A.14c}$$

where the quantities $F_{x,y,z}$ represent the body forces per unit volume due to gravity in the x-, y-, and z-directions. Since coordinate axes are in general direction, gravity-force components may exist in all directions. In Eqns. (6A.14), it would be more useful to express both pressure and gravity forces in one derivative with respect to direction for easy analysis.

This can be done by considering the height h_e of any fluid element above an arbitrary zero datum plane. The potential energy per unit mass of the fluid with respect to the datum is h_e, and the force required to move the mass at constant pressure in any one direction will be given by the derivative of the energy with respect to that direction. Since this force is required to overcome the gravitational body force on the mass, we can write

$$\frac{F_x}{\gamma_w} = -\frac{\partial h_e}{\partial x} \tag{6A.15a}$$

$$\frac{F_y}{\gamma_w} = -\frac{\partial h_e}{\partial y} \tag{6A.15b}$$

$$\frac{F_z}{\gamma_w} = -\frac{\partial h_e}{\partial z} \qquad (6A.15c)$$

where the negative sign follows from the fact that the force acts in the direction opposite to gravity. Then equations (6A.15) can be substituted in equation (6A.14) to give

$$v_x = -k_x \frac{\partial h}{\partial x} \qquad (6A.16a)$$

$$v_y = -k_y \frac{\partial h}{\partial y} \qquad (6A.16b)$$

$$v_z = -k_z \frac{\partial h}{\partial z} \qquad (6A.16c)$$

where h is a total head given by the expression

$$h = h_p + h_e \qquad (6A.17)$$

It may be reiterated that velocity head is neglected in the expression of total head given by Eq. (6A.17) since the velocity of fluid in porous media is very small.

Thus the total head is a potential function similar to the voltage potential V in the flow of electricity where Eq. (6A.16), for example, is written as

$$I_x = -\frac{1}{R_x} \frac{\partial V}{\partial x} \qquad (6A.18)$$

in which I_x is the current in the x-direction (directly analogous to v_x) and R_x is the electrical resistivity of the medium through which electrical flow is taking place $(=1/k_x)$. Another potential function is temperature, θ, in heat flow problems for which the appropriate flux equation is

$$q_x = -C_x \frac{\partial \theta}{\partial x} \qquad (6A.19)$$

where q_x is the heat flow (analogous to v_x) and C_x represents the thermal conductivity of the flow medium (analogous to k_x). These analogies will be useful in further studies of fluid flow. The total head in Eq. (6A.17) can also be considered to represent the energy required to raise unit mass of fluid originally at zero pressure and zero elevation (on the arbitrary datum plane) to elevation h_e and to subject it to a pressure p [given by Eq. (6A.11)].

Theoretical verification
Since Darcy's law has been deduced from experimental studies, many attempts have been made to achieve a theoretical confirmation and to determine values of permeability from theoretical considerations of the size of the soil grains and size and shape of the pore spaces through which flow occurs. Many of these investigations have begun from the Hagen-Poiseuille equation for viscous flow through a small capillary tube of diameter d, which is stated as:

$$\bar{v}_s = -\frac{d^2}{32\mu}\frac{dp}{dl} \qquad (6A.20)$$

where $\bar{v}_s$ is the average flow velocity through the tube and l is measured along the tube.

One of the proposed models consists of a bundle of such tubes arranged in parallel with a ratio of pore cross section to total area of n(porosity), to simulate in which case $\bar{v}_s$ represents a seepage velocity. Noting the similarity of Eqns. (6A.20) and (6A.12), an expression can be written for the coefficient of permeability involving the diameter of the tubes, the porosity of the bundle, and the viscosity of the fluid (Taylor, 1964). This expression can be used to attempt to predict the permeability of a given soil.

However, in practical applications, some differences between the model and soil exist. For example, it is very difficult to specify the "diameter" of the pores, and also measure it, since the pores vary greatly in size and may not, indeed, be continuous throughout the porous medium. Also, the flow in the soil from entrance to exit does not take place in parallel streams.

Although the pore size may depend on the dimensions of the soil grains, there can be no unique relationship since the same collection of grains can be arranged in various assemblages resulting in different pore sizes. Thus, in various studies for correlations between models and natural soil, some form of average pore diameter is used. This diameter may be arrived at through the concept of specific surface (area) or by means of the hydraulic radius as used by Kozeny (Scott, 1963). In addition, since the actual path of fluid flow through the medium is not straight but extremely wavering and twisted, "tortuosity" factors have been proposed to account for the difference in length between the flow path of an "average" water molecule and the distance in a straight line through the medium. Some models suggest changes in the cross-sectional area of the tubes to allow for the varying pore diameter in natural soils.

The assumptions implicit in the various models and their drawbacks are thoroughly discussed by Scheidegger (Scott, 1963).

Limitation

Noting that flow through soils as expressed by Darcy's law is a viscous phenomenon similar to that on which Hagen-Poiseuille's law is based, there are some limitations for its application as in other flow problems. A dimensional analysis of the related parameters indicates that the pressure drops over the length of a tube (or a length of porous medium) is (among other relationships) a function of the dimensionless Reynolds number, R_e defined as:

$$R_e = \frac{\gamma_w}{g}\frac{\bar{v}_s d}{\mu} \qquad (6A.21)$$

242

where $\bar{v}_s$ is the average velocity and d is the diameter of the tube. Reynolds first investigated fluid flow through tubes and found that under certain conditions laminar or smooth flow breaks down and turbulence begins. These conditions are characterized by a limiting value of R_e equal to about 2000, in tube flow, although the value in a particular experiment depends on environmental conditions. Below this value the flow is considered laminar / smooth, and above it, it is considered turbulent.

In these flow studies are carried out in parallel-walled tubes which have continuous and smooth walls without breaks. This is totally different in the case of flow in soils, where the diameter of a continuous pore may vary widely from point to point in the medium. The fluid velocity will also change widely as water flows through the soil. Hence, we may therefore expect, that the transition in flow through soil may not be characterized by the same Reynolds number (since the conditions under which the transition from smooth to turbulent flow takes place in pipes in laboratory investigations are different than the ones in flow through soil media). The same difficultly which occurred in theoretical representation of the soil permeability arises in arriving at a Reynolds number for a given flow through soil, since it involves some unknown values of velocity and pore diameter. If the superficial velocity given by Darcy's law is used and a grain diameter, d, given by the relation

$$d = \sqrt[3]{\sum n_s d_s^3 / \sum n_s}$$ (6A.22)

where n_s is the number of grains of diameter d_s occurring in the soil, then it is found that Darcy's law is valid to values of Reynolds number of the order of unity. The pore diameter is thus not directly expressed, but instead an average grain diameter is used. Even this value of Reynolds number is only very broadly applicable, since there is probably an uncertainty about its accuracy. If, in the determination of Reynolds number, the effective diameter of the soil, d_{10}, is used, the limiting Reynolds number is found to be in the range of 3 to 10. It is pointed out by Muskat that for normal applications in soil mechanics, Darcy's law appears to be valid at least up to the size range of medium to coarse sands. For turbulent flow in coarse soils or under high heads, Scheidegger (Scott, 1963) presents formulae, which may be applied to water flow through coarser materials such as in rock-filled dams. Deviations from Darcy's law can also be expected in extremely fine-grained (cohesive) soils.

It has been suggested that, as a result of physicochemical interactions between the soil and water in clays, flow of water will not occur until a certain limiting gradient i_0 of total head is surpassed (Scott, 63) and in such a case, Eq. (6A.16a), for example, may have to be modified as:

$$v_x = -k_x \left(\frac{\partial h}{\partial x} - i_0 \right)$$ (6A.16d)

The limiting gradient i_0 depends on both the structure and the void ratio of the soil and could be as high as 20 to 30 in very dense clays.

In most soil mechanics applications Darcy's law needs not be modified due to the above factors, although their effect may not be negligible in certain soils. In such specific cases, the gradients in the pore water in laboratory tests are much higher than those generated in the same soil in the field. Non applicability of Darcy's law from such effects even in cohesive soils only occurs when the pore fluid is a gas which is not relevant for seepage of water in soils.

Coefficient of permeability

The coefficient of permeability, k used in Darcy's law, Eq. (6A.16), includes the viscosity and unit weight of the pore fluid. This expression of permeability is applicable when the flow is predominantly due to only one fluid, such as water. The coefficient k is usually determined from field tests (field permeability tests, well pumping tests) or in the laboratory using either constant head permeameter test (preferable for cohesionless soils) and (or) variable / falling head permeameter test (preferable for cohesive soils) (Scott,1963, Taylor, 1964), Evaluations based on the grain size of the soil are, in general, inaccurate due to the reasons mentioned in the above paragraphs. In the various laboratory tests, k_T, the coefficient of permeability obtained at temperature $T\,^oC$, is normally corrected to k_{20}, the value at $20\,^oC$ ($68\,^oF$), by means of the relation (obtained from Eq. (6A.13)) as:

$$k_{20} = \frac{\mu_T}{\mu_{20}} \frac{\gamma_{20}}{\gamma_T} k_T \qquad (6A.23)$$

where the subscripts refer to the temperature.

The results of tests, such as the falling-head permeameter test referred above, carried out on different types of soil, are given in Table 6A-1 showing the range of permeabilities for different soils. The values show a very wide variation with grain size. According to Darcy's law, Eq. (6A.16), permeability is expressed in terms of a velocity (usually centimeters per second or feet per day in soil mechanics) and varies from typical values such as 10^{-3} cm/sec for sands to 10^{-9} cm/sec for clays.

Table 6A.1 Approximate values of permeability of soils

Type of Soil	Grain size, mm	Size at which permeability is measured, mm	Coefficient of permeability	
			cm/sec	ft/yr
Gravel		4	1	10^6
	2.0			
Sand		0.6	10^{-2}	10^4
	0.06	0.06	10^{-4}	10^2
Silt		0.008	10^{-6}	1
	0.002			
Clay		0.001	10^{-8}	10^{-2}

Also, if we carry out a permeability test on a sand whose density is varied, we will find that the permeability decreases with decreasing void ratio, though the relationship cannot be determined uniquely due to large number of factors involved.

For the various clay minerals and different exchangeable cations, some qualitative relations can be given, largely reflecting the different average sizes of the clay platelets in the minerals (Scott. 1963).

For unsaturated soils, the coefficient of permeability will be smaller compared to saturated soils. Experiments indicate (Scott, 1963) that the ratio of the permeability of an unsaturated soil to that of a saturated material *at the same void ratio* varies, However, in the range of degree of saturation which is of most interest in soil engineering, that is, from 80-100%, the ratio of the permeabilities above is nearly a linear function of the degree of saturation (Scott, 1963).

Any analysis of flow under conditions of turbulence (beyond a Reynolds number of about 1) becomes very complicated due to the difficulty of determining the constants used in the appropriate flow equation. In soil mechanics applications, such flows take place through very coarse materials. Under these conditions, use of nonlinear velocity versus gradient equations have been proposed as given below (Scott, 1963)]:

$$v_x = P\left(\frac{\partial h}{\partial x}\right)^n \tag{6A.24}$$

where P and n are constants to be determined from special tests.

6A.4 General Equations of Flow

If we combine the continuity equation (6A.7) and the Darcy's law, Eq. (6A.16), we get the expression for the storage in the element as:

$$\frac{\partial}{\partial x}\left(\gamma_w k_x \frac{\partial h}{\partial x}\right) + \frac{\partial}{\partial y}\left(\gamma_w k_y \frac{\partial h}{\partial y}\right) + \frac{\partial}{\partial z}\left(\gamma_w k_z \frac{\partial h}{\partial z}\right) = \frac{\partial W}{\partial t} \tag{6A.25}$$

The values of $k_{x,y,z}$ depend on the kinematic velocity of the pore fluid as well as on the geometric properties of the soil (Eq. 6A.13). In the range of temperature of interest in soil mechanics, the viscosity does not change appreciably with pressure (Scott, 1963), but there could be changes in the geometric properties of the soil structure. Due to changes of pressure the void ratio or porosity of the soil and therefore the diameter of the pore spaces will change so that $k_{x,y,z}$ cannot, be considered to be constant with pressure. Both the intergranular soil pressure and the fluid pressure will vary from place to place in a soil medium during the flow of water, so that even an initially homogeneous soil may experience spatial variations in permeability as flow continues. Consequently, the permeability of granular media may not change much as flow takes place, but finer-grained cohesive soils may undergo appreciable variations in permeability. However, the directional permeabilities are assumed to be invariant during the flow process in practice. With the assumption of constant coefficients of permeability Eq. (6A.25) can be expanded as:

$$\left(k_x \gamma_w \frac{\partial^2 h}{\partial x^2} + k_y \gamma_w \frac{\partial^2 h}{\partial y^2} + k_z \gamma_w \frac{\partial^2 h}{\partial z^2} \right) + \left(k_x \frac{\partial \gamma_w}{\partial x} \frac{\partial h}{\partial x} + k_y \frac{\partial \gamma_w}{\partial y} \frac{\partial h}{\partial y} + k_z \frac{\partial \gamma_w}{\partial z} \frac{\partial h}{\partial z} \right) = \frac{\partial W}{\partial t} \qquad (6A.26)$$

It is assumed that any gas present in the pores remains immobile, and in effect constitutes part of the soil solid phase. Since derivatives of γ_w are present in Eq. 6A.26), the equation of state for the pore fluid, Eq. (6A.9), is required to expand it further. Differentiating Eq. (6A.9) gives

$$d\gamma_w = \gamma_{w0} \beta dp \qquad (6A.27)$$

However, Eqns. (6A.11b) and (6A.17) relate the total head in the fluid to its pressure and elevation heads, i.e.

$$h = h_e + \int \frac{dp}{\gamma_w} \qquad (6A.28)$$

if the fluid is considered to be compressible. Differentiating Eq. (6A.28), we obtain

$$dp = \gamma_w \left(dh - dh_e \right) \qquad (6A.29)$$

which may be substituted in Eq. (6A.27) to give

$$d\gamma_w = \gamma_{w0} \beta \gamma_w \left(dh - dh_e \right) \qquad (6A.30)$$

Usually the x- and y-axes are considered to lie in the horizontal plane and z-axis to be vertical, with the positive direction taken upward. Then, inserting Eq. (6A.30) into Eq. (6A.26), we obtain

$$\gamma_w \left[k_x \frac{\partial^2 h}{\partial x^2} + k_y \frac{\partial^2 h}{\partial y^2} + k_z \frac{\partial^2 h}{\partial z^2} \right] + \gamma_w \gamma_0 \beta \left[k_x \left(\frac{\partial h}{\partial x} \right)^2 + k_y \left(\frac{\partial h}{\partial y} \right)^2 + k_z \left(\frac{\partial h}{\partial z} \right)^2 - k_z \left(\frac{\partial h}{\partial z} \right) \right] = \frac{\partial W}{\partial t} \qquad (6A.31)$$

Since value of β (compressibility of the fluid) is very small, the second group of terms on the left-hand side of Eq. (6A.31) can be neglected compared with the first group of second derivatives in most soil mechanics problems. Then equation (6A.31) can be simplified as:

$$\gamma_w \left[k_x \frac{\partial^2 h}{\partial x^2} + k_y \frac{\partial^2 h}{\partial y^2} + k_z \frac{\partial^2 h}{\partial z^2} \right] = \frac{\partial W}{\partial t} \qquad (6A.32)$$

6A.5 Steady State Flow

If the soil is saturated and both the soil and the pore fluid are incompressible, the right-hand side of Eq. (6A.32) becomes zero. Then the equation (6A.31) reduces to that for the steady-state flow of fluid through a homogeneous porous medium / soil medium and is expressed as:

$$k_x \frac{\partial^2 h}{\partial x^2} + k_y \frac{\partial^2 h}{\partial y^2} + k_z \frac{\partial^2 h}{\partial z^2} = 0 \tag{6A.33a}$$

This is the well-known **Laplace's equation** for steady state flow of fluids in porous media / soil media.

When the right-hand side of Eq. (6A.32) is not zero due to compressible soil and (or) fluid (or) variable head h, it becomes a non-steady or transient flow problem which has to be dealt with separately. In general, the horizontal permeability coefficients k_x and k_y are greater than vertical permeability coefficient, k_z.

By proceeding on similar lines, identical equations can be derived for electrical current and heat flow, in which the permeabilities $k_{x,y,z}$ for water flow are replaced by the reciprocals of electrical resistivities $R_{x,y,z}$ and by the thermal conductivities $C_{x,y,z}$ respectively. In these analogous processes, electrical potential (voltage) V and thermal potential (temperature) θ take the place of the hydraulic potential h. These equations [see Eqs. (6A.18) and (6A.19)] are expressed as:

$$\frac{1}{R_x}\frac{\partial^2 V}{\partial x^2} + \frac{1}{R_y}\frac{\partial^2 V}{\partial y^2} + \frac{1}{R_z}\frac{\partial^2 V}{\partial z^2} = \text{Rate of storage of electricity} \tag{6A.34}$$

and

$$C_x \frac{\partial^2 \theta}{\partial x^2} + C_y \frac{\partial^2 \theta}{\partial y^2} + C_z \frac{\partial^2 \theta}{\partial z^2} = \text{Rate of storage of heat} \tag{6A.35}$$

For steady state flows of current and heat, the right hand side of Equations (6A.34) and (6A.35) are made equal to zero for analysis.

If, in the steady-state forms of Eqs. (6A.32), (6A.34), and (6A.35) the permeabilities, resistivities, and conductivities are the same in all directions, the flow medium is called isotropic, and the equations are reduced to

$$\frac{\partial^2 h}{\partial x^2} + \frac{\partial^2 h}{\partial y^2} + \frac{\partial^2 h}{\partial z^2} = 0 \tag{6A.33b}$$

with the appropriate substitutions of V and θ for h, in the analogous processes respectively.

6A.6   Two-Dimensional Flow / Laplace's Equation:

In soil mechanics, it may be necessary for practical reasons to simplify the physical problem by assuming it to be two-dimensional. This can be done, for example, when flow takes place through an earth dam which is long (measured along the crest) compared with its height and width or when flow occurs in the previous soil under a sheet-pile wall supporting the sides of a trench etc. Such an assumption cannot be made if the sheet-pile wall forms the four sides of an excavation, say for a bridge pier. In such situations, the full equation (6A.33a) or (6A.33b) must be analysed. In problems amenable for simplification as two-dimensional problems in x and z coordinates, the above Eq.

(6A.33b) reduces to the two dimensional steady state flow equation / Laplace Equation i.e.

$$k_x \frac{\partial^2 h}{\partial x^2} + k_z \frac{\partial^2 h}{\partial z^2} = 0 \qquad (6A.33c)$$

in which the z-direction is usually taken to be vertical.

If flow occurs in two dimensions radially toward or away from a cylindrical sink or source such as a well and the pervious layer is completely confined by impermeable materials, the two-dimensional Laplacian equation (6A.33c) can be expressed in radial coordinates, r and θ. Then the equation becomes one-dimensional in the radial coordinate, r only (since axial symmetry exists along θ direction in such problems i.e. flow is axisymmetric) i.e.

$$\frac{1}{r} \frac{\partial}{\partial r}\left(r \frac{\partial h}{\partial r}\right) = 0 \qquad (6A.36)$$

when the origin of coordinates is taken at the wall axis. In this case, r is measured horizontally, and the soil is considered to be isotropic with permeability k.

Three-dimensional flow may also occur radially in isotropic soil toward or away from a spherical sink or source (thus having symmetry in spherical coordinates), and Laplace's equation in spherical coordinates is applicable in such cases, the origin again being taken at the center of the sink or source. The same is expressed as:

$$\frac{1}{r^2} \frac{\partial}{\partial r}\left(r^2 \frac{\partial h}{\partial r}\right) = 0 \qquad (6A.37)$$

In most steady-state problem, flow takes place in a region with fixed boundaries at which the flow or head conditions are imposed. These are referred to a confined flow problem. The solution of such problems can be obtained by solving the Laplace's equation inside the region which satisfies the boundary conditions imposed along the edges of the region.

Usually, the boundary conditions of such domains may be one of three types: (1) a potential boundary along which the total head is constant, (2) an impervious boundary along which the fluid is constrained to flow, or (3) a free surface boundary along which flow takes place, and where the total head at each point is equal to the height of the point above the assumed datum line since the pressure in the fluid is equal to the external (usually atmospheric) pressure. These boundary conditions can be expressed mathematically as follows.

(1) Along S, $h = H$ (constant),

(2) Along S, $\frac{\partial h}{\partial n} = 0$ (no flow across boundary),

(3) Along S, $\frac{\partial h}{\partial n} = 0$ and $h = f(z)$ (when z-axis is directed vertically)

where S represents the boundary and n is the normal to it at any point.

For example, Figure 6A.2 shows the cross section of an earth dam constructed on an impermeable foundation with a rock drain at the toe (downstream end of the dam) The three boundary conditions mentioned above are shown in the figure. The rock drain is provided so that the water permeating through the dam does not break out on the downstream face, with the resultant possibility of erosion. Water flows from the reservoir to the rock drain within the region of the dam marked *ABCD* and bounded by the flow lines *BC* (boundary condition (2) above) and *AD* (3). The flow line *AD* is called a free surface, or phreatic line, since the pressure in the fluid along this surface equals atmospheric pressure. In practice, there will most probably be some region above this line partially saturated with water as a result of the phenomenon of capillary rise.

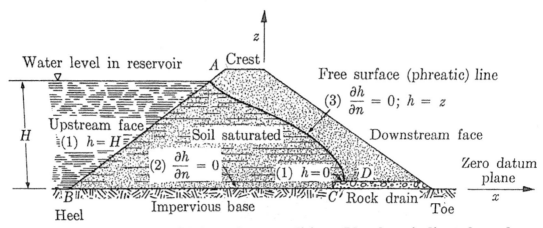

Figure 6A.2 Earth dam with boundary conditions (Numbers indicate boundary conditions as mentioned in the above paragraphs).

There are many such examples of fluid flow as in flow below dams etc. which are analyzed using Laplace equation in one, two or three dimensions. (Taylor, 1964, Scott, 1963).

6A.7 Methods of Solution

Following methods are commonly used to get the solutions for problems of steady state flow in porous media / soil media.
1. Graphical Solution by Sketching – Drawing of Flow nets
2. Mathematical / Analytical Methods / Complex Variables / Conformal Mapping Techniques
3. Numerical Methods – Finite difference / Relaxation and other Numerical approaches
4. Model Studies
5. Electrical Analogy and Capillary Flow Analogy Methods
6. FEM for Flow Problems

Extensive literature is available giving details on the above methods of solution such as Crandall (1956), Scott (1963), Canale and Chapra (1989) etc.

References

Bowles, J.E. (1996) Foundation Analysis and Design, McGraw-Hill International Book Company.

Budynas, R.G. (1999), Advanced Strength and Applied Stress Anaysis, McGraw-Hill Co., U.S.A.

Canale, S.C. and Chapra, R.P. (1989) Numerical Methods for Engineers, McGraw-Hill Book Co., U.S.A.

Chong Chee Siang (2013). Dynamic Response of Beams due to Impact Load (Master thesis). Universiti Malaysia Sabah, Kota Kinabalu, Malaysia.

Chong Chee Siang (2018). Dynamic Responses of Plates and Slabs due to Impact Loads (Doctoral thesis). Universiti Malaysia Sabah, Kota Kinabalu, Malaysia.

Crandall, S.H. (1956) Engineering Analysis, McGraw-Hill Inc., USA.

Crandall, S.H., Dahl, N.C., and Lardener, T.J. (1972) Mechanics of Solids, McGraw-Hill Kogokusha Ltd., New Delhi, India.

Desai, C.S. and Abel, J.F. (1972) Introduction to the Finite Element Method, Affiliated East–West Press Ltd., New Delhi, India.

Hertz, H. (1884) Uber das gleichgewicht schwimmender elasticher platten. Wiedemmanns Annalen der Physik und Chemie, 22, 255.

Hetenyi, M. (1946) Beams on Elastic Foundations, The University of Michigan Press, Ann Arbor, Mich., U.S.A.

Hetenyi, M., "A General Solution for the Bending of Beams on an Elastic Foundation of Arbitrary Continuity," Journal of Applied Physics, Vol. 21, 1950, pp. 55-58

Jones, G. (1997) Analysis of Beams on Elastic Foundations using Finite Difference Theory, Thomas Telford, London, U.K.

Kameswara Rao, N.S.V., Chong Chee Siang, Kenneth Teo Tze Kin, Journal of Technologies in Education, Interactive Learning of Impact Behavior of Structures using Virtual Impact Tests, Volume 11, Issue 4, 2015, ISSN 2381-9243.

Kameswara Rao, N. S. V. (2011) Foundation Design, Theory and Practice, John Wiley & Sons (Asia) Pte Ltd. ISBN: 978-0-470-82534-1.

Lambe, T.W., (1973) Predictions in soil engineering, 13th Rankine Lecture, Geotechnique, 23(2), June, pp. 148–202.

Rajasekaran, S. (1993) Finite Element Analysis in Engineering Design, A.H. Wheeler / Wheeler Publishing, New Delhi, India.

Rao, S.S. (1982) The Finite Element Method in Engineering, Pergamon Press, Oxford, U.K.

Reissner, E. (1937) On the theory of beams resting on yielding foundations. Proceedings of the National Academy of Sciences of the United States of America, 23, 328–333.

Salvadurai, A.P.S. (1979) Elastic analysis of soil-foundation interaction, Developments in Geotechnical Engineering, vol. 17, Elsevier, Amsterdam

Scott, R.F. (1963) Principles of Soil Mechanics, Addison-Wesley Publishing Co., U.S.A.

Strang, G. and Fix, G. (2008) An Analysis of Finite Element Method, Wellesley – Cambrdge Press.

Taylor, D.W. (1964) Fundamentals of Soil Mechanics, Asia Publishing House, New Delhi, India.

Terzaghi, K. (1955) Evaluation of coefficient of sub grade reaction. Geotechnique, 5(4), 297–326.

Timoshenko, S. and Goodier, J.N. (1951) Theory of Elasticity, McGraw-Hill, New York, U.S.A.

Timoshenko, S. and Woinowsky-Krieger, S. (1959) Theory of Plates and Shells, McGraw-Hill Book Co. Inc., New York, U.S.A.

Tomlinson, M.J. (2001) Foundation Design and Construction, Prentice-Hall, Singapore.

Vlasov, V.Z. and Leontev, U.N. (1966) Beams, Plates and Shells on Elastic Foundations, (Translated from Russian), NASA TT F-357.

Winkler, E. (1867) Die Lehre von der elasticitaet und festigkeit, Prag Dominicus, Berlin, p. 182.

Zienkiewicz, O.C. (1971) The Finite Element Method, McGraw-Hill, London, U.K.

Zienkiewicz, O.C. and Taylor, R.L. (1967.1989) The Finite Element Method, McGraw-Hill, London, U.K.

Bibliography

Astley, R.J. 1992 " Finite Elements in Solids and Structures: An Introduction." Chapman & Hall.

Becker, A.A., "The Boundary Element Method in Engineering," Mc.Graw-Hill Book Co., U.S.A, 1992

Bhattarai, P., Chaudary, R., Thapa, C., Dhakal, D.R., Saha, P. 2012. Behavior of Metallic Plates using Finite Element Method and Its Application in Civil Engineering. International Journal of Advance Scientific Research and Technology. 2 (2): 308-316.

Chandrakant, S.D., John, F.A., "Introduction to the Finite Element Method - A Numerical Method for Engineering Analysis", Van Nostrand Reinhold Company, New York, 1972.

Chong, C. S., Kameswara Rao, N.S.V., Mariappan, M., "Finite Element Modelling of Four Edges Simply Supported Steel Plate Under Impact Load", Journal of Fundamental and Applied Sciences, Vol. 9 (3S), 2017, p. 257-278.

Gazetas, G. and Tassoulas, J. L., "Horizontal Stiffness of Arbitrary Shaped Embedded Foundations," J. of GT Div., ASCE, Vol. 113, No. 5, 1987, p.458-475

Gonzalez-Perez, I., Iserte, J.L., Fuentes, A. 2011. Implementation of Hertz Theory and Validation of a Finite Element Model for Stress Analysis of Gear Drives with Localized Bearing Contact. Mechanism and Machine Theory. 46: 765- 783.

Gorbunov-Posadov, M.I., "Beams and Plates on an Elastic Base," (in Russian), Stroizdat, Moscow, U.S.S.R., 1949

Gregory, C.M., Steven D.G. 2009. High-Fidelity Conical Piezoelectric Transducers and Finite Element Models Utilized to Quantify Elastic Waves Generated from Ball Collisions. Proceeding SPIE 7292. Sensors and Smart Structures Technologies for Civil, Mechanical and Aerospace System. 72920S.

Harr, M. E., "Foundations of Theoretical Soil Mechanics," McGraw-Hill Book Co., U.S.A., 1966

Hertz, H., "Uber das Gleichgewicht Schwimmender Elasticher Platten," Wiedemmanns Annalen der Physik und Chemie, Vol. 22, 1884, p. 255.

Hooper, J. A., "Observations on the Behavior of a Pile-Raft Foundation in London Clay," Proceedings of the Institute of Civil Engineers, 55, 1973, pp. 855-877

Hooper, J. A., "Review of the Behavior of Piled Raft Foundations," Construction Industry Research and Information Association, Report No. 83, 1979

IS: 1080 – 1962, Code of Practice for Design and Construction of Simple Spread Foundations, India, 1962

IS: 2974-1966, (Parts I-V), I.S. Codes of Practice for Design and Construction of Machine Foundations, Indian Standards Institute, New Delhi, 1966

IS: 5249-1969, I.S. Method of Test for Determination of in-situ Dynamic Properties of Soils, Indian Standards Institution, New Delhi, First Reprint, July 1975

IS: 8009-1976, Code of Practice for Calculation of Settlement of Foundations Subjected to Symmetrical Vertical Loads: Part I-1976, Shallow Foundations, India, 1976

IS: 2950-1981, Code of Practice for Design and Construction of Raft Foundations, Bureau of Indian Standards, New Delhi, India, 1981

IS: 1888-1982, I.S. Method of Load Test on Soils, Indian Standards Institution, New Delhi, Second Revision, March 1983

IS: 2720 (Parts III, XXVIII, XXXIX, XXXII), I.S. Methods of Tests for Soils, Indian Standards Institution, New Delhi, 1983

IS: 11089-1984, Code of Practice for Design and Construction of Ring Foundations, BIS, New Delhi, India, 1984

IS: 456-2000, Code of Practice for Plain and Reinforce Concrete, Bureau of Indian Standards, New Delhi, 2000

Iyengen, K.T.S.R., and Ramu, S., "Design Tables for Beams on Elastic Foundations and Related Structural Problems," Applied Science Publishers Ltd., London, 1979

Jain, A. K., "Reinforced Concrete – Limit State Design," Fourth Edition, Nem Chand and Bros, Roorkee, India, 1997

Jones Glyn, "Analysis of Beams on Elastic Foundations using Finite Difference Theory," Thomas Telford, U.K., 1997

Kameswara Rao, N.S.V., "Variational Approach to Beams and Plates on Elastic Foundations," Ph.D Thesis, I.I.T., Kanpur, 1969

Kameswara Rao, N.S.V., "Variational Approach to Beams on Elastic Foundations," Journal of EM Division, ASCE, USA, April, 1971, pp. 271-294

Kameswara Rao, N.S.V., "Dynamics of Soil-Structure Systems – A Brief Review," J. of Struct. Engg. India, Vol. 4, 1977, pp. 149-153

Kameswara Rao, N.S.V., "Vibration Analysis and Foundation Dynamics," A. H. Wheeler and Co., New Delhi, India, 1998

Kameswara Rao, N.S.V., "Dynamic Soil Tests and Applications," Wheeler Publishing, New Delhi, 2000

Kameswara Rao, N.S.V., "Mechanical Vibrations of Elastic Systems," Asian Books, New Delhi, India, 2006

Kameswara Rao, N.S.V., Chong C.S., "Virtual Tests for Studies in Impact Engineering and Prototype Testing", IEM an open access journal, Vol. 3, No. 5, 2014, pp. 20.

Kreyszig, E., "Advanced Engineering Mathematics," Wiley Eastern Pvt Ltd, New Delhi, 1967

Kurian, N. P., "Design of Foundation Systems – Principles and Practices," Vol.1, Narosa Publishing House, New Delhi, 1992

Lambe, T. W., "Soil Testing for Engineers," John Wiley and Sons, New York, 1951

Lambe, T. W., and Whitman, R. V., "Soil Mechanics," John Wiley and Sons, USA, 1969 June, 1973, pp. 148-202

Lin, C.L., Chang, Y.H., Liu, P.R. 2008. Multi-factorial Analysis of a Cusp-replacing Adhesive Premolar Restoration: A Finite Element Study. Journal of Dentistry. **36**. 194-203.

Lysmer, J. and Richart, F. E. "Dynamic Response of Footings to Vertical Loading," Journal of SM Division, ASCE, Vol. 92, SM1, January, 1966, pp. 65-91

Malter, H., "Numerical Solutions for Beams on Elastic Foundations," Journal of SM Division, ASCE, 84, 2, Part I, Mar, 1958

Papanikos, G., Gousidou-Koutita, M.C. 2015. A computational Study with Finite Element Method and Finite Difference Method for 2D Elliptic Partial Differential Equations. Applied Mathematics. **6**: 2104-2124.

Prabowo, A.R., Bae, D., Sohn, J., Zakki, A.F. 2016. Evaluating the Parameter Influence in the Event of a Ship Collision based on the Finite Element Method Approach. International Journal of Technology. **4**: 592-602.

Rao, S. S., "The Finite Element Method in Engineering," Pergamon Press, Oxford, U.K., 1982

Rao, S.S., "The Finite Element Method in Engineering, Fourth Edition", Elsevier Science and Technology Books, Oxford, U.K., 2004.

Reddy, J.N., "An Introduction to FinieElement Method." Mcgraw Hill Book.Co., New York, 1984

Reddy, J.N., "An Introduction to Non linear FiniteElementAnalsis," Oxford University Press, UK, 1985

Reddy J.N., Gartling, D.K., "The Finite Element Method in Heat Transfer and Fluid Mechanics," CRC Press, London and New York, 2010

Reese L.C., Isenhower W.M, Wang S.T., "Analysis and Design of Shallow and Deep Foundations," John Wiley, 2005

Richart, F. E., Hall, J. R., and Woods, R. D., "Vibrations of Soils and Foundations," Prentice-Hall Inc., New Jersey, 1970

Rijhsinghani, A., "Plates Subjected to Concentrated Loads and Moments," M.S. Thesis, Illinois Institute of Technology, Chicago, 1961

Rombach, G. A., "Finite Element Design of Concrete Structure," Thomas Telford Publishing, USA, 2004, ISBN: 0727732749.

Taylor, D. W., "Fundamentals of Soil Mechanics," Asia Publishing House, India, 1964

Terzaghi, K., "Theoretical Soil Mechanics," Wiley, New York, 1943

Terzaghi, K., and Peck, R. B., "Soil Mechanics in Engineering Practice," John Wiley, New York, 1967

Tomlinson, M. J., "Foundation Design and Construction," Prentice-Hill, Singapore, 2001

Trivedi, N., Singh, R.K. 2013. Prediction of Impact Induced Failure Modes in Reinforced Concrete Slabs through Nonlinear Transient Dynamic Finite Element Simulation. Annals of Nuclear Energy. 56:109-121.

Vesic, A. S., "Bending of Beams Resting on Isotropic Elastic Solid," Journal of Eng. Mech. Div., ASCE, Vol. 87, EM 2, April, 1961, pp. 35-53

Wang, C.G., Liu, Y.P., Lan, L., Tan, H.F. 2016. Free Transverse Vibration of a Wrinkled Annular Thin Film by Using Finite Difference Method. Journal of Sound and Vibration. **363**: 272-284.

Wieghardt, K., "Uber den Balken auf Nachgiebiger Unterlage," Zeitschrift fur Angewandte Mathematik und Mechanik, Vol. 2, 1922, pp.165-184

Zimmermann, H., "Die Berechnung des Eisenbahnober-baues," Berlin, 1888

Zienkiewicz, O.C., Taylor, R.L., "The Finite Element Method - Fifth Edition, Volume 1: The Basis", Butterworth-Heinemann, Oxford, 2000.

Zienkiewicz, O.C., Taylor, R.L., "The Finite Element Method - Fifth Edition, Volume 2: Solid Mechanics", Butterworth-Heinemann, Oxford, 2000.

Subject Index

ABAQUS 32,168, 172
Adiabatic 237
Aircraft industry 2, 7
Analysis 1-67
Analytical method 1-3, 140-141, 157, 249
ANSYS 32, 168, 172
Approximate method 1-4, 32-33, 144
Area 22-30, 38-48, 64, 155, 194, 198, 209-228, 234
Area moment of inertia 53-55, 59, 198
Artificial Intelligence (AI) 4
Assemblage 7-14, 26, 31,157, 223, 242
Axial stress 38, 55
Axisymmetric 16, 28, 148, 171, 247
Beams 5-6,16, 31-32, 51-60, 106-111, 132, 139, 143-149,152, 160-190, 197,205
Bending moment 31, 51-59,140-141, 149,161, 168, 178, 198
Bending theory 51, 53-55, 144-146, 160-163
Body forces 29, 42, 240
Boundary condition 15, 30, 41, 176, 184, 206-207, 214, 226, 248
Boundary loads 30, 65, 66
Brick 17, 139
buckling 60, 130
Caissons 139
Cartesian coordinate 235
Charpy Impact Test 4-5
Classical bending theory 53, 55, 143,146, 168
Cohesionless 198, 233, 244
Cohesive 198, 233, 243-245
Collocation method 35
Combined footing 137, 185, 192
Compatibility 9-11, 19, 23, 26, 63, 158
Compression 16, 96-105
Concentrated loads 13, 30, 42-44, 54, 65, 166,191
Concrete 6, 139, 143, 173, 191
Consistent force approach 43
Consistent method 14, 30, 56, 57, 65, 66, 164
Constant Strain Triangular (CST) 11, 19-24, 61-67

Constitutive equation 63
Constitutive relation 32, 137,144, 171
Continuous system 1-4, 26
Continuum 1-4, 7, 10, 24, 31-34, 61, 144, 208, 212, 229
Convention 51, 54, 146, 149, 152, 162, 191
Conventional Design 140-143
Conventional method 68-128
Cyclic plate load test 201, 202-203
Dam 247-249
Darcy's law 211, 228, 239-245
Deep foundation 139-140
Deflection 6, 52, 55, 145, 150-154, 172, 194
Degree of freedom 37, 39, 48, 233
Direct stiffness analysis 7, 26, 32-36, 39, 43, 51, 161
Discrete system 1-2, 17, 26
Discretization 2-3, 7, 15, 18, 27, 169, 184, 208, 210, 229
Displacement 7, 11, 19, 29, 37, 149, 164, 183, 195, 203
Displacement method 10, 32, 33, 158
Distributed loads 7, 56, 147
Drilled piers 139
Dynamic 200, 201-203, 234
Eccentric load 140, 143, 184
Elastic foundation 145-147, 154-155, 169, 198
Elastic solids 26, 61-67, 151, 208
Elastic-continuum 153-155
Electrical networks 2
Element stiffness matrix 9, 14, 20, 45, 47, 54, 68-125, 158, 164, 169, 170
Elementary theory 115, 120, 123
Equivalent method 56
Errors 17-18, 200
Euler-Bernoulli bending theory 51, 160
Euler's theory 33, 207, 213
Exact solution 33, 34
Extremum principle 2, 7, 32, 33-34, 36, 205, 206, 208, 213
Finite difference method 1, 7, 218, 249
Finite element method 2, 7, 17, 141, 157, 205, 233
Finite element shapes 16
Flexible footing 141

Flexible plate	199, 201
Flexural rigidity	58, 145, 147, 152, 166, 174
Flow	2, 7, 23, 32, 205-232, 233-249
Fluid dynamic	238
Fluid flow	2, 7, 23, 32, 211, 216, 229, 234, 238, 240-249
Fluid mechanics	33, 205, 238
Fluid Potential	23, 206, 217
Force method	10, 158
Foundation	32, 136-193, 194-204, 248
Foundation-structure interaction	150
Frames	16, 32, 59-60
Free-body diagram	40
Galerkin's method	32, 35, 36
General variational method	154, 155
Geometric	16, 19-20, 33, 39, 41, 154, 245
Geotechnical Engineering	137
Geotechnique	137
Global coordinates	10, 12, 47-50, 159, 219
Governing equations	32-34, 145, 149, 205, 206, 208
Gradient	211, 218, 229, 239, 243, 245
Hagen-Poiseuille law	239, 241, 242
Hammer Drop Test	4, 6
Head	23, 206, 211, 215, 216-228, 239-249
Heat conduction	2, 7, 32, 33, 205, 206
Hinged	91-135
Homogeneity	7, 152, 215
Hooke's Law	155, 194
Hybrid methods	10, 158
Hydraulic radius	242
Impact Engineering	4
Impermeable	214, 247, 248
Independent footing	140
Initial lack of fitness	13, 44-46, 55, 158-163
Initial strain	7, 9, 28, 30, 31, 44-46, 65, 158-160
Internal strain energy	54
Isoparametric element	25
Isothermal	237, 238
Isotropic	19, 28, 144, 206, 218, 229, 247, 248
Judicious approach	65, 67, 166, 185
k matrix	64
Kirchhoff	147

Laplace operator	146, 152-153
Load-displacement relation	59, 154
Local coordinates	10-12, 19, 47-51, 60, 159, 161
Loss	218, 233-236
Lumped force approach / Lumped Load method	42, 44, 56
Mat foundation	137, 140, 156
Material Properties	2, 7, 39, 64, 144
MATLAB	60, 172
Member forces	91-105, 129-131
Method of joints	91-105, 129
Method of Least Squares	35
Method of sections	129
Modulus of elasticity	28, 38, 48, 53-63, 139-156, 194-204
Moment of inertia	53, 58, 106-111, 145, 174, 198
NASTRAN	32, 168, 172
Nodal forces	9-14, 28-31, 37-46, 55, 65-66, 158-166, 169, 213
Numerical method	1, 2, 4, 17-18, 20, 141, 157, 168, 206, 249
Permeability	206, 211-229, 239-247
Pile	136, 139, 197, 247
Pins	129
Plain stress	21
Plane element	25, 61, 62
Plane strain	21, 26, 28, 61, 64, 65, 145, 146, 154
Plane stress	11, 16-17, 21, 26-29, 61, 64, 65, 112-125, 134-135, 154
Plastic settlement	202
Plate	16, 31, 112-135, 143-157, 160, 168-191, 194-204
Plate load test	156, 194-204
Poisson's ratio	28, 59, 63, 139, 145-147, 156, 173, 191, 196-201, 205
Polar coordinate	148, 171
Polynomial	19, 20, 37, 39, 52
Porous media	23, 32, 33, 36, 205-208, 216, 233-249
Potential energy	20, 32, 33, 36, 54, 205, 240
Pressure	137-204
Procedure of FEM	9-10, 17-18, 67, 143, 157-160
Quasi-harmonic differential equation	205, 206
Radius of curvature	55

Rational approach 156, 168
Rational Design 140, 143, 155
Rayleigh-Ritz method 33, 39, 56
Reaction forces 68-130
Residual stress 44, 46
Reverse engineering 3, 4
Reynolds number 242-244
Right-handed coordinate system 145
Rigid footing 139, 141, 183
Ritz process 7, 36
Rod 37-48, 55, 59, 68-90, 126-128
Rollers 115, 120, 123

SAP 32, 172
Seepage 21, 32, 205-232, 233-249
Shallow foundation 137-140
Shear forces 16, 51, 54, 65, 106-111, 141, 146-154, 161-168, 190
Shear modulus 153, 200-201
Shell 25, 143, 144, 169
Simulation 3-4, 157
Simultaneous equation 15, 34-36, 55, 58, 67, 172, 185, 213
Slope 31, 51-58, 132, 143, 146, 160-163, 168, 178, 202
Soil properties 147, 194-204
Soil-structure interaction 2, 7, 143-145, 155-156, 169, 186
Solid mechanics 2, 15, 17, 23, 32, 37, 209, 213
Spread foundation 136-138, 140, 156, 172-182, 191
Spring constant 38, 150, 155, 165, 194, 202-204
Static 56, 65, 234, 235, 239
Static reaction 56
Stone 139
Strain 7, 11, 20, 23, 24, 27, 29, 31, 39, 41, 44, 63, 67, 171
Strength of materials approach 103, 132-133
Stress 7, 31, 38, 44-46, 55, 63, 158, 159, 233, 234, 238
Stress matrix 9, 31, 38, 64, 158
Stress-strain 28, 39, 53, 63, 137
Strip footing 140, 145
Structural analysis 2, 7, 9, 10, 14-15, 59, 141, 149, 157-158, 164, 165, 168, 217
Structural idealization 9, 10, 157
Subdomain method 35
Subgrade reaction 155, 156, 165, 166, 168, 171, 172, 194-203
Substructure 136, 172-182

Superstructure 136-139
Surface forces 43, 45
Symmetric 14, 50, 53, 140, 162, 164, 168, 177, 180, 187, 211, 227, 231
Synthesis 1, 4
System Equilibrium Equations 12-15, 35, 41, 44, 45, 46, 49, 50, 55, 57, 58, 163-165, 176, 184-187, 213
Temperature 44, 55, 77-79, 88-90, 134, 163, 233-246
Tension 16, 91-105, 144, 152, 198
Theory of elasticity 61, 198
Thermal expansion 44, 65, 77-84, 88-90, 126-128, 134
Thermal nodal force 69, 83
Thermal strain 44
Torque 147
Torsional moments 59, 140
Torsional rotations 59
Transformation matrix 21, 48, 50
Transformation of coordinates 10-12, 47, 48, 50, 60, 159
Trial solution 34, 36
Triangular element 21, 23, 24, 61, 63-67, 112-125, 134-135, 209
Truss 16, 37, 47-50, 60, 91-105, 129-130
Two-plane bending 57-59

Virtual Testing 3-4
Vlasov's model 198

Wall reaction 81-84
Wave propagation test 200, 201
Weighted residual methods 2, 35, 36
Well foundation 139
Winkler model 145, 150-160, 165, 168-170, 174, 194, 198
Work potential 54

Author Index

Abel	19, 21, 25, 32, 144, 168, 171, 206, 212, 215
Bowles	32, 139, 166, 170, 171, 198
Budynas	16, 39, 54, 57, 59, 60
Canale	32, 34, 36, 249
Chapra	32, 34, 36, 249
Chong Chee Siang	4
Crandall	32, 33, 34, 36, 51, 55, 160, 207, 213, 249
Dahl	51, 55, 160
Desai	19, 21, 25, 32, 144, 168, 171, 205, 212, 215
Filonenko-Borodich	152, 153, 154
Fix	34, 36
Hertz	150
Hetenyi	151, 152
Jones	198
Kameswara Rao	54, 137-157, 162, 168, 170, 194, 197-202
Kozeny	242
Kreiger	147, 148, 151
Leontev	144, 146, 147, 154, 155, 198
Pasternak	152-154
Rajasekharan	60
Rao, S. S.	144, 206
Reissner	150-153
Scheidegger	242, 243
Scott	211, 237, 242-245, 249
Strang	34, 36
Taylor	2, 7, 11, 25, 32, 34, 36, 39, 43, 45, 46, 60, 63, 64, 65, 144, 149, 168, 171, 198, 239, 242, 244, 249
Timoshenko	10, 147, 148, 151
Tomlinson	139
Vlasov	144, 146, 147, 154, 155, 198
Wieghardt	153, 154
Winkler	150, 151, 194
Zienkiewicz	2, 7, 11, 15, 28, 32, 34, 35, 36, 39, 43, 45, 46, 60, 63, 64, 65, 144, 168, 169, 171, 205, 207, 215

SWASTHI VACHANAM

ॐ द्यौः शान्तिरन्तरिक्षं शान्तिः
पृथिवी शान्तिरापः शान्तिरोषधयः शान्तिः ।
वनस्पतयः शान्तिर्विश्वेदेवाः शान्तिर्ब्रह्म शान्तिः
सर्वं शान्तिः शान्तिरेव शान्तिः सा मा शान्तिरेधि ॥
ॐ शान्तिः शान्तिः शान्तिः ॥

Om Dyauh Shaantir-Antarikssam Shaantih

Prthivii Shaantir-Aapah Shaantir-Ossadhayah Shaantih |

Vanaspatayah Shaantir-Vishvedevaah Shaantir-Brahma Shaantih

Sarvam Shaantih Shaantir-Eva Shaantih Saa Maa Shaantir-Edhi |

Om Shaantih Shaantih Shaantih ||

**May peace radiate herein the whole sky as well as in the vast ethereal space
everywhere.
May peace reign all over the earth, in water and in all herbs, trees, and plants.
May peace pervade everywhere in the whole universe.
May peace be in the supreme being Brahman.
And may there always exist in all, peace and peace alone**

ॐ शान्तिः शान्तिः शान्तिः ॥

Om śhāntiḥ śhāntiḥ śhāntiḥ ॥

Printed in Great Britain
by Amazon

77862439R00163